Energy: A Historical Perspective and 21st Century Forecast

By
Amos Salvador

AAPG Studies in Geology #54

Published by

The American Association of Petroleum Geologists

Tulsa, Oklahoma

AAPG Editor: Ernest A. Mancini
Geoscience Director: James B. Blankenship

This publication is available from:

The AAPG Bookstore
P.O. Box 979
Tulsa, OK U.S.A. 74101-0979
Phone: 1-918-584-2555 or 1-800-364-AAPG (U.S.A. only)
Fax: 1-918-560-2652 or 1-800-898-2274 (U.S.A. only)
E-mail: bookstore@aapg.org
www.aapg.org

About the Author

Amos Salvador received his B.S. in geology from the Universidad Central de Venezuela in 1945 and his Ph.D. in geology from Stanford University in 1950. He worked as a surface geologist for Mene Grande Oil Co. (Gulf) in Venezuela from 1945–1947 and rejoined the company after receiving his Ph.D. and worked for Mene until 1955.

In 1955, Salvador began working for affiliates of the Exxon Corporation, holding several positions with the following affiliates:

Supervisor of Exploration Group for eastern and western Venezuela for the Creole Petroleum Corporation, Venezuela; Vice President of Exploration Research for Jersey Production Research Company, Tulsa, Oklahoma; Manager of the Geological Department, Creole Petroleum Corporation; Assistant Chief Geologist and Manager of Gulf Coast Exploration Division for Humble Oil and Refining Company, Houston, Texas; Executive Vice President of Esso Production Research Company, Houston, Texas; Chief Geologist of Exxon Company U.S.A., Houston, Texas.

In 1980, Salvador became Alexander Deussen Professor of Energy Resources and Morgan J. Davis Professor of Petroleum Geology at The University of Texas at Austin, where his teaching included an undergraduate/graduate course on Energy Resources.

Since 1993, he has been the Morgan J. Davis Emeritus Professor of Petroleum Geology at The University of Texas at Austin.

Salvador served as co-director of the AAPG COSUNA Stratigraphic Charts of North America in the 1980s, was editor and author of the 1991 GSA DNAG *The Geology of North America Volume J, The Gulf of Mexico Basin*, and was editor of the 2nd Edition of the International Stratigraphic Guide in 1994.

AAPG wishes to thank the following
for their generous contributions to

Energy: A Historical Perspective and 21st Century Forecast

By Amos Salvador

ExxonMobil Exploration Company

Contributions are applied toward the production cost of publication,
thus directly reducing the book's purchase price
and making the volume available to a larger readership.

Table of Contents

Preface

Thirteen years of teaching a course on energy resources served to reinforce my conviction that the availability of affordable energy will be a critical element in determining mankind's quality of life everywhere on Earth during the 21st century, second only to the adequate supply of clean fresh water, with both elements essential for the production of food.

Also, I was surprised to find that it was not always easy to obtain information essential for discussing the many aspects of energy consumption and its sources of supply: population and energy consumption statistics; data on the production, reserves, and estimated ultimate recoverable amounts of the nonrenewable sources of energy (oil, natural gas, coal, and oil shales); and on the history of other possible sources of energy (nuclear, hydroelectric, and geothermal power, energy from the sun, the wind, biomass, and hydrogen). This information, I found, is scattered through many separate publications, not all of them readily available.

Once I found and recorded this essential information, I was able to better understand the critical relationships between world human population, energy consumption, and the availability of the various energy sources, and I realized that with this information in hand, I would be in a much better position to look into the energy picture in the 21st century.

Therefore, one of the goals of this study, and perhaps its main value, was to assemble in one place the much needed detailed historical statistical data on human population, energy consumption, and current information about present and possible future sources of energy (production, reserves, and geographic distribution). This information is set forth in the numerous tables, some in the text, others in the appendix and on the CD accompanying this book, and is illustrated in the figures.

The study does not attempt to be an exhaustive discussion of the various subjects covered. It places particular emphasis on the kind of data that would allow one to establish trends that can be projected 100 years—looking back can be the best way to see into the future. To allow readers to enlarge their knowledge and to satisfy their need for additional information, an extensive list of references is included, particularly of recent, up-to-date information judged to be significant in covering current thinking.

The study was not expected to either solve all problems or answer all questions concerning energy during the 21st century. It does not. Many problems remain unsolved, and many questions remain unanswered. However, I am hopeful that this study will serve to clarify some problems and answer some questions, support some concepts and assumptions, and contradict or weaken others. I also hope that it will provide a foundation, a point of departure, for future studies. The field is certainly wide open, and I hope that the publication of this study will encourage new and more extensive investigations. To enable this study to be used as the foundation of further studies, an effort was made to separate factual data from estimates, interpretations, and projections. Efforts were also made whenever possible to clarify the terminology in each of the sections of the study.

The term "energy" as used for the sake of brevity refers to what has been called "primary commercial energy" or just "primary energy." It includes the energy recovered directly from the different sources of oil and natural gas and from coal, oil shales, and tar sands. Primary energy is converted to "secondary energy," fuels and electricity, in refineries and power plants, respectively.

The difference between demand and consumption of energy must be kept in mind. Demand may not always be satisfied; what can be provided and paid for will be the amount consumed. Consumption and supply, of course, will have to match. If, for whatever reasons, there is not enough supply of energy to satisfy the demand, the population of the world or of any region or country will have to make do with what is available.

The major problem (not to call it a nightmare) faced during this study was the many units used in the various data sources to express the volumes of consumption of energy and the production, reserves, and resources of the various sources of energy: British thermal units (Btu, and multiples), joules (j, and multiples), barrels of oil equivalent (BOE), metric (or short) tons of oil equivalent (toe), metric (or short) tons of coal equivalent (tce), cubic feet (ft^3), cubic meters (m^3), watts (W, and multiples), etc. The situation is further complicated by the diversity of conversion factors that have been proposed and used to convert from one unit to another. Particularly bothersome was the conversion of electricity units to other kinds of energy units. This diversity of conversion factors may, in fact, account for some of the differences in the statistical data reported by the difference sources.

Nevertheless, the order of magnitude of the discrepancies was, in all cases, within the degree of accuracy needed for the present study (the historical data are "in the ballpark"). Predictions of the world's population, of the consumption of energy, and of the availability of sources of energy for the rest of this century involve such a large degree of uncertainty that they are not appreciably influenced by relatively small differences in the historic data on which they are based.

An effort was made in this study to use as few different units as possible to avoid confusion and to make easier to assemble and compare the data. The conversion factors used will be discussed as needed in each of the sections dealing with energy consumption and each of the various sources of energy.

The historical data were obtained from the publications of a great number of different governments and organizations worldwide and from several commercial journals. Analogous figures from different sources were in close agreement in some cases but not in others. Even in the case of the publications of the same organization or the same journal, statistics are commonly revised and changed from year to year. Statistics, it has been said, are never current. How true.

The time span covered by the statistical data varies from one subject to another. Whenever possible, I tried to get data from 1950 to as close as possible to the present. I did not always succeed.

The analysis of the historical data and trends of population growth and distribution and of consumption and supply of energy made possible the preparation of five forecasts (scenarios) of how and how much energy will be consumed in the 21st century and what sources will most likely supply it.

Other forecasts of energy consumption and supply during part or all of the 21st century have been published during the last 10 or 15 years. Perhaps the first was presented by John F. Bookout at the 28th International Geological Congress in Washington, D.C., in July 1989, and subsequently published in *Episodes* and *Geotimes* (Bookout, 1989a, b). The time restriction for the presentation of this extremely interesting and timely study made it impossible for Bookout

to fully document the bases for his statements and assumptions. The study, unfortunately, did not receive the attention and recognition it deserved. Since then, several energy forecasts have been published, the most detailed and best documented by the Energy Information Administration of the U.S. Department of Energy (*International Energy Outlook*, several years, with projections through 2020), the International Energy Agency of the Organization for Economic Cooperation and Development (OECD), the International Institute for Applied Systems Analysis, the World Energy Council (*Global Energy Perspectives*, 1998, Nakicenovic et al., eds., with projections to 2050 and some issues presented out to 2100), and J. D. Edwards (1997, 2001, 2002). Much valuable information for the present study was obtained from these publications.

The completion of a study like this would not have been possible without the help and contributions of many individuals and organizations. All, unfortunately, cannot be properly acknowledged here. Special recognition for contributions to the various subjects of this study are due to the Population Division of the United Nations Department of Economic and Social Affairs (population); C. D. Masters, T. S. Ahlbrandt, and British Petroleum (oil); A. Bizon and B. A. Rottenfusser (Canadian oil sands); T. S. Ahlbrandt (natural gas); A. I. Scott (coalbed methane); H. J. Gluskoter, W. M. Bogomazov, and J. L. Qian (coal); J. R. Dyni, Vello Kattai, R. E. M. Novicki, and J. L. Qian (oil shales); Division of Nuclear Power of the International Atomic Energy Agency (nuclear power); World Commission on Dams (hydroelectric power); and M. H. Dickson and M. Fanelli (geothermal energy).

Joel Lardon, of the Bureau of Economic Geology of the University of Texas drafted the excellent figures.

ExxonMobil contributed to the cost of printing all figures in color.

Three individuals deserve exceptional recognition: Dennis Trombatore, the librarian of the Department of Geological Sciences, provided invaluable help in searching the literature and in obtaining publications not available at the University of Texas; Hugh Hay-Roe contributed many good ideas about the organization of this publication and carefully edited the text into clear English; last, but far from least, Betty Kurtz went along with my frequent revisions, additions, and deletions and typed and retyped the many versions of the text.

To all of them, I express my deepest gratitude.

1

Introduction, Brief History, and Chosen Approach

ABSTRACT

Because energy is essential to provide the basic needs of humanity, namely food, clean fresh water, fuel for transportation, the generation of electricity, and the production of heat, it is at the core of economic and social activity; and because energy is consumed by people, estimates of the consumption of energy during the 21st century need to be based on projections of the size, composition, and mode and amount of energy use of the human population of the world. Estimates of the possible supply of energy are based on the availability of the presently known sources: principally oil, natural gas, coal, and hydroelectric and nuclear power. The development of these estimates is the objective of this study. They support the conclusion that, although the global distribution of energy sources is obviously uneven, during the 21st century, there will be sufficient sources to adequately satisfy the demand for energy of a human population whose growth rate shows signs of starting to level off.

INTRODUCTION

The ascent of mankind, from its appearance on Earth to its present condition, has depended on the availability of food, clean fresh water, certain materials, and energy. These are not independent factors; they are interrelated in many ways, and the circumstances of their availability and efficacy of their use have established important milestones in the progress of humanity. Each and every one of these factors is essential for the prosperity and growth of every country and every region of the world; sustaining and enhancing our quality of life depend on them. To speculate about the future of the human family and to attempt to predict the circumstances under which it will live or perish, we must consider whether these essentials of human life will be available in the future in the quantities necessary to maintain a growing world population.

Although the global distribution of energy sources is obviously uneven, there will probably be adequate energy resources for a world population whose growth rate shows signs of starting to level off. The inevitable price fluctuations and transportation problems will mean periodic shortages in some areas of the globe. Continuing technological progress in many fields, from petrochemistry and electronics to more efficient energy transformation and use, will stretch the available resources while lessening negative impacts on the environment. Much depends on the ability of the world's political leaders to fashion realistic long-range policies while continuing to support the scientists and technical specialists who work toward cleaner energy sources at reasonable cost.

A BRIEF HISTORY

The availability and use of energy has had a fundamental influence throughout the history of mankind. The "discovery" and control of fire completely changed people's everyday life. It allowed them to keep warm, to drive off predators, to cook, to dry and harden wood, and to heat and split bones and gave them access to

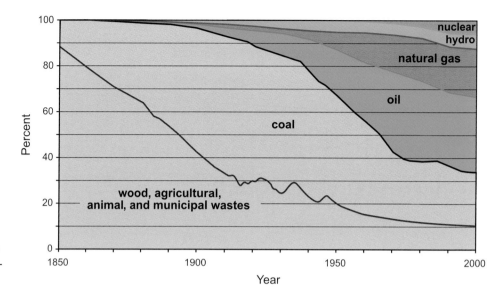

Figure 1. World primary energy sources (generalized after Nakicenovic et al., 1998).

an entirely new class of materials: the metals. Similar changes were brought by the subsequent use of the wind, running water, coal, the steam engine, oil and natural gas, and nuclear power. Each new source of energy provided new benefits and physical comforts. Energy is at the core of economic and social activity. The conditions governing supply, transportation, distribution, and consumption of energy affect all individuals everywhere in the industrialized countries as well as in the Third World. An adequate supply of energy is essential for the achievement of political, social, and economic stability and world peace, a matter of life or death for all humanity.

Energy sources are needed to fight hunger and poverty. Tractors, other agriculture equipment, trucks, trains, and cargo ships that increase the efficiency and mechanization of agriculture and make possible the transportation and distribution of food all need energy to run. The availability of energy therefore will be critical to ensure enough food for the world's growing population. Similarly, availability of a source of energy will determine the eventual feasibility of desalinating sea water as clean fresh water becomes insufficient for the needs of an increasing world population. Fossil fuels (oil and coal) are used in the production of plastics, which have replaced some metals and other materials in innumerable products; and energy is necessary to produce the materials crucial in an advanced civilization.

In discussing the consumption and sources of supply of energy in the 21st century, it is essential to keep in mind that energy is used, in order of importance,

• to fuel land, water, and air transport (cars, trucks, tractors, trains, ships, and airplanes)
• to generate electricity
• to produce heat

Before the Industrial Revolution, during the last half of the 18th century, the main sources of energy to provide mankind with heat, light, and work were wood, agricultural residues, and animal wastes (what have been called the "traditional renewables") and, to a small extent, wind and running water.

The introduction of the steam engine during the Industrial Revolution opened a new chapter in the use of energy and brought a considerable increase in the production and use of coal. During the second half of the 19th century and the first two decades of the 20th century, coal and the traditional renewables supplied more than 90% of the energy consumed in the world. By the end of the 19th century, coal had replaced the traditional renewables as the principal energy source over much of the world (Figure 1).

The beginning of the 20th century witnessed the introduction of two revolutionary innovations in energy use: electricity and the internal combustion engine. Electricity provided new and safer sources of light, heat, and work; the internal combustion engine completely transformed transportation. Other new sources of energy were put into use: oil, hydropower, and eventually, natural gas (see Nakicenovic et al., 1998, p. 11–14).

Crude oil, "discovered" in 1859 in the United States, progressively replaced coal during the 20th century as the principal source of energy, and during the middle part of the century, 60–70% of world energy demand was supplied by coal and oil (Figure 1). Most of this demand was for transportation fuels and for the generation of electricity, the demand for which increased rapidly during the 20th century. The main fuels for generating electricity during the early and middle parts of the 20th century were coal, hydropower, and oil. Demand for natural gas increased during the second half of the 20th century, particularly as a cleaner fuel for the generation of electricity.

At the start of the 21st century, more than 80% of the demand for energy is supplied by oil, coal, and natural gas; they provide all the fuels for transport and 65% of the primary energy for the generation of electricity. Traditional renewables are still used, particularly in many developing countries, and hydropower and nuclear power provide about 35% of the input for the generation of electricity, but only about 10% of the total energy supply.

Other sources of energy, namely geothermal, solar (photovoltaic), and wind, make only a minute contribution to the total supply of energy in the early years of the 21st century. In considering their potential contribution during the rest of this century, we must keep in mind that solar and wind power, as well as hydropower and nuclear power, will be used mainly in the generation of electricity. Geothermal power can be and has been used as a source of heat in addition to its contribution to electricity generation; wind can be used to pump water and grind grain, and solar power is also used directly to heat water. However, the enormous demand for land, water, and air transportation fuels can only be supplied at this time (and for a long time in the future) by the fossil fuels: oil, natural gas, and coal.

The quantity of energy economically available during the 21st century will determine, to a considerable extent, how people live and what their standard of living will be. What will the future bring? As the human population of the world continues to increase and as people aspire to a higher standard of living, will energy sources be available to supply the growing demand? What sources of energy will supply the demand?

The present study tries to answer these questions.

ANALYTICAL APPROACH

To estimate the rates of energy consumption during the 21st century, I experimented with several possible approaches. I finally came to the conclusion that the most direct and best approach was to first compile the past trends of human population size and rates of growth and of consumption of energy. These trends could then be combined to obtain the past trends of consumption of energy per capita, not only for the whole world, but also for specific countries, groups of countries, and regions of the world. These trends could then be combined as needed and extrapolated into the 21st century.

The next most promising approach would have been to attempt to establish for these countries or groups of countries the historical trends of energy consumption per unit of economic activity, as expressed by the gross domestic product (GDP) or some other economic indicator — what has been called "energy intensity" (the amount of primary energy needed per unit of economic output). Estimates of the consumption of energy in the 21st century would then be based on the extrapolations of these trends of energy intensity.

The first approach proved to be more direct and preferable. To compile historical statistics on population and on energy consumption (two fairly objective pieces of information) and to compute past trends of energy consumption per capita on the basis of this information was reasonably easy and involved only two variables. Statistics on economic activity, on the other hand, proved much more difficult to obtain and were found to be considerably less well documented and commonly inconsistent, particularly in the case of the developing countries. As stated by Nakicenovic et al. (1998, p. 16), "A major difficulty in comparing GDPs across countries is the need to translate everything into a common currency. Most often, this is done using market exchange rates, with the United States dollar as the common currency. Problems arise for several reasons. First, not all economies have a free market for foreign currency exchange. Second, the use of market exchange rates implicitly assumes that domestic prices are comparable with international prices. This is not the case for most developing countries, where prices for food and basic services, for example, are substantially below international levels. Third, many transactions are not accounted for in the formal economy, especially in less developed economies." In addition, many economists have come to doubt that there is a necessary direct relationship between GDP and energy consumption, and statistics show that this certainly seems to be the case. Finally, to attempt to use trends of economic activity would involve three variables: population, consumption of energy, and economic activity, making the process considerably more complicated.

Economic indicators (conservation and efficiency in the use of energy, cost of production and transportation of energy and its effect on the price of energy, and possible economic improvements resulting from new and better technology) were, of course, used in projecting the consumption of energy into the 21st century.

Following the chosen approach, the results of this study will be presented in five sections.

The first section deals with the historical trends of the size and growth rates of human population and of energy consumption and energy consumption per capita since the middle of the 20th century. It is critical for the purpose of this study to know how many people have lived on Earth during the 20th century, how this population has been distributed in the various regions of the world, and what the trends of population growth were. The volumes of energy consumption and the levels of energy consumption per capita in the various regions provide essential historical data. Without that, it would be futile to try to estimate the multiple aspects of energy consumption in the 21st century.

The second section attempts to extrapolate the historical population and energy consumption trends into the 21st century.

The third section deals with the sources that have supplied energy in the past and that most likely will supply it in the future.

The fourth section deals with the generation and consumption of electricity historically and in projections for the future.

The fifth and final section of the study discusses five scenarios of estimated energy consumption during the 21st century and the possible sources of energy that may supply it. These five forecasts are based on the historical data discussed in the previous four sections; the key to predicting the future is understanding the past. For each scenario, the choices of basic information, assumptions, and predictions of possible future developments will be specified, documented, and discussed.

The first, third, and fourth sections are reasonably objective; the second and fifth necessarily involve numerous assumptions and a considerable level of uncertainty. The predictions and estimates of future developments and events range from fairly well-documented inferences to guesses, some of relatively minor significance, others critical to the results of the study.

Multiple projections within ranges believed to be realistic have been made in some cases: population growth, increase in the efficiency of energy use, ultimate recovery of some energy sources, and others. Obviously, assumptions, predictions, and estimates had to be attempted to generate a picture of the consumption and possible sources of energy in the 21st century, the objective of this study. Like gemstones, predictions need not be perfect to be valuable.

Readers are, of course, free to disagree with these assumptions, predictions, and estimates. However, they will have available the historical statistical data assembled in this study, on the basis of which they can make their own assumptions and derive their own predictions of the probable consumption and sources of energy in the 21st century.

The five forecasts of supply and use of energy during the 21st century do not include energy provided by wood, charcoal, and agricultural, animal, or other wastes, important as they may have been and still are in some regions of the world (Figure 1). It has been said that as late as about 1940, more energy was obtained worldwide from agricultural wastes than from hydrocarbon fuels, and that in 1959, farm wastes contributed between 10 and 15% of world energy requirements or three times that provided by wood. This may or may not be true. These sources of energy are the most poorly documented, and their consumption is very difficult to estimate. The percentage of the total world energy consumption supplied by wood, charcoal, and the various types of wastes, as shown in Figure 1, is highly conjectural. In the developed countries, both wood and farm wastes have progressively been replaced by more convenient sources of energy, but in many developing countries, wood, charcoal, and agricultural wastes still provide an important percentage of the energy requirements.

It has been mentioned earlier, but it may be worthwhile to state again, that the overall energy picture for the 21st century presented in this study is generally optimistic for three reasons:

- There will be plenty of energy sources to supply the demand, although their uneven geographic distribution will, in some cases, result in long-distance transport and price fluctuations, which may restrict at certain times their availability and affordability in certain countries and regions of the world.
- The growth rate of the world's population is decreasing and may, perhaps, nearly reach stability toward the end of the 21st century.
- Advances in present technologies will make possible a better, more efficient, and environmentally more benign supply and use of energy.

The demand for energy by the world's population, for the most part, will be satisfied. However, this optimistic outlook should not breed overconfidence. The future supply of energy is perhaps the most urgent question facing the world, and it is possible that a misguided choice of energy policies will create problems of many kinds, including critical environmental problems. Energy policies will need early and careful consideration and action.

2

Human Population and Energy Consumption: The Past

ABSTRACT

The rate of growth of the world's human population, alarmingly rapid during the last couple of centuries, slowed down appreciably in the late 20th century in the developing and developed countries alike. However, the human population continues to grow slowly in the developed countries and faster in the Third World, particularly in sub-Saharan Africa. The developing countries increased their share of the total world population from 67% in 1950 to 79% in 2000.

Total energy consumption, as a result, also increased faster in the developing than in the developed countries. The fossil fuels (oil, natural gas, and coal) provided the bulk of the energy consumed in the world in the last 50 years. Only hydroelectric power and, since 1950, nuclear power contributed additional, albeit smaller, amounts of energy.

Energy consumption per capita rose worldwide since the mid-20th century until the early 1970s. Since then, it has grown at a much lower rate everywhere, even remaining essentially unchanged in some countries. Energy consumption per capita in the developed countries in the last 50 years has been much higher (7–10 times higher) than that in the developing countries and 2–4 times higher in the United States than in the rest of the developed countries.

HISTORICAL DATA ON THE WORLD'S HUMAN POPULATION

The world's human population grew very slowly from the first appearance of *Homo sapiens* on Earth until about the 18th century; growth accelerated in the 18th and 19th centuries and became relentless in the 20th century. In the 1960s, the growth of the world's population had accelerated to more than 2.5% per year and to more than 3% per year in some developing countries (Figures 2, 3). As has been pointed out by many authors, it took more than a million years for the world's population to reach 1 billion, around the year 1815. The second billion was added in the next 112 years, by 1927, and the third in another 33 years, by 1960.

This accelerating rate of growth of the world's population has stabilized in the last 40 years; it took 14 years to reach 4 billion, in 1974, another 13 years to

reach 5 billion, in 1987, and 12 years to reach 6 billion in 1999. It appears that the growth of the world population is becoming linear instead of exponential, with the addition of each of the last 3 billion taking 12–14 years (Appendix Table 1; Figure 3). That the growth of the world's population is slowing down is also indicated by the fact that annual population additions, which increased consistently from 20 to 25 million per year in the 1930s and 1940s, peaked at 75 million per year in 1972–1974, decreased to 72 million per year in 1978, increased again to 87 million per year in 1988, and have decreased since then to 78 million per year by 2000 (Appendix Table 1; Figure 4). In the last 10 or 15 years, the worldwide trend has clearly been toward a slower rate of population growth.

Information on the total population of the world, however, is not sufficient for the purpose of the present study. Because of the great diversity of the

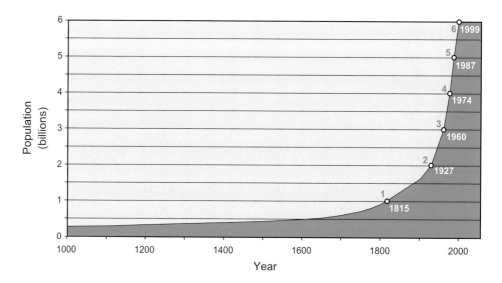

Figure 2. World population.

world's population, the rate of population growth has been far from uniform in the different countries or groups of countries of the world; although in the developed countries, the population has been growing very slowly, if at all, and in most developing countries, the population has been growing at a very fast rate, in some cases doubling every 20–25 years. Population growth in the different developing countries also varies widely. Similar disparities between the developed and the developing countries and between different developed and developing countries should be expected in the future.

It clearly makes sense, therefore, to acknowledge this diversity by not treating the whole world as one unit in attempting to make population projections into the 21st century. Regions are different not only from a demographic standpoint, but also in social, political, economic, and cultural terms. These differences clearly affect population trends. Aggregating certain different countries and regions would only serve to obscure the picture and make projections meaningless.

The growth rate and size of the population have, of course, a close connection with the consumption of energy. Energy is consumed by people, and in general, it should be expected that energy consumption will increase as the population increases. However, different countries consume energy at very different rates: as will be discussed in detail in the following chapters,

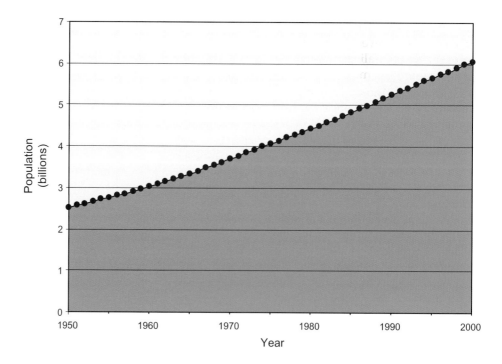

Figure 3. World population.

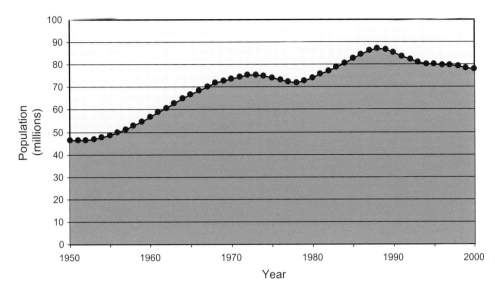

Figure 4. Annual increases to the world population.

the industrialized, developed countries have a much higher consumption per capita than the developing countries, and even in these two general groups of countries, consumption of energy varies considerably.

It is therefore necessary to estimate not only how many people will inhabit the Earth during the 21st century, but how the population will be distributed between the various countries or regions of the world and what the consumption of energy per capita has been and will likely be in these different countries and regions.

As a first breakdown, on the basis of the stage of their economic development, the countries of the world have been grouped into two major categories: the "developed" and the "developing" (Third World) countries.

This grouping into developed and developing countries is necessarily somewhat arbitrary. Some countries included in the developing category are in advanced stages of industrialization and considerably more developed than some of the countries included in the developed category. In addition, the division of the countries of the world into developed and developing countries is not precisely the same in all publications or in the product of the work of all authors or organizations. This inconsistency, however, is not believed to introduce serious problems or influence, in any important way, the results of this study. The population of some of the countries inconsistently assigned to one or the other of the two categories represents only a small part of the total population of the world.

The present study includes, as developed countries, the United States, Canada, Australia, New Zealand, Japan, Western Europe, Eastern Europe, and the 15 presently separate countries that made up the former Soviet Union (FSU). The developing countries include all the rest.

Western Europe comprises Austria, Belgium, Denmark, Finland, France, Germany, Greece, Iceland, Ireland, Italy, Luxembourg, the Netherlands, Norway, Portugal, Spain, Sweden, Switzerland, and the United Kingdom.

In Eastern Europe, those included are Albania, Bulgaria, the Czech Republic, the former East Germany, Hungary, Poland, Romania, Slovakia, and the former Yugoslavia (Serbia and Montenegro).

Adjustments had to be made for the reunification of Germany: East Germany was grouped with the rest of the countries of Eastern Europe before 1991 and as part of the reunited Germany since then (Appendix Tables 2, 3; Figures 6, 7).

Appendix Table 1 shows the total population of the world and its subdivision into the developed and the developing countries from 1950 to 2000. It shows that the portion of the population of the world represented by the developing countries increased from 67% in 1950 to 79% in 2000, indicating a much faster population growth of the developing countries than that of the developed countries. In fact, about 97% of the yearly growth of the world's population occurs now in the developing countries, and the population of 50 of the least developed countries will triple by the middle of the 21st century. However, in 1997, 51 developed countries containing 44% of the world's population had fertility rates (the number of children a woman bears during her lifetime) below population replacement (2.1 children per woman). Figure 5 shows graphically the contrasting rates of population growth in the developed and in the developing countries.

More specific information on the population of the developed countries is shown in Appendix Tables 2 and 3 and in Figures 6 and 7.

Appendix Table 2 and Figure 6 show the population of the developed countries from 1950 to 2000 broken down into nine countries or groups of countries: the

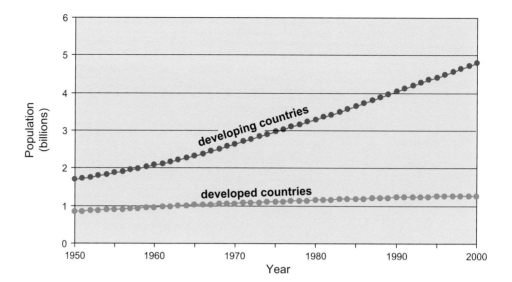

Figure 5. Population of the developed and developing countries.

United States, the total population of the developed countries minus the United States, Canada, Australia, New Zealand, Japan, Western Europe, Eastern Europe, and the FSU.

As can be seen in Appendix Tables 2 and 3 and in Figures 6 and 7, the rate of growth of the population

of the developed countries is very low, having slowed down appreciably and begun to flatten during the last 35 years. The population of the countries that made up the FSU and of the countries of Eastern Europe, in fact, has not changed appreciably in the 1990s. The population of the countries of Western

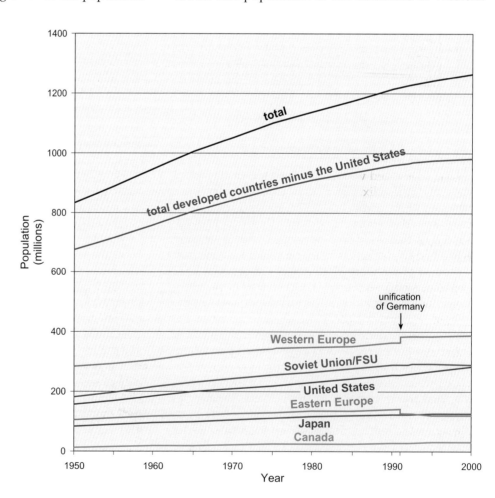

Figure 6. Population of the developed countries.

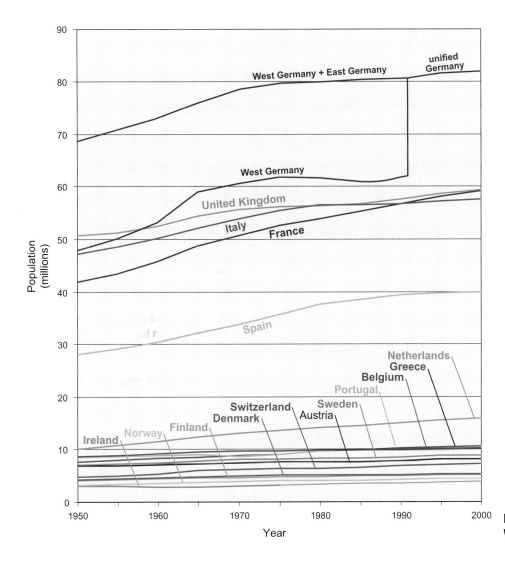

Figure 7. Population of the Western European countries.

Europe has changed very little during the last 30 years, as shown in Appendix Tables 2 and 3 and in Figures 6 and 7.

Fertility rates in the Western European countries averaged 2.6 in the early 1960s. They have been decreasing steadily since then to the present 1.4, below the replacement level. In Japan, the fertility rate is 1.3. Only three developed countries have a little more than doubled their populations since 1950: Canada, Australia, and New Zealand.

The reasons behind the stable or slow-growing population of the developed countries differ from country to country, as a result of a complex combination of social, economic, political, and cultural elements. The United States population, for example, has grown at a somewhat faster rate than that of the countries of Western Europe; half of this growth is caused by a large influx of immigrants who generally favor larger families. The United States fertility rate is now just a little less than 2.1.

The situation is quite different in the developing countries. Appendix Tables 1 and 4 and Figures 5 and 8 show the growth of the population of this group of countries from 1950 to 2000. During this 50-year interval, the population of the developing countries has increased from 1.706 to 4.865 billion, a 185% increase. However, exponential growth during the 1950s, 1960s, and 1970s seems to have slowed down and, since 1980, seemed to approach a more linear growth. The annual population additions in the developing countries peaked at 80 million people per year in 1988–1989 and has decreased to 75–76 million per year from 1993 to 2000.

Appendix Table 4 and Figure 8 show the historic growth of the population of China, India, Africa, and Latin America (including Mexico and the Caribbean countries). The combined population of these four countries or groups of countries has represented between 77% of the total population of the developing countries in 1950 and 75% in 2000. The range of their

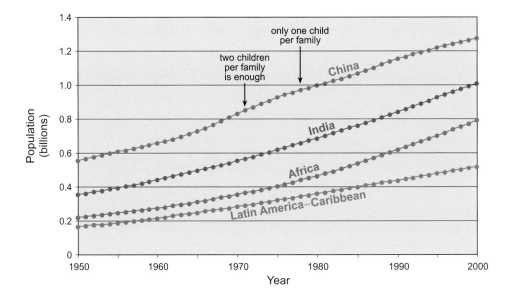

Figure 8. Population of the developing countries.

rates of population growth is representative of the growth rates of the remaining developing countries and can serve as the foundation for estimating future trends.

Population growth in the developing countries, as mentioned before, has been disparate: population has been growing at a much lower rate in some countries than in others, but in general, the rate of growth of the population of most developing countries is definitely slowing down. As shown in Appendix Table 4 and Figure 8, the rate of population growth in China, exponential during the 1950s and 1960s, has slowed down considerably after the government instituted in the 1970s a strict policy concerning the desired number of children per family: First, in 1971, the "two children per family is enough" policy; then, in 1978, the desired number of children per family reduced to one. As a result, the fertility rate in China decreased from 6 in the 1950s to 1.8 in the late 1990s, and the annual population additions, which reached 22 million in 1970, have decreased to a little more than 10 million in 2000. Similar reductions were attained in South Korea and Taiwan. The rate of population growth of Latin America has also shown a slowing trend: the annual population additions have remained at 7–8 million since 1970. The annual population additions in India increased from 7 million in the early 1950s to 17 million in 1992 and have remained at the 16–17 million level since then. Only the population of the countries of Africa (despite high mortality rates because of starvation, diseases spawned by malnutrition, hunger, and the acquired immune deficiency syndrome [AIDS]) continues to increase at an exponential rate, particularly in those of sub-Saharan Africa. The rate of the population growth in the rest of the devel-

oping world varied but showed, in general, a slowing trend.

The slowing down in the rate of population growth in many of the developing countries can be attributed to several reasons: 80% of the population of these countries lives now in countries where the government has felt in the last few years that the growth in population has been excessive and where steps are being taken to slow it down by initiating and implementing, and in some cases subsidizing, family-planning programs; by providing economic incentives for the choice of having fewer children; and by making available much improved public health care. As health improves and children are more likely to survive, families commonly show less desire to have many children. Better health, however, results in a decline in mortality, which undermines the desired reduction in population and favors an older demographic distribution.

Also contributing to the slowing down of the population growth of many developing countries has been better education and the improved status of women. Education, it has been said, is the best contraceptive. As a result, in the last 40 years, the percentage of couples in the developing countries using family planning has increased from 10% in the 1960s to 55% at the start of the 21st century. Fertility rates have started to drop in many Third World countries; their women are now having an average of 3.8 children, down from 6 in the mid-1960s. Decline of fertility during the last 30 years is now the rule instead of the exception in these countries. Only in sub-Saharan Africa has the fertility rate remained at about 6.5 children per woman during the last 30–35 years. However, in all developing countries, the average number of children a woman bears during her lifetime is still

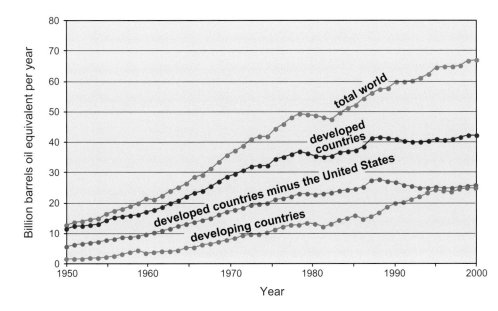

Figure 9. Energy consumption: world, developed and developing countries.

well above 2.1 children per woman, the number that would keep the population roughly stable.

Finally, urbanization and industrialization also have had a strong influence on the reduction of the rate of population growth of the developing countries. City dwellers generally have more of a desire, as well as the necessity, for fewer children than rural families. At the start of the 21st century, half of the world's population (more than 3 billion people) are living in megacities, and only 3 of the 20 largest cities in the world—Tokyo, New York, and Los Angeles—are in the developed world.

At the dawn of the 21st century, the population of the world is clearly growing at an appreciably lower rate. However, it is continuing to grow. Whereas each mother in the developing as well as in the developed countries is having fewer children, there are still more mothers.

HISTORICAL DATA ON THE WORLD'S ENERGY CONSUMPTION

As in the case of world population, statistical data on past world energy consumption were compiled for the world as a whole and for the developed and the developing countries, as well as for several countries or groups of countries in these two broad categories. The countries placed in the developed and developing categories are the same as those used in the discussion of the world's population.

Appendix Table 5 and Figure 9 show the consumption of energy from 1950 to 2000 for the world as a whole and for the two groups of developed and developing countries.

World energy consumption can be seen to have increased steadily from 1950 to 2000, interrupted only by the two "oil shocks" of 1973–1974 and 1979–1983. World energy consumption in 2000 was almost five-and-a-half times as much as in 1950.

The energy consumption of the developed countries also increased from 1950 to the late 1990s. Their energy consumption, however, flattened at about 40 billion BOE per year after 1989–1990, reflecting the collapse of the Soviet Union. A possible economic recovery of the former Soviet Union and the countries of Eastern Europe in the late 1990s could be responsible for a small increase in the total energy consumption of the developed countries during the late 1990s.

The developing countries, however, were not as strongly affected by the "oil shocks" or the collapse of the FSU. Their overall energy consumption has increased more than 20 times during the last 50 years. In 1950, they accounted for about 9% of the energy consumed in the world. This percentage has increased steadily to 37% in 2000.

Appendix Table 6 and Figure 10 show the energy consumption from 1950 to 2000 of the developed countries individually or in groups of countries: the United States, Canada, Japan, Western Europe, Eastern Europe, and the former Soviet Union. Appendix Table 6 also shows the energy consumption of the developed countries minus the United States, which is graphically shown in Figure 9.

The oil shocks of 1973–1974 and 1979–1983 are clearly reflected in the energy consumption of the United States and Western Europe. The former Soviet Union, the Eastern European countries, and Canada were not affected, as they did not depend on the oil from the producers of the Persian Gulf area.

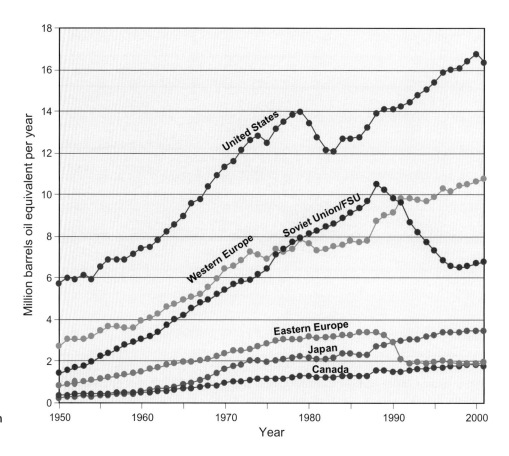

Figure 10. Energy consumption of the developed countries.

The collapse of the former Soviet Union, on the other hand, was responsible for the sharp drop in the consumption of energy of the countries of the FSU and those of Eastern Europe. Also responsible for the sharp drop in the energy consumption of the Eastern European countries was the unification of Germany in 1991 and the transfer of East German data from Eastern to Western Europe.

Appendix Table 7 and Figure 11 show the consumption of energy from 1950 to 2000 in China, India, Africa, and Latin America. These four countries or groups of countries, as mentioned earlier, have made up between 75 and 77% of the total population of the developing countries during the years of record.

As noted before, the consumption of energy of the developing countries increased 20-fold from 1950 to 2000. However, although their population increased during this time span from 67 to 79% of the world's population, their share of the world's energy consumption represented only 9% in 1950 and 37% in 2000. China, India, Africa, and Latin America accounted for 73–75% of the energy consumed by the developing world during the early 1950s and 56–58% in the 1990s.

PAST SOURCES OF ENERGY

Appendix Table 8 and Figure 12A and B show the world's consumption of energy by source since 1950. Figure 12A shows the amounts of the contribution of each source of energy, and Figure 12B shows their percentages of the total energy consumption in the world. They both clearly bring out that fossil fuels (oil, natural gas, and coal) have provided the bulk of the energy consumed in the world in the last 50 years, as they had done since the beginning of the 20th century. Only hydroelectric power and, since the 1970s, nuclear power contributed additional, albeit small, amounts of energy. The share of hydroelectric power has remained essentially unchanged at about 5–7% of the total world energy consumption, whereas the share of nuclear power increased slowly from its first commercial use in the 1960s to about 6% in 1990. It has remained at that level since then.

Oil and coal provided more than 80% of the energy consumed worldwide in the 1950s and early 1960s. The share of natural gas, however, has increased steadily since 1950 and now stands at about 22%. Oil's share, however, has decreased from a peak of 49% in 1974 to about 37% at present, as has coal's share, from about 55% in 1950 to about 26% now.

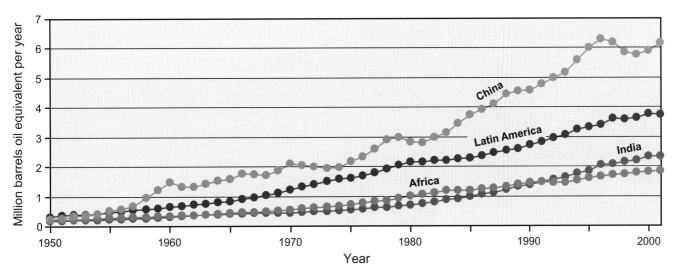

Figure 11. Energy consumption of the developing countries.

Other sources of energy (geothermal, shale oil, solar [photovoltaic], and wind) have so far made only an insignificant contribution to the energy consumed in the world. Biomass (wood, and combustible agricultural, animal, and municipal waste) has not been included as a source of energy, because although it is the oldest source of energy and is still in wide use in some developing countries (see Figure 1), it is the most poorly documented of all sources of energy.

HISTORICAL DATA ON ENERGY CONSUMPTION PER CAPITA

The result of combining historical data on human population and on energy consumption in the past half century to obtain energy consumption per capita (for the world as a whole, for the developed and developing countries, and for some countries or groups of countries) brings up the following significant overall trends (Appendix Tables 9–24; Figure 13):

1) The total world energy consumption per capita, expressed in barrels of oil equivalent per person per year (BOE/p/yr), doubled from about 5 BOE/p/yr in 1950 to 10 BOE/p/yr in the early 1970s, remaining in the 10–11-BOE/p/yr range since then (Appendix Table 9).

2) Energy consumption per capita has remained, as expected, much higher in the developed countries than in the developing countries, but this difference has diminished from 15–16 times higher in the 1950s to less than half that in the late 1990s (Appendix Tables 10, 11). This decrease in the dis-

parity in per-capita energy consumption between the developed and developing countries in the decade of the 1990s is in part caused by the collapse of the former Soviet Union in 1991.

3) The per-capita consumption of energy in the developed countries, like that of the world as a whole, increased relatively slowly from 1950 to the early 1970s and has remained fairly constant since then: 14 BOE/p/yr in 1950, 29–30 BOE/p/yr in the early 1970s, and 30–33 BOE/p/yr in the 1980s and 1990s (Appendix Table 10).

4) In the developing countries, per-capita energy consumption has been extremely limited in the past half century. It increased slowly over this period but showed some actual year-to-year decreases during the early 1980s and late 1990s (Appendix Table 11).

5) Individual countries or groups of developed countries showed a range in per-capita energy consumption.

The countries of Western Europe, New Zealand, and the total of the developed countries minus the United States had patterns of per-capita energy consumption similar to those of the whole of the developed countries, albeit at lower levels.

The energy consumption per capita of the former Soviet Union followed that of the countries of Western Europe from 1950 until the early 1970s but continued to increase at about the same rate during the late 1970s and 1980s probably because of the generally recognized inefficiency of energy use in the former Soviet Union. Per-capita energy consumption reached 36 BOE/p/yr in the late 1980s. After its collapse in 1991, the per-capita consumption of energy in the countries of the

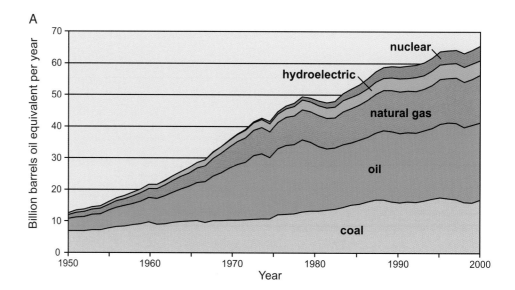

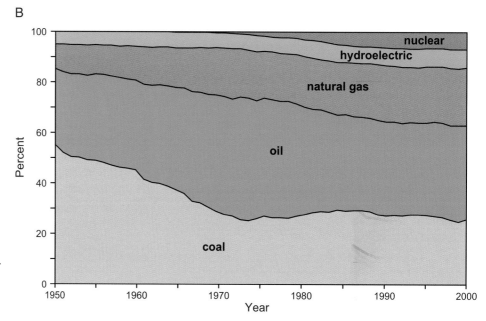

Figure 12. (A) World energy consumption by source. (B) World energy consumption by source (percent).

FSU plummeted to 22–23 BOE/p/yr in the late 1990s but showed a slight increase in the last 2 years of the century.

The pattern of per-capita energy consumption in the countries of Eastern Europe more closely followed that of the Western European countries until 1990. Then, as in the case of the former Soviet Union, per-capita energy consumption dropped sharply from about 24 BOE/p/yr in the late 1980s to 16 BOE/p/yr in the early 1990s and remained at that level during the rest of the 1990s.

The beginning of a possible economic recovery in the FSU and Eastern Europe may be presaged by the leveling and slight increase in the consumption of energy and per-capita consumption during the closing years of the 1990s in these countries.

Japan's per-capita energy consumption recovered from the effects of the Second World War during the 1950s, 1960s, and early 1970s from a low of about 3 BOE/p/yr in 1950 to 18 BOE/p/yr in the early 1970s. It has increased at a much slower rate since then, reaching about 27 BOE/p/yr in the late 1990s.

The largest departure from the level of per-capita energy consumption in the developed countries is represented by the United States and, in the last 20 years, by Canada.

The United States is by far the most lavish user of energy in the world. In 1950, with a little more than 6% of the population of the world, it consumed almost 46% of the total energy consumed in the world. This high percentage has progressively decreased since

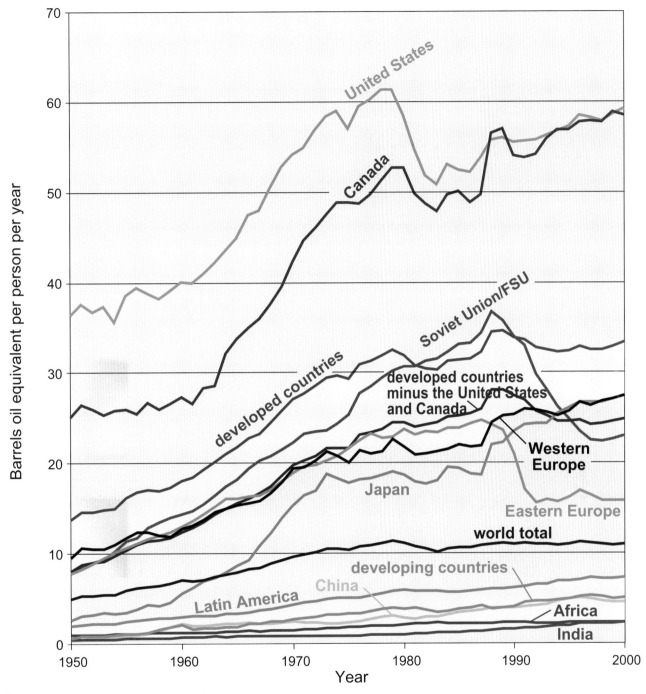

Figure 13. Per-capita energy consumption.

then, but in 2000, with 4.7% of the world's population, the United States still used a little more than 25% of the energy consumed in the world.

The large amounts of energy consumed in the United States are reflected in the per-capita energy-consumption levels. Per-capita energy consumption increased from 36 BOE/p/yr in 1950 to 61 BOE/p/yr in the late 1970s, dropped to 51–52 BOE/p/yr in the early 1980s, and has

increased slowly since then to reach about 58–59 BOE/p/yr in the late 1990s. The ups and downs in the late 1970s and early 1980s reflect in the United States, as in many other countries, the oil shocks during those years.

The high levels of per-capita energy consumption in the United States were three to four times those of all of the rest of the developed countries and even of those of

Western Europe during the 1950s and early 1960s. Despite significant increases in the consumption of energy attained in other developed countries since 1950, the per-capita consumption of energy in the United States since the mid-1960s has remained 2 to 3 times as high as that of the rest of the developed countries and has varied from 12 to 20 times that of the total of the developing countries.

Canada's per-capita energy consumption during the 1950s, 1960s, and 1970s increased from 70 to 85% of that of the United States and reached parity during the 1990s at 55–59 BOE/p/yr.

6) The per-capita energy consumption of the combined developing countries is extremely low and has seen little increase in the last 50 years. It increased very slowly from 0.68 BOE/p/yr in 1950 to 3.86 BOE/p/yr in 1980 but reached only about 5 BOE/p/yr in the next 20 years and remained essentially unchanged during the second half of the 1990s. China followed a very similar path, increasing from 0.45 BOE/p/yr in 1950 to 4.63 BOE/p/yr in 2000. India and Africa had even lower rates of per-capita energy consumption: from 0.5 and 0.94 BOE/p/yr, respectively, in 1950 to 2.32 and 2.3 BOE/p/yr, respectively, in 2000. Latin America had somewhat higher per-capita energy consumption: from 1.93 BOE/p/yr in 1950 to a little more than 7 BOE/p/yr in the late 1990s but still much lower than the average of the developed countries.

7) The above historical data about the per-capita consumption of energy in the world and in certain countries or groups of countries clearly indicate that per-capita energy consumption rose worldwide since 1950 until the early 1970s, and that since then, it has increased everywhere at a much lower rate or remained essentially unchanged. The countries of the FSU and Eastern Europe are the exception because of the shattering events that followed the collapse of the former Soviet Union.

This marked slowing down of the per-capita consumption of energy since the early 1970s is believed to be caused by definite improvements in the efficiency of energy use brought by technological advances and perhaps by efforts to conserve energy and protect the environment. As will be discussed later, the future prolongation of this trend will have a fundamental influence on the estimation of energy consumption in the 21st century. Inhabitants of both the developed and the developing countries seem not to be increasing their individual consumption of energy, but their numbers will steadily increase during the 21st century.

SOURCES OF INFORMATION

The main sources of data on energy consumption were the publications of the United Nations [*World Energy Supplies, 1950–1978* (United Nations Department of Economic and Social Affairs, Statistical Office, 1976–1979); *Yearbook of World Energy Statistics, 1979–1981* (United Nations Department of Economic and Social Affairs, Statistical Office, 1981–1983); *Energy Statistics Yearbook, 1982–2001*(United Nations, Department of Economic and Social Affairs, Statistical Office/Statistics Division, 1984–2002)]; those of the Energy Information Administration of the U.S. Department of Energy [*International Energy Annual* (U.S. Department of Energy and Energy Information Administration, 1979); *International Energy Outlook* (U.S. Department of Energy and Energy Information Administration, 1982); British Petroleum's (1988–2003) *Statistical Review of World Energy*, and PennWell's *Oil & Gas Journal Data Book*.

The data obtained from these different publications did not always precisely correspond, but as mentioned in the Preface, the magnitude of the discrepancies was, in all cases, within the range of accuracy needed for the present study. The fact that the amounts of energy consumption were given in different units in different publications (metric tons of oil equivalent [mtoe] and metric tons of coal equivalent [mtce], both units of weight; barrels of oil equivalent [BOE] and cubic meters [m^3], both units of volume; quadrillions of British thermal units [Quads], a unit of heat; and joules, a unit of energy) may account for some of the discrepancies, because different organizations apparently use different factors to convert from one unit to another. To convert from metric tons of oil equivalent to barrels of oil equivalent, for instance, the conversion factor used by different organizations ranges from 7.2 to 7.57 BOE per mtoe, evidently depending on the estimated average specific gravity of the oil produced in the world or in different regions of the world.

In the tables and figures of this study, the consumption of energy has been expressed in barrels of oil equivalent (BOE). The following conversion factors were used:

$$1 \text{ mtoe} = 7.33 \text{ BOE}$$
$$1 \text{ mtce} = 5.04 \text{ BOE}$$
$$1 \text{ Quad} = 172.4 \text{ BOE}$$
$$1 \text{ joule} = 163.4 \times 10^{-12} \text{ BOE}$$

Some adjustments also had to be made because the amounts of the different sources of energy obtained from different publications sometimes did not add up to 100% of the total amount of energy consumed.

3

Human Population and Energy Consumption: The Future

ABSTRACT

The growth rate of the world's human population decreased during the late 20th century. It should be expected to continue to decrease during the 21st century. Three projections are favored that show the world's population increasing from 6 billion at the end of the 20th century to either 10, 11, or 12 billion by 2100. In all three projections, it is estimated that the percentage of the world's population living in the developing countries will reach about 90% in 2100.

Energy consumption per capita in the 21st century is predicted to decrease in the developed countries, probably to increase moderately in the developing countries, and, as a result, to increase in the world as a whole. Given that the developing countries will contain the great majority of the world's people during the 21st century, the correct estimation of their future energy consumption will be critical in forecasting the total energy consumption in the world. Even a modest increase in the energy consumption per capita in the developing countries will result in their total energy consumption surpassing that of the developed countries during the first half of the 21st century; by 2100, the developing countries could be consuming three times as much energy as the developed countries. Will this be possible?

ESTIMATION OF THE WORLD'S HUMAN POPULATION DURING THE 21ST CENTURY

The rate of growth of the world's human population during the 21st century has been the subject of extensive and factious controversy. Equally controversial has been the maximum size of the population that the world would be able to support.

Concern about the growth of the world's population apparently goes back to Niccolo Machiavelli (1469–1527) and Sir Walter Raleigh (1552–1618) (Neurath, 1994). However, this issue was most forcefully brought to the attention of the world by the publication in 1798 of Thomas Robert Malthus' *An Essay on the Principle of Population* and, about 174 years later, by Meadows et al.'s (1972) *The Limits of Growth*.

Malthus (1798, and following editions) claimed that man, if not restrained from time to time by Mother Nature with what he called "positive checks," such as famine, war, or pestilence, has a tendency to multiply faster than he can increase his food supply. Meadows et al.'s (1972) book, sponsored by the Club of Rome, similarly warned that if mankind were to continue to grow at the rapid rate at which it had been growing during the previous 200 years, a collapse of the life-supporting system on Earth would have to be expected.

Both Malthus and Meadows et al. treated the growth of the world's population on a global basis, without reference to the differences between certain countries or special regional groups of countries. For this, they were justifiably criticized. Because of the great heterogeneity of the growth patterns of the different regions and countries of the world, it clearly does not make any sense to base the reasoning used in estimating the world's future population growth on assumptions related to the world as a whole. This deficiency was corrected in the Second Club of Rome Report, *Mankind at the Turning Point* (Mesarovic and Pestel, 1974).

Meadows et al., as well as Mesarovic and Pestel, indicated that the breakdown of the life-supporting system could be avoided only if both the growth of the human population and the per-capita consumption of nonrenewable resources, including sources of energy, were to slow down considerably.

Malthus' book first, and then Meadows et al.'s *The Limits of Growth* became, during their time, the subjects of exhaustive debate, pro and con, in hundreds of articles in newspapers, magazines, journals, and books the world over. Meadows et al. updated their work in 1992, and interesting contributions to this controversy have been provided during the last 20 years or so by Ehrlich (1968), Marchetti (1978), Ehrlich and Ehrlich (1990), Keyfitz and Flieger (1990), Lutz (1994), Neurath (1994), and many others.

Several specific projections of the growth of the world's population during the 21st century have been ventured during the last two or three decades. They range widely from those that predict an implausible unrestrained continuation of the high rates of growth during the early and middle 20th century with the world's population reaching 20 or 25 billion before the end of the 21st century to those (equally implausible and unrealistic) that optimistically forecast that the population of the world will grow at an increasingly slower rate and stabilize at about 7 billion during the second half of the 21st century.

Current information indicates that the rate of growth of the world's population is slowing down. As mentioned before, the human population grew exponentially for 200 or 300 years before the late 20th century, but in the last three decades, the growth has been linear (Figures 2, 3). The annual rate of growth peaked in the late 1960s at about 2.1%, a rate that would double the world's population in 33 years. The rate decreased during the 1970s but increased again in the 1980s, reaching another peak of about 1.75% in the late 1980s. It has decreased progressively since then to 1.4% in the late 1990s. The annual population additions, naturally, show a similar pattern (Appendix Table 1; Figure 4).

The late 1980s may indeed have been a turning point in the world's demographic trends; the start of an increase in the global awareness that the population of the world cannot continue to grow indefinitely, and that there must be limits to its growth.

To assume that the decrease of the growth rate of the world's population will continue to some degree during the 21st century is reasonable in view of the previously discussed significant declines of the fertility in most countries of the world during the last 30 or 40 years, a decrease that has surprised most demographers. The countries of sub-Saharan Africa are the most notable exception.

This steady decline in fertility throughout most of the world has probably been the reason why projections of the world's population during the 21st century put forth in the last two decades by several organizations, notably the United Nations, the World Bank, and the U.S. Bureau of the Census, have assumed progressively lower future rates of growth and have predicted lower numbers for the population of the world at the end of the century. Their central, base, or medium variant cases have the world population reaching 8–13 billion in the year 2100 and stabilizing at between 11 and 14 billion sometime during the 22nd century.

It would be unrealistic to believe that the population of the world will grow beyond these limits, or that it will stabilize sometime during the 21st century. Population growth takes a long time to slow down just as it takes a long distance to slow down and stop a fully loaded oil supertanker. Demographic momentum, also referred to as population momentum, determined largely by the age distribution of the present population of the world, particularly that of the developing countries with a large proportion of their population in the reproductive years, will make unlikely a persistent and rapid decrease of the growth rate of the world population.

For the purpose of this study, I have chosen three possible projections of the growth of the world's human population: they predict that the size of the population will reach 10, 11, or 12 billion by the year 2100 (the low, middle, and high cases, respectively), the order of magnitude favored by most recent estimates (Appendix Table 25; Figure 14). In the 10-billion case, the world's population is assumed to stabilize at about this level around 2120; in the 11-billion case, the world's population is expected to stabilize in 2160–2170 at about 11.5 billion; and in the 12-billion case, leveling off will occur at about 13–13.5 billion late in the 22nd century. Three projections are used because when dealing with the future growth of the world's population, it is necessary to think in terms of ranges and avoid single numbers whenever possible.

The rate at which the population of the world has been growing is not the same everywhere. As discussed before, the population of the developing countries has been growing at a much faster rate than that of the developed countries. While the population in some developing countries is doubling every 20 years,

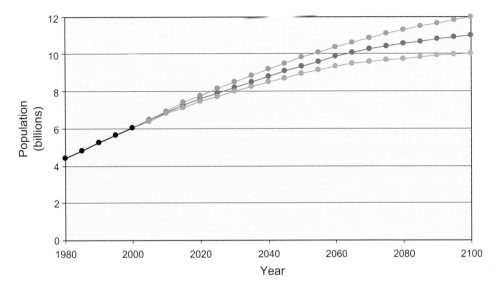

Figure 14. World population projections.

in some countries of Western Europe, the population has been essentially stable or has even decreased slightly during the last few years (see Figures 5–8). Because the population of the developing countries is growing so much faster than that of the developed countries, the percentage of the world's population living in the developing countries has been increasing steadily and is expected to continue to do so during the 21st century. In 1950, the Third World population represented 67% of the world in total; this percentage increased to about 79% in 2000 and is estimated to reach 90% in the year 2100 (Appendix Table 26; Figure 15).

Because the developing countries will contain the bulk of the world's population during the next century, it goes without saying that correctly estimating the future consumption of energy in these countries will be critical in arriving at a realistic assessment of the possible total world consumption of energy in the 21st century.

Appendix Table 26 and Figure 15 show the percentage of the world population in the developed and the developing countries from 1950 to 2000 and the estimated percentages during the 21st century. The percentage for the developing countries has been broken

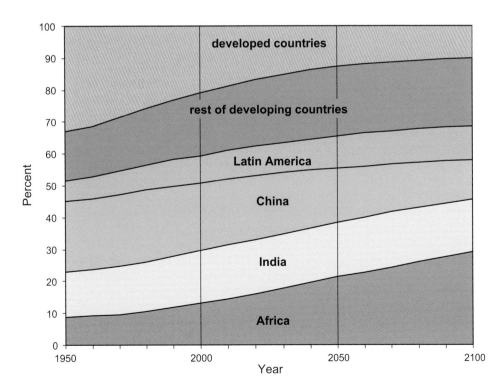

Figure 15. Percent of the world population in the developed and developing countries.

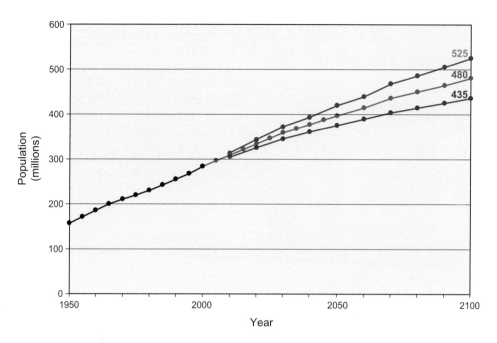

Figure 16. United States population projections.

down to show estimates for Africa, India, China, Latin America, and the rest of the developing countries. Africa is shown to increase its share of the world's population from 13% in 2000 to 29% in 2100; India will represent 16–17% throughout the 21st century; China will decrease its share from 21% in 2000 to 12.5% in 2100, and Latin America's share will increase from 8.6% in 2000 to 10.5% in 2100. The rest of the developing countries represent 19.7% of the world's population in 2000 and will increase their share to 21.5% in 2100. These percentages reflect the historic trends of the population growth during the 20th century of these countries or groups of countries as discussed in the previous section of this study: the rapid growth of the population in Africa, particularly in sub-Saharan Africa; the more moderate growth in India and Latin America; and the slowdown of the population growth in China as a result of the government's demographic policies.

Appendix Tables 27–29 show the projected population of the world in the 21st century for the developed and the developing countries and for the countries or groups of countries of the Third World for the three possible cases used in this study.

The population of the United States, which grew from 158 million in 1950 to 283 million in 2000, at a somewhat faster rate than that of the other developed countries mainly because of significant additions caused by immigration (see Appendix Table 2; Figure 16), is forecasted in the middle case to increase to 400 million in 2050 and approaching 500 million in 2100 (Appendix Table 30; Figure 16). In 1950, the population of the

United States represented 19% of the population of the developed countries. This percentage increased to 22.4% in 2000, and it has been estimated that in the middle case, it will reach about 33% in 2050 and 43–44% in 2100.

Appendix Table 28 and Figure 17 show the projected total population of the world and the share of the total for the developed and developing countries for the middle case. The increase of the world's population is projected to take place almost entirely in the developing countries. The population of the developed countries is estimated to decrease from 1.26 billion in 2000 to 1.1 billion by 2100 and that of the developing countries to grow from 4.79 to 9.9 billion. In 2100, Africa's population is estimated to reach 13.9 billion; India will reach 1.81 billion, having replaced China as the most populous country in the world by 2050; China's population is shown to reach a peak of about 1.6 billion in about 2040 and then decrease during the second part of the 21st century to 4 billion in 2100; Latin America will increase its population from 520 million in 2000 to 1.15 billion in 2100; and the rest of the developing countries will reach 2.36 billion by the end of the century.

Appendix Tables 27 and 29 show world population projections for the low and high cases, respectively.

The future evolution of the world's population will, of course, be different from these projections. This is unavoidable; it is very difficult to model uncertainty. Unpredictable events (demographic, socioeconomic, cultural, political, medical, and climatic) will take place. Some may restrain the rate of growth of the world's

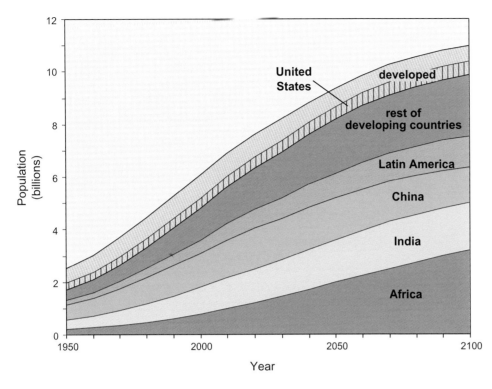

Figure 17. World population projection: Middle case.

human population; others may increase it; combined, they may compensate each other, making the favored projections reasonably realistic. Not considered, for example, is the possible massive migration from the developing to the developed countries, something very likely to take place, perhaps unavoidable, but extremely difficult to predict and quantify, something to which most governments and individuals in the developed countries have not yet given serious and realistic consideration. How long will the restraints on the population growth in China, effective in the last three decades, remain in effect? What long-range effect will AIDS have on the population of some countries? How extensive will the migration from rural to urban areas be?

If these projections are of the proper order of magnitude, as I hope, Malthus' positive checks may be avoided, at least in their most drastic and widespread form. Wars, old and new pestilences, famines, and massacres, still painfully evident in some parts of the world today, will undoubtedly recur, but if the people of the world persevere in their efforts to slow down the growth of the population, these banes of humanity will be kept from spreading over large regions of the Earth; they can, in many cases, be blamed on crowding, scarcity of food, fresh water, and energy sources, all of them commonly a direct consequence of overpopulation.

ESTIMATION OF THE ENERGY CONSUMPTION PER CAPITA DURING THE 21ST CENTURY

Introduction

The wide differences in the past consumption of energy per capita among the various countries and groups of countries, particularly between the developed and developing countries (Appendix Tables 9–24; Figure 13), make it essential to estimate separately the probable consumption of energy per capita in these various countries and groups of countries to estimate the worldwide consumption of energy in the 21st century.

Appendix Tables 9–24 and Figure 13 indicate that

1) Consumption of energy per capita from at least 1950 to the late 1970s increased progressively, to a greater or lesser extent, in all countries and groups of countries for which information was computed:

- The world as a whole
- The developed and the developing countries
- The United States
- Canada
- The developed countries minus the United States and Canada
- Australia
- New Zealand

- Japan
- Western Europe
- Eastern Europe
- The former Soviet Union (FSU)
- China
- India
- Africa
- Latin America

Since the late 1970s, the energy consumption per capita in all these countries and groups of countries has either leveled off or grown at a much slower rate. In the FSU and the Eastern European countries, the beginning of the slowdown in the rate of energy consumption per capita did not take place until the late 1980s.

2) Based on their rate of energy consumption per capita, the countries and groups of countries examined in this study fall into three categories, clearly distinguished from each other (Appendix Tables 9–24; Figure 13):

- The United States and Canada, with by far the highest rate of energy consumption per capita (59 BOE/p/yr in 2000)
- The remaining developed countries: Australia, New Zealand, Japan, Western Europe, the FSU, and the Eastern European countries, with intermediate rates of energy consumption per capita (25 BOE/p/yr in 2000)
- The developing countries, with very low energy consumption per capita (5 BOE/p/yr in 2000); among these countries, India and the African countries have had the lowest rate of energy consumption, and Latin America has had the highest (2.3 and 7 BOE/p/yr, respectively, in 2000)

3) The rate of energy consumption per capita of the world as a whole is closer to that of the developing countries, as would be expected, because these countries account for the major part of the human population of the world: 67% in 1950, increasing to 79% in 2000.

In estimating probable rates of consumption of energy per capita in the 21st century, therefore, it would be most significant to make estimates for (1) the United States and Canada, as the most lavish users of energy; (2) the remaining developed countries; (3) the total of the developed countries; (4) the total of the developing countries; and (5) China, India, Africa, and Latin America individually. Combining these partial estimates, estimates have also been made for the world as a whole.

The assumptions made in estimating these future rates of energy consumption per capita, particularly those for the developing countries, are of fundamental importance in determining the total amount of energy consumption in the 21st century. These assumptions must therefore be discussed in some detail.

The United States and Canada

Appendix Tables 12 and 13 and Figure 13 show that the past rate of energy consumption per capita in the United States increased from about 36–37 BOE/p/yr in the early 1950s to as much as 61 BOE/p/yr in 1978–1979. The United States per-capita consumption decreased to about 50 BOE/p/yr in the mid-1980s and has leveled off at about 55–60 BOE/p/yr in the 1990s.

Canada's annual energy consumption per capita was lower than that of the United States in the 1950s, 1960s, and 1970s, increasing from 25 BOE/p/yr in 1950 to 53 BOE/p/yr in 1980. It decreased to about 50 BOE/p/yr in the mid-1980s and, as in the United States, leveled off at about 55–60 BOE/p/yr during the 1990s.

For the purpose of this study, it is optimistically predicted that Americans and Canadians will progressively become increasingly efficient and frugal in their use of energy. Two cases have been considered: one that assumes that by 2100, the energy consumption per capita in the United States and Canada will have decreased to 40 BOE/p/yr and a second that assumes a decrease to 30 BOE/p/yr. Contributing to such a decrease would be better technology, more efficient use of electricity, more efficient transportation vehicles, changes in traditional patterns of behavior, and the inevitable increase in the price of energy.

The Remaining Developed Countries

Appendix Table 14 and Figure 13 show that the past rate of energy consumption per capita in the developed countries other than the United States and Canada increased from 8 BOE/p/yr in 1950 to almost 25 BOE/p/yr in 1980, peaking at 28 BOE/p/yr in 1988–1989. After that, reflecting the collapse of the communist regimes in the FSU and Eastern Europe, the consumption of energy per capita of the developed countries other than the United States and Canada has decreased, leveling off at 24–25 BOE/p/yr in 1990s.

Individual countries of this group have shown somewhat different patterns in the consumption of energy per capita. Australia's rate has been consistently higher than that of the rest of the developed countries other than the United States and Canada (Appendix Table 15) but lower than that of those two countries; New Zealand has had rates of energy consumption more in line with the North American developed countries (Appendix Table 16).

The consumption of energy per capita in Western Europe has increased more or less gradually from 9.5 BOE/p/yr in 1950 to almost 22.5 BOE/p/yr in 1979, remaining at about this level during the 1980s and gradually increasing again to 26–27 BOE/p/yr during the late 1990s (Appendix Table 18; Figure 13).

As mentioned previously, it took Japan more than two decades (the 1950s, 1960s, and early 1970s) to recover from the effects of the Second World War. Its consumption of energy per capita in 1950 was only 2.6 BOE/p/yr, but by the mid-1970s, it was only a little lower than that of the rest of the developing countries other than the United States and Canada. In the late 1990s, Japan's energy consumption per capita was comparable to that of the countries of Western Europe (Appendix Table 17; Figure 13).

The former Soviet Union and the countries of Eastern Europe had rates of consumption of energy per capita comparable to those of the other developed countries excluding the United States and Canada from 1950 until 1991, somewhat higher in the former Soviet Union (unaffected by the oil shocks of the 1970s) and somewhat lower in Eastern Europe. After the collapse of the former Soviet Union, the energy consumption per capita of both groups of countries dropped sharply during the early 1990s, leveled off during the mid-1990s, and appeared to start a recovery in the late 1990s. By 2000, the rate in the Eastern European countries was close to 16 BOE/p/yr (Appendix Table 19; Figure 13), and that of the FSU was about 23 BOE/p/yr (Appendix Table 20; Figure 13).

Despite steady increases in their energy consumption per capita during the 1950s and 1960s, the developed countries other than the United States and Canada have used energy during the last 20 years at a rate per capita that is only about 60–65% of the rate in the two North American countries. This considerable difference in their past rates of energy consumption per capita makes it necessary, as mentioned above, to develop separate estimates of energy consumption per capita during the next century for (1) the United States and Canada and (2) the other developed countries. It is not essential, however, to consider separately certain other developed countries or groups of countries, because the differences in their rates of energy consumption per capita are relatively minor and certainly within the degree of accuracy needed for the present study.

As in the case of the United States and Canada, it can be predicted that the rest of the developed countries will become more efficient in their consumption

of energy during the 21st century. However, because these countries, as a whole and individually, are and have always been considerably more efficient in the consumption of energy than the United States and Canada, the decrease in their rate of energy consumption per capita will necessarily be less sharp than that predicted above for the two North American countries. It is estimated that the energy consumption per capita of the developed countries other than the United States and Canada, which decreased from about 28 BOE/p/yr during the late 1980s to about 25 BOE/p/yr in 2000, will decrease to 20 BOE/p/yr by the year 2100. The sharp drop in the energy consumption per capita in the FSU and the Eastern European countries after the collapse of the communist system is probably temporary; it is presumed that these countries, after they overcome the present period of political and economic instability and go through the difficult restructuring from a centrally planned to a free-market economic system, will probably return to consuming energy at a per-capita rate similar to that of the other non-North American developed countries. The beginning of such a recovery has been evident in the late 1990s.

The Total of the Developed Countries

The past rates of energy consumption per capita for the developed countries, including the United States and Canada, were computed as a point of reference and as a test of the corresponding computations for the United States, Canada, and for the remaining developed countries. They are shown in Appendix Table 10 and Figure 13. As should be expected, the rates of energy consumption per capita for the total of the developed countries are higher than those of the group of developed countries that exclude the United States and Canada.

The consumption of energy per capita in the total of the developed countries increased steadily from 13.7 BOE/p/yr in 1950 to about 32.5 BOE/p/yr in 1979 and has remained at 31.5–34.5 BOE/p/yr since then.

Combining the assumptions made above for the energy consumption per capita in the 21st century for the United States and Canada and for the remaining developed countries, it is estimated that the energy consumption per capita for the total of the developed countries will decrease from the present 33–34 BOE/p/yr to between 25 and 30 BOE/p/yr in 2100.

The Developing Countries

For successful estimates of the consumption of energy in the world during the 21st century, the correct estimation of the rate of energy consumption per capita in the developing countries during that period is critical, because by the year 2050, about 87% of the world's population will live in the developing countries, probably reaching 90% in 2100. The amount of energy consumed by these countries undoubtedly will be the most significant factor in the energy equation of the 21st century. How much energy the Third World countries will consume or be able to afford to consume also will be crucial in determining their fate, i.e., their standard of living and their economic and social activities. The rate of energy consumption per capita of the Third World countries (in total, in certain groups, and individually) is, however, extremely difficult to estimate, if for no other reason than that it is not easy to estimate the future growth rate of the population of many of these countries.

Some of the countries included in this study under the general label of developing countries undoubtedly will do much better than others; some, like South Korea and Taiwan, already do; others are expected to improve their economies during the next century. China and some South American countries have commonly been mentioned in this respect.

The historical rate of energy consumption per capita of the total of the developing countries, as previously discussed, has been extremely low; it was about 0.7 BOE/p/yr in 1950 and, in the next 50 years, increased to only about 4 BOE/p/yr. India and the African countries, particularly the sub-Saharan countries, had per-capita energy consumption below the average for the developing countries; Latin America was above the average, whereas China's energy consumption per capita has been close to the average (Appendix Tables 21–24; Figure 13).

The rate of energy consumption per capita in the developing countries since 1950 has been only 10–12% of that in the developed countries as a whole and 5–8% of that in the United States. Are the developing countries going to be able to increase their energy consumption per capita and, consequently, their standard of living in the 21st century? Or are they going to remain in their present status as developing or underdeveloped countries for many years to come? These are very thorny questions.

As could be expected, answers offered to these questions differ greatly. Optimists believe that the per-capita energy consumption in Asian, African, and Latin American countries might increase as much as three to five times in the next 50 years if expected social and economic improvements are realized (and this, of course, is a big "if"). Pessimists hold that the developing countries, burdened by their large and fast-growing human populations, will never be able to increase their economic standing appreciably and will therefore not be able to afford a higher consumption of energy per capita,

particularly when energy inevitably becomes increasingly expensive, as what is generally believed will be the case when oil becomes scarcer and in high demand as a source for transportation fuels.

Out of a population of about 4.8 billion living in the developing countries in 2000, 2 billion are still reported to live in poverty and hunger. Half a billion, 180 million of them children, live in abject poverty. One billion people are suffering from critical shortages of energy. This situation may not get much better. For about 2 billion human beings in the developing countries, the main sources of energy are still wood, agricultural waste, and other noncommercial sources. Even if the rates of population growth in most developing countries decrease appreciably, as discussed above, it is possible that by the middle of the 21st century, the inhabitants of many developing countries may run out of wood as their forests are progressively cut down. The environmental degradation caused by the destruction of the forests will also contribute to other damage to the environment (erosion of soils, reduction in the rate of ground-water recharge, and floods) and will impair, in many ways, the economic and social development of many developing countries and thus reduce the financial means necessary for the procurement of energy. Poverty, population growth, and environmental degradation are elements in a chain reaction of problems in which numerous developing countries are caught, contributing to economic, social, and political instability from which it will be very difficult to escape.

To increase their consumption of energy per capita, most developing countries will need to overcome three very serious obstacles: their inefficient and wasteful consumption of energy, the scarcity of capital, and, perhaps the most important, the growth of their population. These obstacles will not be easily or quickly surmounted.

If the energy consumption per capita in the developing countries were to increase during the 21st century at the same rate it has increased from 1950 to 2000 (from 0.7 to about 5 BOE/p/yr), it would reach about 9–10 BOE/p/yr in 2050 and about 15 BOE/p/yr in 2100. At that rate, as will be discussed in the next section, at the end of the 21st century, the developing countries would be consuming roughly three times as much energy as all the developed countries put together. Although some of the developing countries, as discussed above, will be able to increase their energy consumption per capita substantially, it does not seem realistic to predict that the developing countries as a whole will be able to afford the procurement of such large amounts of energy, particularly in view of the increases in the price of energy that are forecasted for the 21st century.

Also to be considered is the fact that the consumption of energy per capita in the developing countries, like that in the developed countries, seems to have leveled off during the 1990s and may not increase in the 21st century at the rate it increased from 1950 to the late 1980s.

In estimating the future rate of energy consumption per capita of the developing countries for this study, I have therefore considered three cases:

- The most pessimistic case assumes that the energy consumption per capita of the developing countries during the 21st century will remain at about the same rate as it was in the 1990s (about 5 BOE/p/yr).
- The other two cases assume increases in the energy consumption per capita that will result in it reaching 10 and 15 BOE/p/yr, respectively, in 2100. These increases, particularly the higher one, may be unrealistic, unless the growth of the population of most developing countries is radically controlled during the coming decades. Greater increases in the energy consumption per capita of the developing countries, with concomitant major improvements in their standard of living, have not been contemplated. However, even if the energy consumption per capita in the developing countries remains low during the 21st century, their total consumption of energy could increase considerably as their population will increase, even in the most optimistic projections.

The World

The average energy consumption per capita of the world has remained essentially unchanged at 10–11 BOE/p/yr since 1970 (Appendix Table 9; Figure 13). As the rate of consumption hopefully decreases in the developed countries and increases in the developing countries, the world's energy consumption per capita is expected to increase moderately during the 21st century to 12–15 BOE/p/yr by 2100.

ESTIMATION OF THE WORLD'S ENERGY CONSUMPTION DURING THE 21ST CENTURY

The historical information provided in the previous sections of this study, on human population, energy consumption, and energy consumption per capita from 1950 to 2000, along with forecasts of world population and energy consumption per capita during the 21st century, should provide the best available foundation for estimating the world's energy consumption during the coming 100 years. The established historical trends cannot be ignored in estimating energy consumption in the 21st century.

The following are separate energy-consumption forecasts for the 21st century for (1) the developed countries, (2) the United States and Canada combined, (3) the developed countries outside North America, (4) the developing countries, and (5) the world as a whole.

These are, of course, not the only estimates that could be derived from the historical data of previous sections. Population and energy consumption statistics can be combined in multiple other ways, leading to a profusion of forecasts. Many possibilities exist: selection of different regions or combinations of countries; different choices of future population trends; and use of different trends of per-capita energy consumption.

Those discussed in the following pages and presented in Figure 18 (based on Appendix Tables 31–35) are believed to be based on sound available historical data and to be, for the most part, reasonably realistic. They primarily serve to illustrate and provide examples of how historical data on human population and energy consumption can be combined to forecast the future consumption of energy and to construct scenarios of energy consumption and supply for the 21st century, as will be attempted in the final section of this study.

Realistic estimates, however, need to take several other factors into consideration: availability of energy sources, economic strength, prices of energy sources, technological progress, and possible efforts to protect the environment. The relevancy of these factors to the consumption of energy in the 21st century will be discussed in connection with the proposed energy consumption and supply scenarios.

Table 1 provides the basic information (historical data on population and energy consumption per capita in 2000 and estimates of these factors in 2100) used in making the following estimates. These assumptions and the resulting estimates need not, and will not, be accepted by all, but as mentioned before, all the basic information to make other estimates is available in the tables and figures in previous sections of this study.

The Developed Countries

The combination of the three possible sizes of the population and the two possible rates of energy consumption per capita of the developed countries in the year 2100 produced six estimates of the consumption of energy of this group of countries in 2100 shown in Table 2.

The average of the six estimates of energy consumption in the developed countries in 2100 is 30.25 billion BOE/yr.

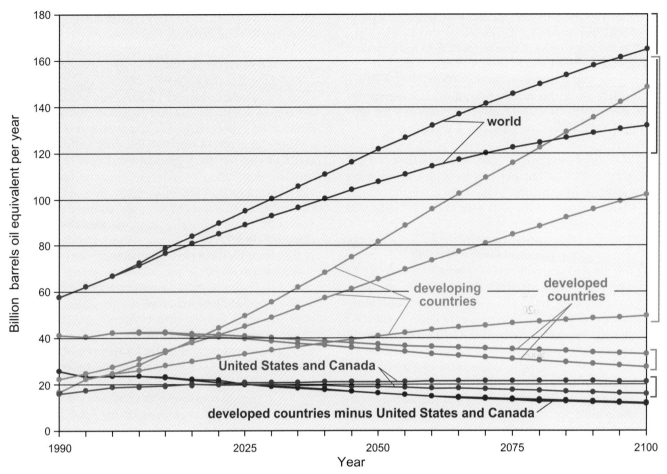

Figure 18. Energy consumption projections.

The computation by 5-year intervals of the middle two cases for the developed countries (population of 1.1 billion and consumption per capita of 25 or 30 BOE/p/yr by 2100) is shown in Appendix Table 31 and Figure 18. Also shown in Figure 18 is the full range of the estimated energy consumption of the developed countries in 2100.

In the two middle cases, the population of the developed countries is projected to decrease from 1264 million in 2000 to 1100 million by 2100, and the rate of energy consumption per capita is likewise projected to decrease from 33.4 to 25 or 30 BOE/p/yr. If these projections are judged to assume an unrealistic population reduction, the high case may be favored in estimating the energy consumption of the developed countries during the 21st century. This case shows the population growing slowly during the first few decades of the 21st century and then decreasing to 1200 million by 2100, essentially the same amount as in 2000, and the rate of energy consumption per capita also decreasing to 25 or 30 BOE/p/yr by the end of the 21st

century. The difference in total energy consumption between the two cases is not large: 27.5 and 33 billion BOE/yr in the middle case vs. 30 and 36 billion BOE/yr in the high case or less than 20% difference. Even using the population projection of the high case (1200 million in 2100) and an increase in energy consumption per capita to 35 BOE/p/yr by 2100, the total consumption of energy of the developed countries would reach only 42 billion BOE/yr by the end of the 21st century, about 16% more than the higher amount of the high case.

To assume that the total population of the developed countries is going to remain fairly even during the 21st century and that their rate of energy consumption per capita is going to decrease or, at most, increase at a small rate is not considered unrealistic, considering that the data on past population and energy consumption trends in the developed countries suggest that the population of the majority of these countries is growing very slowly or, in some cases, decreasing, and that they are becoming increasingly

Table 1. Different cases used in estimating the world's energy consumption during the 21st century.

Population (millions)		Energy Consumption Per Capita (BOE/p/yr)	
2000	2100	2000	2100
Developed Countries			
1264	1000	33.4	25
	1100		30
	1200		
United States and Canada			
314	474	59	30
	526		40
	580		
Developed Countries Minus the United States and Canada			
950	526	25	20.5
	574		
	620		
Developing Countries			
4792	9000	5	5
	9900		10
	10,800		15
World			
6057	10,000	11	12
	11,000		15
	12,000		

efficient and frugal in their use of energy. If this proves to be true, the total consumption of energy in the developed countries should decrease gradually during the 21st century from a little more than 40 billion BOE/yr to about 30–35 billion BOE/yr. It may decrease to 25 billion BOE/yr or remain at about 40 billion BOE/yr by 2100, but these differences in the estimates are

Table 2. Consumption of energy in the developed countries in 2100.

Population (millions)	Consumption Per Capita (BOE/p/yr)	Energy Consumption (million BOE/yr)
1000	25	25,000
	30	30,000
1100	25	27,500
	30	33,000
1200	25	30,000
	30	36,000

Table 3. Consumption of energy in the United States and Canada in 2100.

Population (millions)	Consumption Per Capita (BOE/p/yr)	Energy Consumption (million BOE/yr)
474	30	14,220
	40	18,960
526	30	15,780
	40	21,040
580	30	17,400
	40	23,200

insignificant when compared with the magnitude and range of the estimates of the potential total energy consumption in the developing countries.

The United States and Canada

The six cases resulting from combining the three estimates of the population of the United States and Canada in 2100 (474, 526, and 580 million) with the two estimates of the possible decrease in the rate of energy consumption per capita (from 59 BOE/p/yr in 2000 to either 30 or 40 BOE/p/yr in 2100) indicate a relatively small range in the amount of energy consumption in the United States and Canada during the 21st century, as shown in Table 3.

The average of the six estimates is 18,433 million BOE/yr.

The computations by 5-year intervals of the middle two cases (population of 526 million and consumption per capita of 30 or 40 BOE/p/yr by 2100) are shown in Appendix Table 32 and Figure 18. The full range of the estimated energy consumption of the United States and Canada in 2100 is also shown in Figure 18.

In the two middle cases, the population of the United States and Canada is projected to increase from 314 million in 2000 to 526 million in 2100, and the energy consumption per capita to decrease from 59 to 30 or 40 BOE/p/yr. If the consumption of energy per capita decreases to 30 BOE/p/yr by 2100, the consumption of energy would increase from 18,570 million BOE/yr in 2000 to about 19,600 million BOE/yr in 2030 and then decrease to about 15,800 million BOE/yr in 2100. If the energy consumption per capita decreases to only 40 BOE/p/yr, the consumption will reach 21,000 million BOE/yr in 2045 and remain at that level during the rest of the 21st century.

In all cases, at any time, the total combined consumption of energy of the United States and Canada represents more than half of the energy consumption of the total combined developed countries.

Table 4. Consumption of energy in the developed countries minus the United States and Canada in 2100.

Population (millions)	Consumption Per Capita (BOE/p/yr)	Energy Consumption (million BOE/yr)
526	20.5	10,783
574	20.5	11,767
620	20.5	12,710

The Developed Countries Minus the United States and Canada

Three estimates of the population of the developing countries other than the United States and Canada and a single estimate of the consumption of energy per capita of this group of countries have been combined to come up with three estimates of their energy consumption in 2100 (Table 4).

The average of the three estimates is 11,753 million BOE/yr.

In Appendix Table 33, the consumption of energy of the developed countries other than the United States and Canada has been estimated in two ways:

1) By subtracting the energy consumption of the United States and Canada in the middle case and using an energy consumption per capita of 30 BOE/p/yr (Appendix Table 32) from the energy consumption of the total developed countries in the middle case and an energy consumption per capita of 25 BOE/p/yr (Appendix Table 31), and

2) By subtracting the energy consumption of the United States and Canada in the middle case and an energy consumption per capita of 40 BOE/p/yr (Appendix Table 32) from the energy consumption of the total developed countries in the middle case and using an energy consumption per capita of 30 BOE/p/yr (Appendix Table 31).

Both combinations of basic data result in very similar figures for the consumption of energy and for the energy consumption per capita of the developed countries other than United States and Canada. They are close to calculations that assume a decrease of the energy consumption per capita from 25 BOE/p/yr in 2000 to 20.5 BOE/p/yr by the end of the 21st century.

The Developing Countries

In the case of the developing countries, three estimates of the population in 2100 have been combined with three estimates of the energy consumption per capita

to calculate the following estimates of energy consumption in the developing countries in 2100 (Table 5).

The average of the nine estimates is 99,000 million BOE/yr.

The computation by 5-year intervals of the three estimates of the middle case is shown in Appendix Table 34 and Figure 18. The full range of the estimated energy consumption of the developing countries in 2100 is also shown in Figure 18.

As discussed earlier, to estimate what the energy consumption per capita in the developing countries will be during the 21st century is far from easy. How much the standard of living of the developing countries will improve, if at all, during the 21st century and, consequently, how much their consumption of energy will increase are very difficult and controversial subjects about which widely disparate predictions have been made. Of the two main factors that determine the consumption of energy, the growth of the population can probably be estimated within not-too-widely-divergent limits based on reasonably good data; future trends in the energy consumption per capita, on the other hand, are extremely difficult to predict in the case of most developing countries.

One of the three estimates of the middle case, a constant rate of energy consumption per capita of 5 BOE/p/yr, may appear to be too pessimistic but reflects the stable level of energy consumption during the 1990s; 10 BOE/p/yr may seem more realistic, as it corresponds to the long-range increase from 1950 to 2000, whereas 15 BOE/p/yr may be too optimistic, possibly even unrealistic, because some developing countries may reach that rate of energy consumption per capita during the 21st century, but many others may not do that well. If the average of the developing countries reached that level of energy consumption, they would be consuming five to six times the amount

Table 5. Consumption of energy in the developing countries in 2100.

Population (millions)	Consumption Per Capita (BOE/p/yr)	Energy Consumption (million BOE/yr)
9000	5	45,000
	10	90,000
	15	135,000
9900	5	49,500
	10	99,000
	15	148,500
10,800	5	54,000
	10	108,000
	15	162,000

of energy consumed by the developed countries by 2100. Reality, however, may prove to be closer to the lower estimate. In all three estimates, the developing countries would surpass the developed countries in total energy consumption during the first half of the 21st century.

The range of estimates of the energy consumption of the developing countries in 2100 is extremely wide, from a low of 45,000 million BOE/yr to a high of 162,000 million BOE/yr. However, even if the range of estimates of the energy consumption per capita is narrowed to 5, 7.5, and 10 BOE/p/yr, the size of their human population is so large that the range of estimates of their consumption of energy in 2100 would remain very wide: 45,000 to 148,000 million BOE/yr. It is evident, therefore, that to properly estimate the world's consumption of energy in the 21st century, a realistic estimate of the future energy consumption of the developing countries is fundamental.

The World as a Whole

Three estimates of the global population and two estimates of the energy consumption per capita in 2100 were used to forecast the total amount of energy that will be consumed in the world at the end of the 21st century (Table 6).

The average of the six estimates is 148,500 million BOE/yr.

Appendix Table 35 and Figure 18 show the computation of the middle case by 5-year intervals. Figure 18 also shows the full range of the estimated energy consumption of the world in 2100. The two estimates of the energy consumption per capita for the whole world assume a gradual but small increase from 11 BOE/p/yr, sustained from the mid-1970s to 2000, to 12 and 15 BOE/p/yr in 2100. This small increase considers that although the developed countries may reduce their energy consumption per capita during the 21st century, the developing countries, which will account for the majority of the population of the world, will probably increase theirs somewhat.

Table 6. Consumption of energy in the world in 2100.

Population (millions)	Consumption Per Capita (BOE/p/yr)	Energy Consumption (million BOE/yr)
10,000	12	120,000
	15	150,000
11,000	12	132,000
	15	165,000
12,000	12	144,000
	15	180,000

Any estimate of the total world energy consumption should, of course, be the sum of certain estimates of the energy consumption of the developed and the developing countries. The estimate of the energy consumption of the middle case using energy consumption per capita of 12 BOE/p/yr (132,000 million BOE/yr) fits the sum of the middle case for the developed countries with an energy consumption per capita of 30 BOE/p/yr (33,000 million BOE/yr) and the middle case for the developing countries with an energy consumption per capita of 10 BOE/p/yr (99,000 million BOE/yr). In this case, the developing countries in 2100 would be consuming three times the amount of energy consumed by the developed countries and 75% of the total energy consumed in the world.

Numerous other combinations of the estimates of the energy consumption of the developed and the developing countries are possible in arriving at an estimate of the total world energy consumption at the end of the 21st century, from as low as 70,000 million BOE/yr (by combining the lowest cases for the developed and the developing countries) to 198,000 million BOE/yr (by combining the highest cases). Both of these extreme estimates are considered unrealistic; a range between 100,000 and 150,000 million BOE/yr is considered as better founded. The choice of the estimates for the energy consumption of the developing countries will have a much greater significance in the elaboration of the estimates for the whole world, because the probable range of the estimates for the developed countries, as previously discussed, is relatively narrow.

Probable Deviations from Chosen Estimates

The estimates of the world's energy consumption during the 21st century shown in Appendix Tables 30–35 and Figure 18 are just that— estimates. We know they will be wrong; the question is, how wrong? Too high? Too low? The curves representing the increases or decreases in the rates of energy consumption will not be as smooth as those shown in Figure 18. As in the past, there will be peaks, plateaus, and valleys caused by wars, oil shocks, embargoes, changes in the energy policies of energy-producing or energy-consuming countries or groups of countries, changes in the attitudes and habits of energy consumers, and perhaps, regrettably, implacable famines and plagues, not to mention possible major breakthroughs in the technology of energy production and consumption.

Unpredictable as the factors on which the above estimates were based (human population and consumption of energy per capita) can be, the incidence of potential local or worldwide innovation, changes, and turmoil mentioned above is even more unpredictable.

We must also consider that the existence of abundant sources of primary energy (oil, natural gas, coal, and others) does not necessarily mean that they will be available, and that all the inhabitants of all of the countries of the world will have access to them. International political conflicts leading to oil embargoes, for example, may restrict considerably the availability of certain energy sources to certain countries or regions. Who controls some of the largest sources of energy may be critical in this respect. Inevitable increases in the cost of energy will likewise make certain sources of energy unaffordable to the population of many countries, particularly in the Third World. The abundance of the sources of energy, therefore, may not be the principal factor in determining the availability and consumption of energy in the 21st century, and actual consumption may be much smaller than the possible demand.

As the various sources of primary energy are discussed in the following sections of this study, the factors that may control the availability and consumption of energy, irrespective of the abundance of its sources, will become more evident. They will be discussed again in greater detail in the closing section of this study.

4

Sources of Energy

ABSTRACT

During the second half of the 20th century, fossil fuels (oil, natural gas, and coal) provided 85–95% of the total energy consumed in the world. The other 5–15% was supplied mainly by hydroelectric and nuclear power. The fossil fuels can be used to power land vehicles, ships, and aircraft and to generate electricity. Hydroelectric and nuclear power can be used only to generate electricity. To predict what sources will supply the demand for energy during the 21st century, this chapter reviews the essential information about past production and consumption, estimated reserves and resources, ultimate recovery, and potential future availability of all presently known sources of energy. This review indicates that fossil fuels will remain as the principal source of the energy consumed in the world during the 21st century and certainly during its first half.

INTRODUCTION

As discussed in the Introduction to this study, over the centuries, the sources of energy used by mankind have changed widely. Similar changes should be expected in the future.

Wood and other plant materials were first used by early humans as fuel for fire. Many centuries later, they added wind and running water as new sources of energy. The use of coal as a fuel, limited until the mid-19th century, grew dramatically during the Industrial Revolution, with coal replacing wood as the main source of energy in many regions of the world. Coal, in turn, was progressively replaced by oil during the first half of the 20th century, and by the mid-1960s, oil became the main source of energy. During the last decade of the 20th century, oil has provided 37% of the world's energy supply. At the same time, the contribution of natural gas to the total world supply of energy, which was minor in the first half of the 20th century, increased consistently during the second half, and at present, gas provides about 22% of the world's consumption of energy. The energy supplied by running water has been used for many centuries but has never provided more than

6 or 7% of the total energy consumed in the world. The contribution of nuclear power, first used in the 1960s, reached 6–7% about 1990 but has remained at that level during the 1990s (Appendix Table 8; Figure 12A, B).

Fossil fuels (oil, natural gas, and coal) can be the source of transportation fuels and be used to generate electricity. Other primary sources of energy (nuclear, hydroelectric, and geothermal power, sunshine, and wind) can only be used to generate electricity. Biomass can fuel electricity generation and be the source of gaseous and liquid fuels but only in very small amounts. Hydrogen is not a primary source of energy; it needs to be generated by some process.

What sources will supply the consumption of energy during the next century? Will oil be replaced as the main source of energy, as wood and coal were replaced before? What will replace oil, and more importantly, will there be sufficient sources of energy available during the 21st century to satisfy the worldwide demand for energy, which, as discussed before, may well increase considerably?

In trying to answer these questions, the main sources of energy, nonrenewable (oil, natural gas, coal, tar sands, and oil shales) as well as renewable (nuclear, hydroelectric, geothermal, and solar), are reviewed in the following sections. For each, essential information about their past production and consumption will be discussed along with their present and potential future availability.

Wood, peat, and agricultural and other wastes are not discussed in detail. They are mainly noncommercial local fuels used for cooking and domestic heating, and their consumption is so poorly documented that it is not included in most energy statistics. Although their use has decreased in the last 150 years, they are still important sources of energy in many developing countries.

The use of terms such as "reserves" and "resources" to describe the level of certainty of the existence of some naturally occurring materials, such as oil, gas, and coal, has been far from consistent. In the following sections of this study, these terms are used according to the definitions of the U.S. Geological Survey (McKelvey, 1973; modified by Miller et al., 1975):

- Resources: concentrations of naturally occurring solid, liquid, or gaseous materials in or on the Earth's crust in such form that economic extraction of a commodity is currently or potentially feasible; resources include both the discovered (identified) and undiscovered occurrences of the particular material.

- Reserves: that portion of the identified resource that can be economically extracted at the time of classification or estimation using existing technology and at the then prevailing prices.

Refinement of these terms to express, for example, further degree of certainty about the existence of the reserves (proved, probable, and possible or the equivalent terms measured, indicated, and inferred) will not be generally used, because they represent degrees of accuracy unnecessary for the present study.

An additional term used in this study is "estimated ultimate recovery," also called "ultimate resource" and "resource base," and defined as the total amount of the material expected to be produced during its lifetime; it is the sum of the amount of the material already produced (cumulative production), the current reserves, and the amount expected to be discovered and produced in the future.

The expression "in place" is used in the case of some energy sources, principally very heavy oil in the tar sands and coal, to refer to the amount of a material or commodity in the ground prior to any production. Part of it is recoverable, but another part will be unrecoverable under any predictable technological or economic conditions.

OIL

ABSTRACT

Commercial production of oil started in the late 19th century, and since the late 1960s, oil and natural-gas liquids (NGL) have been the world's principal source of energy. At present, the oil-producing countries of the Persian Gulf region provide 28% of the world's production and contain 65% of the world's reserves of oil. The Organization of Petroleum Exporting Countries (OPEC), which now includes the countries of the Persian Gulf region plus Algeria, Indonesia, Libya, Nigeria, and Venezuela, produces 40% of the world's oil and contains 79% of the oil reserves.

Three estimates of the ultimate worldwide oil recovery (3, 3.5, and 4 trillion bbl) include cumulative production, current reserves, natural-gas liquids, reserve growth, and enhanced oil recovery (EOR) additions to known oil fields and future discoveries. If any one of these estimates is correct, oil production will peak between 2025 and 2040 and then decline during the rest of the 21st century. Oil and its products will continue to be an important component of the world energy mix through the end of the century. In addition, conversion of natural gas to liquids (GTL) will provide a considerable increment to the supply of liquid fuels. The world will not run out of oil for a good many years yet.

INTRODUCTION

Oil is, at present, the principal source of energy throughout the world (see Appendix Table 8; Figure 12A, B). This is not surprising because oil is easier to produce,

transport, store, and use than natural gas or coal, the other two fossil fuels that, with oil, now supply more than 80% of the world's energy consumption.

The following discussion, like that of other sources of energy dealt with in the following pages, will first cover historical data (production and reserves) and will then attempt to predict the future production of oil based on ultimate recovery estimates.

Statistics on oil production and reserves are most commonly given in barrels (42 gal). Some sources, however, use metric tons and, less commonly, cubic meters. Because the relation between 1 t (weight) and 1 bbl of oil (volume) is not constant (it depends on the specific gravity of the oil), many different conversion factors can be found in the various sources of information; they most commonly range from 7.2 to 7.57 bbl of oil/t. In the present study, I have used a conversion factor of 7.33 bbl of oil/t, which seems to correspond to an average of crude oils from throughout the world.

Whenever possible, the terms "oil" and "crude oil" will be used for what is sometimes called "conventional crude oil," which constitutes the great bulk of the liquid petroleum produced today, including that produced using enhanced oil recovery techniques. The following statistics of production and reserves include lease condensate (the liquids recovered at natural-gas wells or at small gas/oil separators in the field) but exclude natural-gas plant liquids (those obtained from natural gas at processing plants). Both are referred as "natural-gas liquids." Also excluded are the bitumen of the "Canadian Tar Sands" and the extra-heavy oil (oil with a specific gravity of less than 10° API) in the Orinoco Oil Belt in Venezuela, which will be discussed separately. In some cases, however, statistics for oil or crude oil include natural-gas liquids, and the boundary between heavy and extra-heavy oil, as will be discussed in the section on tar sands, is not always clearly defined.

PRODUCTION

Commercial production of oil is considered to have started in 1859 in the United States with the discovery of oil near Titusville, western Pennsylvania, by "Colonel" Edwin L. Drake. Several countries soon joined the United States as oil producers, and by the turn of the century, the world's production had reached 350 million bbl/yr. Production of oil reached 1 billion bbl in 1923, 2 billion bbl in 1939, and 3 billion bbl in 1947. Production grew at an increasing rate in the following decades, reaching 20.3 billion bbl annually in 1973. The rapid increase in production that characterized the 1950s, 1960s, and early 1970s was interrupted in 1973 by the "first oil shock," the consequence of the Yom Kippur War, and the resulting Arab oil embargo, which brought a sharp increase in the price of oil. Annual oil production dropped sharply to 19.3 billion bbl.

After 1975, annual oil production resumed its increase until 1979, when it reached 22.8 billion bbl.

The Iranian Revolution brought the "second oil shock;" oil exports from Iran ceased, the price of oil increased drastically, and world oil production declined during the next 4 years to about 19.3 billion bbl in 1983. Production increased slowly afterward to about 22 billion bbl in 1990 and remained at that level until 1994. Production has increased slowly since 1994 to close to 25 billion bbl/yr in 2000–2001. This more stable production of oil has been attributed to higher prices, more efficient use of energy, a comeback of coal, the entry of nuclear power, and particularly, natural gas in the generation of electricity and to the realization by the governments of many industrial countries that, for security reasons, it is dangerous to depend too heavily on a single source of energy. The share of the energy market held by oil has declined from 49.4% in 1974 to 36–37% in the late 1990s (see Appendix Table 8; Figure 12B).

(A detailed and extremely interesting history of the oil industry can be found in Daniel Yergin's *The Prize*. It brings up how oil has influenced and will continue to influence world events for many years to come.)

World yearly oil production from 1950 to 2002 is shown in Appendix Table 36 and Figure 19. Cumulative oil production during that same time span is shown in Appendix Table 36 and Figure 20. Both sets of data are as reported by the *Oil & Gas Journal*. As mentioned above, they include lease condensate.

At the start of the 21st century, the 10 principal oil producers were, in descending order (with approximate percentages of the total world oil production):

Saudi Arabia (11%)
Russia (10.5%)
United States (8.8%)
Iran (5%)
China (5%)
Norway (4.8%)
Mexico (4.7%)
Venezuela (4%)
United Kingdom (3.5%)
Canada (3.2%)

Another 17% of the world's oil production is being provided by Iraq, Nigeria, Abu Dhabi, Kuwait, Brazil, Libya, and Indonesia, for a total of 77.5%.

Oil production in the United States increased from 1972 million bbl/yr in 1950 to 3522 million bbl/yr in 1971, an average growth of almost 74 million bbl annually. It decreased until the initiation of oil production from the Prudhoe Bay field in the North Slope region of Alaska but has resumed its decrease since then (Appendix Table 36; Figure 19).

Production in the countries of the former Soviet Union (FSU) from 1952 to 2002 is shown in Appendix

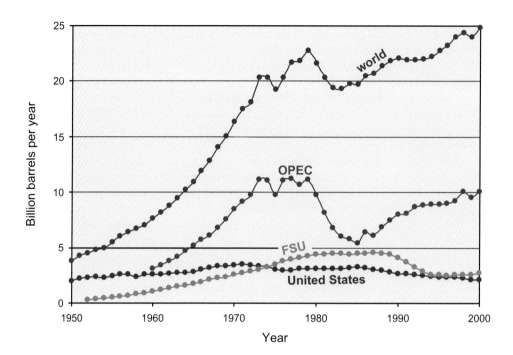

Figure 19. Oil production.

Table 36 and Figure 19. It increased rapidly from about 340–600 million bbl in the mid-1950s to 4500 million bbl in the late 1980s, dropped sharply in the early 1990s after the collapse of the former Soviet Union but began to recover in the late 1990s and the early years of the 21st century.

Since its creation in 1960, the Organization of Petroleum Exporting Countries (OPEC) has represented an important part of the total world oil production. The OPEC now includes the oil-producing countries of the Persian Gulf region (Iran, Iraq, Kuwait, Qatar, Saudi Arabia, and the United Arab Emirates) plus Algeria, Indonesia, Libya, Nigeria, and Venezuela. Ecuador and Gabon, one-time members, withdrew in 1992 and 1994, respectively. As shown in Appendix Table 36 and Figure 19, OPEC's oil production increased steadily from 3166 million bbl in 1960 to 11,180 million bbl in 1973. Production went down to 9771 million bbl in 1975 as a result of the first oil shock, reached around 11,000 million bbl in 1976–1979, and then declined again

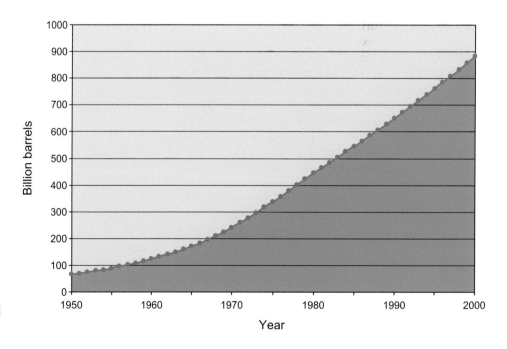

Figure 20. World cumulative oil production.

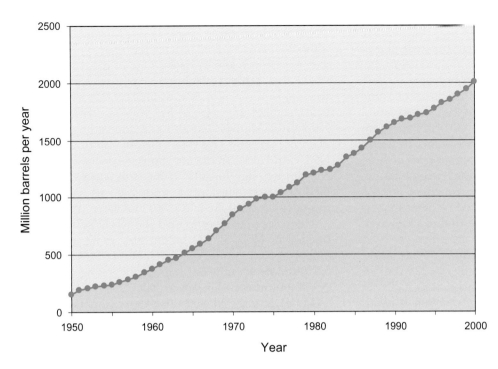

Figure 21. World natural-gas liquids production.

to 5448 million bbl in 1985 in the aftermath of the second oil shock. It has increased steadily since then to about 9000–10,000 million bbl in the late 1990s and the early 2000s.

The percentage of world oil production supplied by the OPEC members increased from 41% in 1960 to 55% in 1973, decreased to 27.6% in 1985, and then increased again to about 39–40% in the late 1990s (Appendix Table 36). At the start of the 21st century, the oil-producing countries of the Persian Gulf region supplied 28% of the world's oil production.

The distribution of world oil production is certain to change radically during the 21st century. As will be discussed in the next section, 66–67% of the present world oil reserves are in the Persian Gulf region. Production is irrevocably declining in the United States, as well as in several other countries that are now producing at capacity, with little hope of increases in the future. Even if the recently discovered major deepwater oil fields in the Gulf of Mexico, off Brazil, and off the west coast of Africa, as well as the new giant fields in the Caspian Sea region, add substantially to future world oil production, the void left by the eventual decrease of the production in other countries will have to be filled in the 21st century by the Persian Gulf producers.

NATURAL-GAS LIQUIDS

Natural-gas liquids are hydrocarbons that exist as gas in the reservoir but become liquid at surface pressure and temperature conditions. They are the liquid content of natural gas that is recovered separately at the surface in lease separators and field facilities or in natural-gas processing plants. The liquids recovered at the well or at small gas-oil separators in the field are called "lease condensate" (or "field condensate") and consist primarily of pentanes and heavier hydrocarbons. Those obtained from processing natural gas at natural-gas processing plants are called "natural-gas plant liquids" or "liquefied petroleum gases" and consist primarily of ethane, propane, butane, and isobutane.

The content of natural-gas liquids (NGL) of natural gas varies widely from essentially none in "dry gas" to more than 200 bbl of natural-gas liquids/mmcf of gas in rich "wet gas."

Figures for the world's production of NGL are not consistently reported and are therefore not easy to obtain. The total amounts of NGL or of the lease condensate production or both are commonly reported as part of the crude oil production, and whereas production volumes of natural-gas plant liquids are sometimes reported, separate figures for lease condensate production can seldom be obtained. As mentioned before, the statistics of oil production and reserves used in this study, as reported by the *Oil & Gas Journal*, include lease condensate. The corresponding figures for the natural-gas plant liquids are reported separately, commonly under the name of natural-gas liquids.

Because of the difficulty for obtaining figures for NGL production, the volumes of the world's total NGL production shown in Appendix Table 37 and

Figure 21 had to be estimated, assuming an average liquids-to-gas ratio of 22.5 bbl of NGL/mmcf of gas. These estimated volumes parallel quite closely the volumes of world NGL production reported by several sources; they correspond to the statistics of *World Oil* from the late 1950s to the mid-1970s; they are a little higher than those provided by British Petroleum from the mid-1970s to the mid-1990s but correspond to those for the late 1990s; and they are lower than those reported for the 1980s and 1990s by the Energy Information Administration of the U.S. Department of Energy. Volumes of NGL production reported by other sources for certain years also fall very close to those shown in Appendix Table 37 and Figure 21.

For recording the NGL production of the world as a whole, the liquids-to-gas ratio of 22.5 bbl of NGL/mmcf of gas is believed to be reasonable. In the United States, where better information is available, the average NGL yield of natural gas is higher: 35–40 bbl/mmcf. The ratio, of course, varies from field to field and from country to country, and the overall world average can only be calculated based on reliable statistics for gas and NGL, statistics that are not always readily available.

The estimates of the reserves and ultimate recovery of NGL are obviously tied directly to those for natural gas, because NGL are a by-product of the production of natural gas. The estimates of the volumes of NGL ultimately to be produced also depend on the estimated yield of NGL of the natural gas to be produced in the future. The future's potentially recoverable NGL needs to be calculated, therefore, on the basis of the estimates of future natural-gas production and estimates of the most likely liquids-to-gas ratios.

As natural-gas production has increased at a faster rate than oil production since about 1970, the percentage of total petroleum liquids represented by NGL has increased steadily from about 4% in 1970 to about 9% in 1995. In the United States, the percentage of the total petroleum liquids represented by NGL has ranged in the last 20 years between 15 and 20%. As the world's crude oil production continues to increase at a slow rate and eventually begins to decrease in the 21st century (see next sections) and as natural-gas production continues to increase at a much faster rate for yet a good many decades, the share of the total petroleum liquids represented by NGL will continue to increase. During the 21st century, NGL will represent a very important share of the total petroleum-liquids production and needs to be given serious consideration in attempting to estimate the sources of supply available to satisfy future world energy demand.

Since, as will be discussed in the section on natural gas, the United States was the only important producer of gas until the early 1960s, the early estimates of the ultimate recovery of NGL were based on North American experience and statistics. One of the earliest estimates was made by Hubbert (1962), who placed the world's future production of NGL at 220 billion bbl, using a yield of about 30 bbl of NGL/mmcf of gas. Hubbert's estimate, like those of succeeding authors, does not include any NGL already produced. Such NGL cumulative production, however, would not have contributed significantly to the ultimate recovery of NGL, because substantial production of natural gas did not start until the late 1940s and early 1950s.

Hendricks (1965) estimated the world's future production of NGL as 520 billion bbl, using a yield of 37 bbl of NGL/mmcf of gas. Hubbert (1969) increased his estimate to 250–410 billion bbl, using yields of 28–32 bbl of NGL/mmcf of gas.

More recently, Meyerhoff (1980) estimated the future production of NGL at 174 billion bbl using a yield of 25 bbl/mmcf of gas, and Parent (1982) reports an estimate of remaining recoverable volume of NGL ranging from 195.9 to 279.5 billion bbl based on a yield of 15–20 bbl/mmcf of gas.

Masters et al. (1987, 1990, 1991, 1992, 1994, 1997) increased their estimates of future production of NGL from 121.6 in 1987 to 192.2 billion bbl in 1997 as their estimates of the ultimate recovery of natural gas increased. They used in all their estimates a liquids-to-gas ratio of only 15 bbl/mmcf of gas. The estimates of Masters et al. include only liquids extracted at NGL plants and exclude condensate recovered at the wellhead. The low liquids-to-gas ratio used and the exclusion of the lease condensate production probably make the estimates of Masters et al. somewhat conservative. Other estimates fall within the range of those mentioned above.

The *U.S. Geological Survey World Petroleum Assessment 2000* estimate of the ultimate recovery of NGL for the world excluding the United States is 324 billion bbl, including 7 billion bbl of cumulative production, representing a yield of 24 bbl of NGL/mmcf of gas.

The estimates of future natural-gas production that will be used in this study, as will be discussed in the section on natural gas, are similar to those of the *U.S. Geological Survey World Petroleum Assessment 2000*. For estimated future conventional natural-gas production of 13,500 and 17,500 tcf and using liquids-to-gas ratios of 20, 22.5, and 25 bbl/mmcf of gas, the following cases for the future production of NGL have been calculated (Table 7). Leaving out the highest and lowest figures, it is estimated that the future production of NGL will range between approximately 300 and 400 billion bbl.

Table 7. Future natural-gas production.

Future Natural-gas Production (tcf)	Future Natural-gas Liquids Production at the Following Liquids-to-gas Ratios (billion bbl)		
	20	22.5	25
13,500	270	304	337
17,500	350	393	437

RESERVES

The estimation of oil reserves is a somewhat subjective exercise, commonly a creative practice not always guided by the facts. Published reserves figures are commonly seasoned with political expediency, economic ulterior motives, and self-serving objectives. Different sources commonly provide different and biased assessments of the honesty of the oil reserves of particular countries or of the world as a whole, commonly made in response to economic or political contingencies and crises.

In addition, it is not always clear what meaning the terms used to report the various kinds of oil reserves have in the different countries or regions of petroleum occurrence. Although in the United States and some other countries the use of the term "proved reserves" as

against "probable reserves" and "possible reserves" is reasonably consistent, in other countries, these and other terms applied to reserves have a range of different meanings. "Identified reserves" or "discovered reserves," for instance, may mean oil recoverable at the time of their estimation under existing price and operating conditions (proved reserves) or may refer to recoverable oil during an indefinite future interval of time (proved, probable, and possible reserves of United States use) (see Masters et al., 1992, 1994). Changes in oil prices also influence the estimation of oil reserves (past, present, and future) and are extremely difficult to predict. Nevertheless, as mentioned in the Preface, the order of magnitude of the discrepancies, particularly concerning the total world reserves, is well within the degree of accuracy needed for this study.

Appendix Table 38 and Figure 22 show the estimates of world proved oil reserves since 1950 as reported by the *Oil & Gas Journal*. Other sources consulted report comparable figures. In using these estimates, the capriciousness of the processes of reserve calculation, as mentioned above, should be kept in mind. The *Oil & Gas Journal* requests from oil producers data on proved reserves, but the responses inevitably include volumes that strict systems for oil-reserves estimation would classify as "indicated" or "inferred." Some countries neither update their reserves estimates every year nor respond annually to the *Oil & Gas Journal*. For these and other reasons, the reserves amounts as reported may be misleading.

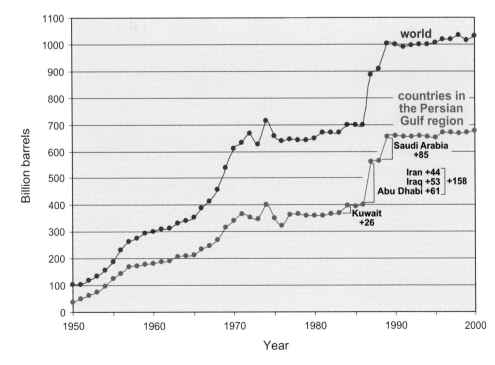

Figure 22. Oil reserves of the world and the Persian Gulf region countries.

All estimates of world oil reserves show four distinct stages:

1) a period of rapid increase from 1950 to about 1974
2) a period of slower growth from 1974 to about 1986, including minor yearly decreases
3) a second period of rapid increase from 1986 to 1989
4) a period from 1989 to the present, during which there has been little or no change in the estimates of the world's oil reserves

The first period, from 1950 to 1974, corresponds to the years during which major oil fields were being discovered in many parts of the world (Persian Gulf region, Nigeria, Libya, the North Sea, southern Mexico, and the former Soviet Union) or when the full magnitude of previous discoveries was being recognized after additional development drilling and the availability of some production history.

The second period, from 1974 to 1986, includes the time of the two oil shocks, a time of considerable confusion and adjustment in the petroleum industry, during which not many new major discoveries were made or announced.

The third period, from 1986 to 1989, marks the time when the oil reserves of the oil-producing countries of the Persian Gulf region and Venezuela (OPEC members), as reported by the *Oil & Gas Journal*, showed large increases (Appendix Table 39; Figure 22). This is an excellent example of what had been called "political reserves." Before 1986, some of the Persian Gulf producers had already boosted their reserves; in 1982, the reported reserves of Iraq rose from 29.7 to 41 billion bbl; and in 1984, the oil reserves of Kuwait were reported to have increased from 64 to 90 billion bbl. The major reserves increases were reported in 1987 by Iran (doubled from 49 to 93 billion bbl), Iraq (also doubled from 47 to 100 billion bbl), Abu Dhabi (tripled from 31 to 92 billion bbl), and Venezuela (more than doubled from 25 to 56 billion bbl). Finally, in 1989, the oil reserves of Saudi Arabia were increased by 50% from 170 to 255 billion bbl (Appendix Table 39). *World Oil* and other sources, including OPEC spokesmen (Miremadi and Ismail, 1993), reported similar increases. During this period, therefore, the oil reserves of both the world and the Persian Gulf region showed an addition of more than 300 billion bbl, which represented an increase of 81% for the Persian Gulf region and 49% for the world (Figure 22). Because it is generally accepted that the Persian Gulf region contains huge oil accumulations, the large reserves additions were not seriously questioned. After these reserves additions were announced, the reserves of the Persian Gulf region have remained essentially the same (as

65–66% of the world's reserves) year after year despite the substantial oil production and the fact that not enough wells have been drilled and no major new discoveries have been announced since the late 1980s to justify such constancy of reserves. The reserves of Venezuela have progressively increased from 56 billion bbl in 1987 to 77.7 billion bbl in 2001 (Appendix Table 39).

Given the economic and political implications of proved oil reserves estimations, it is difficult to say whether the reserves of the OPEC members had previously been underestimated, if these increases were the result of more careful estimation of the reserves of the supergiant oil fields of the Persian Gulf region and Venezuela, or whether they were the product of some creative bookkeeping on the part of these oil-producing countries to justify their requests for larger production quotas, which are related to the size of the proven reserves.

The oil reserves of the prolific Persian Gulf region have, for many years, accounted for the bulk of the reserves of the world (Appendix Table 38; Figure 22). During the mid-1950s and 1960s, the reserves of the Persian Gulf region represented about 60–63% of the total reserves of the world. After 1969, the percentage of the world's oil reserves located in the Persian Gulf region decreased to about 53–56% in the mid-1980s, because no major discoveries were announced in this region, whereas important reserve additions were being made elsewhere, particularly in the North Sea and in southern Mexico. When the large increases of the reserves of Kuwait, Iran, Iraq, Abu Dhabi, and Saudi Arabia were announced in the late 1980s, the percentage of the world oil reserves in the Persian Gulf region increased again to about 65% by 1989. They remain at that level today. As mentioned earlier, the world's oil reserves, at least as reported, have also remained essentially unchanged since 1989 (Appendix Table 38; Figure 22). At the beginning of the 21st century, they stand at a little more than 1 trillion bbl.

The countries belonging to the OPEC have contributed an important part of the total world oil reserves (Appendix Table 40): 70–75% during the 1960s; 65–69% during the 1970s and early 1980s; and about 74–79% since 1987. Three other countries contribute an additional 9.4%: Russia (4.7%), Mexico (2.6%), and the United States (2.1%).

RESERVE GROWTH

Because the magnitude of the announced new oil field discoveries since 1989 cannot account for the replacement of the volumes of oil of about 22–24 billion bbl/yr produced during those years, it is necessary to assume that upward revisions of the reserves of existing fields (what has been called "field growth" or "reserve growth") are responsible for the magnitude of the world oil reserves remaining essentially constant during the last 10–12 years. Either that, or suppose that reserves increases are due to some oil producers' reluctance to acknowledge the decline of their reserves.

Upward revision of the reserves of known oil fields (reserve growth) has unquestionably been a major and probably intensifying factor in the estimation of the world's proved oil reserves. The original estimates of the reserves of a field, commonly conservative because of lack of understanding of the geological, engineering, and production characteristics of the reservoir or the field, grow over time as the estimates are updated when the field is developed and produced. Reserve growth needs to be recognized as a significant source of additions to the proved reserves of a field or of an oil-producing region. Unfortunately, it has regularly been underestimated (see Adelman and Lynch, 1997).

A review of the reported reserves of many known oil fields over the years in different petroleum provinces of the world indicates that the proved reserves have invariably and consistently grown substantially throughout the life of the fields; most oil fields eventually produce much more oil than they had originally been predicted to produce. In some mature oil provinces, more reserves are added in older fields than those provided by new field discoveries. In the United States, for example, reserve growth of known fields accounted for 90% of the annual additions to proved reserves from 1978 to 1990 (Root and Attanasi, 1993).

Many factors contribute to overcoming the original conservative approach to the estimation of the proved reserves of a newly discovered oil field and to make necessary the progressive increases of successive estimates:

- New and better drilling and production technology. Horizontal and extended-reach drilling and subsea production systems have contributed to speed up, to increase, and, in some cases, to make possible the recovery from certain types of reservoirs, to allow the depletion of small accumulations that would otherwise not have been commercial, and to revive economic production from old fields. Injection of water, gas, or other fluids, and enhanced oil recovery procedures have made it possible to increase the oil recovery of many oil fields.
- Infill drilling (vertical and horizontal), recompletions of previously existing wells to produce from additional producing intervals, discovery of new and deeper reservoirs, development of previously known but undeveloped reservoirs, and extensions of the previously identified limits of the field.
- Better understanding of the local geology and more precise reservoir characterization that make it possible to reestimate basic reservoir parameters and to increase the recovery factors of the reservoirs.

Production history, operational experience, and oil-market conditions can, of course, change the reserves estimates of a field downward as well as upward, but this has seldom been the case.

Although reserve growth of oil fields is an unquestionable fact, the different approaches to its estimation and prediction (Arrington, 1960; Hubbert, 1967; Marsh, 1971; White et al., 1975; Root, 1981; Attanasi and Root, 1993, 1994; Root and Attanasi, 1993; Root and Mast, 1993; Schmoker and Dyman, 1998; Attanasi et al., 1999; Masters et al., 1984, 1987, 1991, 1994; Schmoker and Klett, 2000) indicate that the rate and amount of the growth of the reserves of oil fields appear to vary widely. Upward reserves revisions, for instance, seem to be higher and more common in older fields, in which wells were drilled, completed, and produced without access to more effective technology developed since their discovery. Higher oil prices may also make possible the production of oil once considered uneconomic.

However, fields discovered more recently, developed carefully, and studied early in their development with the help of up-to-date technology are much less susceptible to future recurrent increases in their estimated reserves. Moreover, fields requiring very high expenses for their development and production, as, for example, offshore fields in very deep water or fields in hostile environments, may not be able to maintain production beyond a certain minimum rate that allows profitable operation, therefore limiting reserve growth that would have been able to increase their ultimate recovery under different conditions. Finally, an important factor to be considered in estimating the reserve growth of known and yet undiscovered fields is the credibility of the reported reserves. As mentioned before, published reserves are not uncommonly understated or overstated for political or economic reasons and should be expected, therefore, to show large, small, or no increases in the future.

But, however reserve growth is determined, it should be a crucial component of any estimate of the ultimate recovery of oil from any field or any region anywhere in the world. Although not always acknowledged in the past, the importance of reserve growth is fortunately fully recognized now in most recent discussions of the future availability of oil as a source of energy.

ENHANCED OIL RECOVERY (EOR)

As mentioned in the previous section, the oil recovered in many oil fields and, consequently, its estimated reserves can be increased by enhanced oil recovery practices.

The definition of what are considered as enhanced oil recovery methods varies. For the purpose of this study, "primary recovery" is defined as the oil production that uses the natural drive mechanisms of the reservoir (water drive, dissolved gas drive, gas-cap drive); "secondary recovery" is defined as the additional oil production by added energy, usually injection of "natural" reservoir fluids, such as water or gas; and "enhanced oil recovery" (sometimes called "tertiary recovery") is defined as oil production by any technique applied after secondary recovery, principally by injection of materials not normally present in the reservoir such as thermal methods (steam injection or in-situ combustion), miscible or solvent drive (flooding by carbon dioxide or other solvents), or chemical methods (surfactant or polymer injection).

Until now, only the thermal methods, particularly the injection of steam into heavy-oil reservoirs, and carbon dioxide injection have proved to be economically successful. Other enhanced oil recovery methods may become equally profitable in the future.

The volumes of oil estimated to be recoverable by enhanced oil recovery methods vary widely: British Petroleum's Basil Butler is quoted (*Oil & Gas Journal*, October 20, 1986, p. 20) as having stated at the 1986 World Energy Congress in Cannes, France, that enhanced oil recovery has the potential to raise recoverable reserves by only about 2% (35–45 billion bbl, if the amount of remaining oil to be produced, current reserves plus additional oil to be discovered or added to the reserves by revisions, is estimated to range from 1750 to 2250 billion bbl). W. H. Dorsett of Chevron, on the other hand, has estimated (*Oil & Gas Journal*, April 29, 1991, p. 75–76) that as much as 1600 billion bbl of oil may be recoverable in the future by enhanced oil recovery methods if the price of oil increases to about $40 per barrel. DeHaan (1995) estimated that a little more than 300 billion bbl of oil might be recovered during the 21st century by enhanced oil recovery methods, about half by thermal methods, and the rest by various other methods. He estimated that the enhanced oil recovery contribution to total world oil production will be mostly during the first half of the century, when production by primary and secondary methods will probably begin to decrease and the price of oil will increase.

In the present study, it has been assumed that enhanced oil recovery methods will increase the ultimate oil recovery by an additional 8% of the amount estimated to be recovered by primary and secondary methods—current reserves plus oil to be discovered or added to the reserves by revisions—roughly between 160 and 240 billion bbl. This estimate is based on the following considerations:

1) Oil fields producing from good-quality reservoirs (high porosity and permeability, thick continuous producing intervals, and water drive) will attain very high recovery by primary and secondary methods, leaving small amounts of oil to be recovered by enhanced oil recovery methods. Fields producing from poor-quality reservoirs (low porosity and/or permeability, discontinuous or lenticular reservoirs, dissolved-gas or gas-cap drive) do not lend themselves to effective enhanced oil recovery either.

2) Oil fields put on production in the last two or three decades are produced in a much more efficient manner than older fields as a result of the application of improved drilling and production technology. Much more detailed geological description of the reservoir, better log interpretation, more core analysis investigations, and computer simulation programs have been used to study alternative recovery processes and to evaluate injection programs as well as infill and recompletion options. Water or gas injection programs have been initiated much earlier in the life of many fields. As a result, the recovery by primary and secondary methods will reach high percentages of the original oil in place, leaving less oil to be recovered by enhanced oil recovery.

3) Oil produced by enhanced oil recovery methods accounts now for about 3–4% of the world's oil production. This percentage will undoubtedly increase in the future as production by primary and secondary methods decreases and the price of oil increases, but it is likely that oil production by enhanced oil recovery methods will not reach much above 10% of the total remaining world oil production. Much of this enhanced oil recovery production undoubtedly will come from heavy-oil reservoirs.

RESERVES-TO-PRODUCTION RATIOS

The computation of the world oil reserves-to-production ratios (R/PR), the years it will take to deplete the current reserves at the present rate of production, is recorded in Appendix Table 41 and shown graphically in Figure 23. The R/PR, naturally, reflects the history of oil production and reserve additions discussed in previous sections. This ratio is not nearly as significant as popularly considered, because it is not a simple measure of how many more years oil will last. Instead, it is

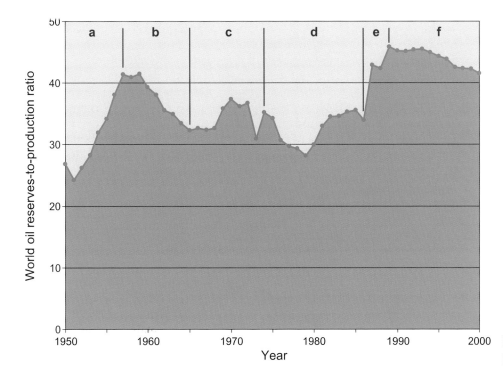

Figure 23. World oil reserves-to-production ratios.

influenced considerably by the intensity of future exploration, the price of oil, the amount of reserve growth, the definition of "reserves," etc., and how honest the estimation of the reserves has been. The oil R/PR has remained at between about 30 and 45 years during the last half century. Six stages can be distinguished (Figure 23).

a) From 1951 to 1957, the additions to the world's oil reserves exceeded the yearly increases in oil production, resulting in a rapid increase in the R/PR from 25 to 41.

b) From 1957 to 1965, the two factors were reversed; a very rapid increase in oil production and a lower rate of increase in additions to reserves resulted in a decrease of the R/PR to about 32.

c) From 1965 to 1974, both production and oil reserves increased at a rapid rate, the R/PR rising to a high of 37 in 1970 and dropping to a low of 31 in 1973.

These first three stages correspond to the first period of the development of estimates of world oil reserves described in a preceding section—a period of steady and generally rapid increase in the discovery of new reserves (see Figure 22).

d) From 1974 to 1986, the overall increase of oil production and of world reserves estimates slowed considerably; both production rates and reserves estimates fluctuated up and down. This period includes the two oil shocks discussed earlier; it corresponds to the second stage of the development of reserves estimates. During this period, R/PR ranged between 28 and almost 36 years.

e) From 1986 to 1989, oil production increased very slowly, whereas world oil reserves dramatically increased principally because of the reporting of much higher reserves estimates by the oil-producing countries of the Persian Gulf region and Venezuela (see Figure 22). This increase in reserves estimates resulted in a consequent increase in the R/PR from about 35 to about 45 (Figure 23).

f) Since 1989, the rate of world oil production (about 22–25 billion bbl/yr), the estimates of the reserves (about 1 trillion bbl), and therefore, the R/PR (about 42–45) have shown little change, indicating that the volumes of oil produced were replaced by corresponding increases in the reserves estimates resulting from both new discoveries and, to a greater extent, from upward revisions (reserve growth) to the reserves of known fields.

ESTIMATION OF THE ULTIMATE RECOVERY OF OIL

For the purpose of predicting the volumes of future oil production, the most critical component is the estimate of the ultimate amount of oil that will be recovered from all oil accumulations in the world. This amount includes

- the cumulative production
- the reserves still recoverable from known (discovered) fields

- the additions to current reserves (reserve growth)
- the reserves from fields to be discovered
- the currently marginal (uneconomic) reserves that may be recoverable in the future under new technology and higher prices.

Eugene Stebinger, chief of the Foreign Mineral Section of the U.S. Geological Survey at the time, was probably the first geologist to attempt such an estimate. Stebinger's estimates of what was called the oil resources of the regions of the world about which information was then available, were first published by David White in his classic paper *Petroleum Resources of the World* (White, 1920). Stebinger's estimates, prepared in the early part of 1920, when world oil production was about 700 million bbl/yr, and broken down by regions, totaled about 43 billion bbl of oil, almost equally divided between the eastern and western hemispheres, but with "a great preponderance of tonnage north of the equator." White remarked that he regarded Stebinger's estimates as conservative, particularly with respect to some of the more promising regions, and ventured the opinion that there was probably another 20 billion bbl of oil available in the world in addition to the 43 billion bbl estimated by Stebinger "or, in round numbers, as much as 60,000,000,000 barrels in all."

No similar attempts to predict the world's ultimate recovery of oil were published until 1942, when Wallace Pratt in his *Oil in the Earth* referred (p. 41 and 67) to a study by Stebinger [with the Standard Oil Company (New Jersey) by then] and coworker Lewis Weeks, indicating that the regions of the world containing a favorable sedimentary section "would ultimately produce some 600-billion barrels of oil, including the 37-billion barrels already found." Pratt adds that "such figures suggest the size of our ultimate reserve of oil in the earth!"

Numerous estimates of the world's ultimate recoverable volume of oil have been published since these early calculations were made. Those reported in readily accessible publications, and which I was able to document, are listed chronologically in Table 8 and shown in Figure 24. The author and/or organization and the amount of the estimate are indicated in each case. An excellent review and discussion of the estimates of the ultimate recovery of oil made from the end of World War II to 1980 has been contributed by Grenon (1982a, b).

One must keep in mind that the different estimates, all by knowledgeable petroleum geologists, have been made by using somewhat different methods, perhaps for different purposes, commonly based on data gathered by others, and with variable degrees of knowledge of the various oil provinces of the world. Undoubtedly,

there is a certain lack of independence, with quoting and requoting of the amounts arrived at by the estimators considered particularly authoritative, and an inevitable reluctance to propose estimates greatly different from those favored by the majority of the experts. A natural bias to join the crowd thus exists. An estimate favored by numerous experts provides an aura of credibility. However, we must remember that even when experts agree, they may be mistaken.

The basic data and the methods used in making the estimates of the ultimate recovery of oil listed in Table 8 and shown in Figure 24 are not always described by the authors, but it is reasonable to assume that they are different for the various estimators, ranging from very subjective and empirical to reasonably rigorous and precise.

A weakness common to most estimates of the ultimate recovery of oil is the uncertainty involved in assessing the fraction of the total oil-in-place that may ultimately be produced, because this fraction depends on additional fragile assumptions concerning future economic conditions, the future price of oil, and possible advances in the technology of producing the oil. It is not always specified, for instance, whether the estimates include the amount of oil expected to be recovered by the application of enhanced oil recovery methods. Some estimators attest that they include the amounts of oil to be recovered by enhanced oil recovery methods; others exclude them, and still others do not specify whether they included them. Not always mentioned either is the magnitude of oil volumes to be recovered by enhanced oil recovery methods after primary and secondary recovery production has been exhausted.

Finally, some of the estimates include all liquids (oil as well as natural-gas liquids), whereas others do not include the natural-gas liquids, and still others do not specify whether the natural-gas liquids are included. Similarly, how much of the very heavy crudes are included is not generally specified.

Nevertheless, the general trends in the order of magnitude of the estimates provide information adequate for the purposes of this study; the estimates of the ultimate amount of oil to be recovered, as listed in Table 8 and shown in Figure 24, increased progressively from 500 to 600 billion bbl in the mid-1940s, when little was known about many potentially petroliferous basins of the world, and next to nothing was known about the offshore regions, to about 2000 billion bbl by 1960, as recognition of the size of the discoveries in the Persian Gulf region improved and as important offshore discoveries were made. The numerous estimates made in the 1960s, 1970s, and early 1980s ranged from a little less than 1500 to about

3000 billion bbl, most of them in close proximity to 2000 billion bbl. Since the mid-1980s, the lower limit of the estimates has slowly increased. Most recent estimates range between 3000 and 3600 billion bbl, implying that 1000–1600 billion bbl of oil still remain to be discovered or to be added to the reserves of known fields through reserve growth. The most notable exceptions among recent estimates are an estimate by British Petroleum's Richard Miller (1992) that places the global ultimate recoverable reserves at a minimum of 4000 billion bbl (see discussion by Ulmishek et al. [U.S. Geological Survey], 1993, objecting to this estimate as being too high) and the low estimates of Campbell (1991, 1995, 1996, 1997), Laherrère (1994), and Campbell and Laherrère (1998) that range between 1650 and 1800 billion bbl.

Undoubtedly the most detailed, realistic, and best supported of recent estimates of the world's ultimate oil recovery is the 2000 assessment of the U.S. Geological Survey. The total 3021 billion bbl include a cumulative production of 710 billion bbl, remaining reserves of 891 billion bbl, 732 billion bbl yet to be discovered, and, most significant, 688 billion bbl of expected reserve growth. An additional 324 billion bbl of estimated ultimate recovery of natural-gas liquids gives a total for oil plus natural-gas liquids of 3345 billion bbl. Previous estimates by the U.S. Geological Survey (Masters et al., 1983, 1984, 1987, 1990, 1991, 1992, 1994, 1997) had shown repeated increases from 1718 billion bbl in 1983 to 2272 billion bbl in 1997. These estimates did not include reserve growth.

For the purpose of this study, three estimates of the world's ultimate oil recovery have been considered: 3000, 3500, and 4000 billion bbl. They cover, in addition to a cumulative oil production of 900 billion bbl and reserves of about 1100 billion bbl, volumes of oil to be discovered in the future, future production of natural-gas liquids, probable large reserve growth in fields already discovered, and oil to be recovered by enhanced oil recovery methods.

The components of the three estimates are shown in Table 9.

This range of estimates is believed to be realistic in view of what is now known about potential petroliferous basins in the world and about the results of exploration for oil throughout the world in the last few decades. The low estimate, in fact, may be too conservative.

Concerning new discoveries, I think it is correct to say that no major sedimentary basin in the world capable of containing sizable oil accumulations is entirely unknown (see, for example, the valuable publications of Masters et al. of the U. S. Geological Survey and the *U.S. Geological Survey World Petroleum Assessment 2000*). For the amount of oil to be ultimately

recovered to reach above 4000 billion bbl, it would be necessary for future oil discoveries to amount to more than 800 or 1000 billion bbl of oil, something very unlikely, unless somewhere in the world, an entirely new oil province is discovered matching the richness of the Persian Gulf region. I do not believe that such a discovery is likely. Oil reserves undoubtedly will be added in the decades to come by the discovery of many small- and medium-size fields, but as has been pointed out by many authors (Moody, 1970; Klemme, 1971, 1977; Holmgren et al., 1975; Nehring, 1978; Ulmishek et al., 1993), major reserves additions are only made by the discovery of enormously rich new plays containing many giant and supergiant fields. The bulk of the reserves of most petroliferous basins are generally contained in a few giant and supergiant fields. Unfortunately, the great majority of the sedimentary basins in the world are now sufficiently well known to preclude the discovery of many such super-rich new plays or fields. The rate of discovery of giant and supergiant oil fields, in fact, has decreased in the last two decades, with fewer of these fields having been discovered during recent years. Important oil fields, however, have been discovered in the Caspian Sea region and in deep-water offshore basins in the United States, Brazil, and along the west coast of Africa, and additional significant discoveries should be expected in these regions, as well in the Persian Gulf area.

However, although there are fewer sedimentary basins in the world still unexplored for oil, there is now vastly more and much better information about most of these sedimentary basins, both onshore and offshore, than there was 20 or 30 years ago, and there is much more advanced exploration technology. Finally, there is the pressing need to discover more oil, which, in turn, creates the incentive to come up with new ideas and new approaches to explore for it. This is very important, because as Wallace Pratt so wisely told us, "unless men believe that there is more oil to be discovered, they will not drill for oil... when no man any longer believes more oil is left to be found, no more oil fields will be discovered" (Pratt, 1942, p. 85).

The amounts of natural-gas liquids estimated to be produced in the future, 300, 350, and 400 billion bbl, are also believed to be plausible. As mentioned before and as will be discussed in more detail in the section on natural gas, production of natural-gas liquids has risen greatly in the last few decades as production of natural gas continues to increase (Appendix Table 37; Figure 21). Serious interest in exploring for natural gas did not develop outside the United States until the late 1960s. Now, natural gas is the fuel of choice, the preferred fuel for the generation of electricity. Many very large natural-gas fields are still stranded for lack

Table 8. Estimates of the ultimate worldwide recovery of oil (estimates are shown in Figure 24 by listed number; for publication, see list of References).

	Year of Publication	Author and/or Organization	Estimated Ultimate Recovery (billion bbl)
1	1946	Duce (Aramco)	500[a]
2	1946	Pogue (Chase Manhattan Bank)	605[b]
3	1948	Weeks (Standard Oil Co., New Jersey)	610[c]
4	1949	Levorsen (Stanford)	1635[d]
5	1950	Levorsen (Stanford)	1635
6	1950a	Weeks (Standard Oil Co., New Jersey)	1010[e]
7	1950b	Weeks (Standard Oil Co., New Jersey)	1100[f]
8	1953	MacNaughton, personal communication	1000
9	1956	Hubbert (Shell)	1250
10	1958	Weeks (Standard Oil Co., New Jersey)	3000[g]
11	1959	Weeks (Standard Oil Co., New Jersey)	3500[h]
12	1961	Weeks (Weeks Petroleum Corp.)	3500[h]
13	1962	Hubbert (Shell)	1250
14	1963	Weeks (Weeks Petroleum Corp.)	2000
15	1965	Hendricks (U.S. Geological Survey)	1984–2480[i]
16	1967	Ryman (Standard Oil Co., New Jersey)	2090[j]
17	1967	Royal Dutch Shell	1800
18	1968	Weeks (Weeks Petroleum Corp.)	3550[k]
19	1969	Hubbert (U.S. Geological Survey)	1350–2100
20	1970	Weeks (Weeks Petroleum Corp.)	3550[k]
21	1970	Moody (Mobil)	1800
22	1971a, b	Warman (British Petroleum)	2000
23	1971	Weeks (Weeks Petroleum Corp.)	3650[l]
24	1972	Warman (British Petroleum)	1800
25	1972	Linden (Institute of Gas Technology)	2945
26	1972	Moody and Emmerich (Mobil)	1800–1900[m]
27	1974	Kirby and Adams (British Petroleum)	1600–2000
28	1974	Parent and Linden (Institute of Gas Technology)	3000–4000
29	1975	MacKay (Bank of Montreal, Calgary) and North (Carleton University, Ottawa)	1000–1050
30	1975	Weeks (Weeks Petroleum Corp.)	3180[n]
31	1975a, b	Moody (consultant) and Esser (Mobil)	2000–2030
32	1975	Moody (consultant) and Geiger (Mobil)	2000[o]
33	1975	Linden and Parent (Institute of Gas Technology)	2685[p]
34	1975	Moody (consultant)	1800–1900[q]
35	1975	National Academy of Sciences	2326
36	1976	Grossling (U.S. Geological Survey)	2200–3000
37	1976	Barthel et al. (West Germany Geological Survey)	2500
38	1977	Parent and Linden (Institute of Gas Technology)	2000[r]
39	1977	World Energy Conference	1889
40	1977	Klemme (Weeks Petroleum Corp.)	1550[s]
41	1978	Desprairies (Institut Francais du Petrole)	2220–2520[t]
42	1978	Moody (consultant)	2030
43	1978	Nehring (Rand Corp.)	1700–2300
44	1979a, b	Wood (Cities Service)	2163[u]
45	1980	Halbouty and Moody (consultant)	2288
46	1980	Schubert (World Energy Conference)	2600
47	1980	Nehring (Rand Corp.)	1600–2000[v]
48	1980	Desprairies and Tissot (Institut Francais du Petrole)	1830–2200
49	1980	Roorda (Shell)	2400
50	1981	Hubbert and Root (U.S. Geological Survey)	2000
51	1982	Nehring (Rand Corp.)	1600–2000[w]
52	1982	Bois (Institut Francais du Petrole)	2600
53	1983	Masters et al. (U.S. Geological Survey)	1718
54	1983	Riva (Library of Congress)	1953
55	1984	Burollet (Total)	2213[x]
56	1984	Masters et al. (U.S. Geological Survey)	1818
57	1985	Tanzil (consultant)	2594
58	1986	Masters (U.S. Geological Survey)	1718

Table 8. Estimates of the ultimate worldwide recovery of oil (estimates are shown in Figure 24 by listed number; for publication, see list of References) (cont.).

	Year of Publication	Author and/or Organization	Estimated Ultimate Recovery (billion bbl)
59	1986	Ivanhoe (consultant)	1700
60	1987	Masters et al. (U.S. Geological Survey)	1744
61	1987	Pecqueur (Elf Aquitaine)	2200[y]
62	1987	Roadifer (Mobil)	2000
63	1988	Riva (Library of Congress)	1765
64	1989a, b	Bookout (Shell)	2000
65	1990	Masters et al. (U.S. Geological Survey)	2074
66	1991	Masters et al. (U.S. Geological Survey)	2079
67	1991	Campbell (consultant)	1650
68	1991	Riva (Library of Congress)	2215
69	1992	Masters et al. (U.S. Geological Survey)	2171
70	1992	Miller (British Petroleum)	>4000
71	1993	Townes (independent petroleum geologist)	2600–3000
72	1993	Miremadi and Ismail (OPEC)	2200
73	1994	Masters et al. (U.S. Geological Survey)	2272
74	1994	Laherrère (Petroconsultants)	1800
75	1995	Campbell (consultant)	1650
76	1996	MacKenzie (World Resources Institute)	1800–2600
77	1996	Campbell (consultant)	1750
78	1997	Campbell (consultant)	1800
79	1997	Edwards (University of Colorado)	2836
80	1997	Masters et al. (U.S. Geological Survey)	2272
81	1997	Al-Jarri and Startzman (Texas A&M)	1760
82	1998	Campbell and Laherrère (consultants)	1800
83	1998	Hiller (Hanover, Germany)	1800–2570
84	1998	Linden (Illinois Institute of Technology)	4000
85	1998	Schollnberger (Amoco)	3300
86	2000	U.S. Geological Survey 2000	3021
87	2001	Deffeyes (Princeton)	2100–2120
88	2001	Odell	3000
89	2001	Edwards (University of Colorado)	2750–3670
90	2002	Edwards (University of Colorado)	3251

[a]100 in the United States, 400 abroad.

[b]49.2 cumulative production, plus 65.8 proved reserves, plus 490 future discoveries.

[c]487 future discoveries.

[d]65 cumulative production, plus 65 discovered reserves, plus 1500 undiscovered reserves. Based on estimates by Pogue (1946), and Weeks (1948) for onshore, and Pratt (1947) for offshore.

[e]Discussion to Levorsen (1950).

[f]610 onshore, 400 offshore shelves.

[g]1500 primary recovery, plus 1500 secondary recovery; includes natural-gas liquids.

[h]2000 primary recovery, plus 1500 secondary recovery; includes natural-gas liquids.

[i]6162–6200 oil in place, 40% recovery.

[j]According to Hubbert (1969).

[k]2200 primary recovery, plus 1350 secondary recovery.

[l]2290 primary recovery, plus 1360 secondary recovery.

[m]1000 discovered, 800–900 yet to be discovered.

[n]1900 onshore, 1280 offshore.

[o]280 (90%), 2000 (50%), and 2200 (10%).

[p]Estimated total remaining recoverable.

[q]1000 discovered, 800–900 yet to be discovered.

[r]Includes natural-gas liquids.

[s]350 cumulative production, 600 proven reserves, 600 undiscovered.

[t]Delphy pool: 1620–2200–2870.

[u]1038 already discovered, 1125 to be discovered; 1500 (95%), 3300 (50%), 3100 (10%).

[v]Includes natural-gas liquids.

[w]Includes natural-gas liquids.

[x]524 billion cumulative production, plus 2313 reserves and to be discovered.

[y]Includes enhanced oil recovery.

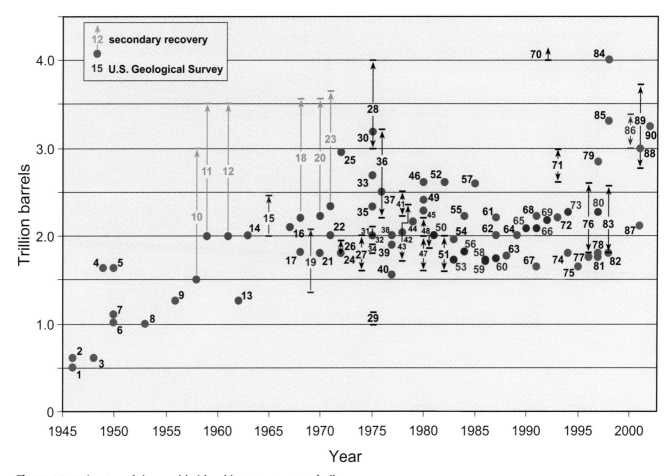

Figure 24. Estimates of the worldwide ultimate recovery of oil.

of means to transport the gas to centers of consumption. This is changing. Natural-gas exploration and production will grow fast in the future and, with it, the production of natural-gas liquids.

Finally, we can also expect that the contribution of the growth of oil field reserves, known now or to be discovered in the future, to the ultimate amount of oil to be produced in the world will be important. As discussed earlier, reserve growth has been commonly underestimated, despite the fact that in many coun-

tries and in many regions, it has made a more important contribution to the appreciation of the world's oil reserves than that provided by the discovery of new fields. Future economic, technological, and political factors will continue to favor this contribution.

Additional oil recovery is also expected from enhanced oil recovery projects, which, as mentioned earlier, may contribute between 160 and 240 billion bbl, principally in heavy-oil reservoirs.

ESTIMATION OF OIL PRODUCTION DURING THE 21ST CENTURY

As discussed in the previous sections, the amount of oil ultimately to be recovered in the world has been estimated for the purpose of this study to range most probably from 3000 to 4000 billion bbl, including natural-gas liquids supplied by the production of natural gas; oil to be recovered by primary and secondary production methods; growth of the estimated reserves;

Table 9. Estimates of the world's ultimate oil recovery.

	Billion Barrels		
Cumulative production and reserves	2000	2000	2000
To be discovered	300	550	800
Natural-gas liquids production	300	350	400
Reserve growth and enhanced oil recovery	400	600	800
	3000	3500	4000

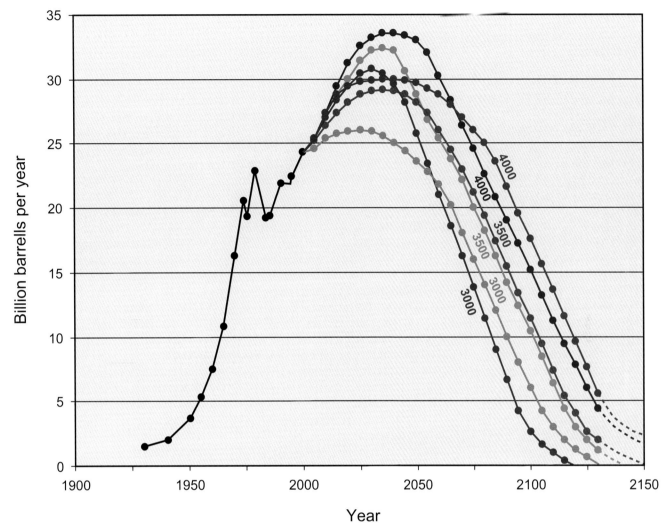

Figure 25. Oil production projections in the 21st century.

and an additional 160–240 billion bbl of oil to be produced by enhanced oil recovery methods.

On this basis, six possible projections of world oil production, two for each of the three estimates of ultimate oil recovery, were prepared by first assuming different trends of oil production during the next 25 or 30 years and then completing the projected production to represent the estimated amounts of ultimate oil recovery (Appendix Table 42; Figure 25). The estimates are based on my assumptions of the possible trends in world oil production and not on any mathematical formulation. The six projections chosen are presented as reasonable scenarios and as a means to illustrate the process of estimating the possible rates of oil production during the 21st century.

Of the two projections based on an ultimate oil recovery of 3000 billion bbl, one (purple curve) assumes a continuing increase in oil production from 24.3 billion

bbl in 2000 to a maximum of 30.8 billion bbl in 2030, whereas the other (gray curve) assumes a much slower increase to only 26 billion bbl in 2025. Both involve a steady production decrease during the rest of the 21st century.

The two projections based on an ultimate oil recovery of 3500 billion bbl also reflect two different assumptions of the increase of oil production during the early years of the 21st century: a faster increase to 32.4 billion bbl in 2035 (blue curve) and a slower one to only 29.2 billion bbl, also in 2035 (green curve). Production is projected to decrease after the peaks in 2035.

Finally, both projections based on an ultimate oil recovery of 4000 billion bbl show peak production in 2035–2040 at 30 billion bbl (red curve) and 33.6 billion bbl (brown curve).

In all six cases, oil production is projected to peak between 2025 and 2035–2040 and to decline during the rest of the 21st century. Faster production increases in the early years of the century will inevitably result in earlier depletion of the world's oil reserves, and even if higher, more optimistic estimates of the ultimate oil recovery are supported, and if it is assumed that oil consumption will rise moderately or that it will stabilize in the first few decades of the 21st century, world oil production will probably peak in midcentury, not much beyond 2035–2040. An important part of the energy demand during the 21st century will be able to be supplied by oil to a greater or lesser extent. The world will not run out of oil for a good many years yet.

Many other projections can be made by assuming (1) different trends in the oil production rates during the next few decades, (2) different estimates of the ultimate amount of oil to be recovered, (3) the possible replacement of oil with some other source of energy, or (4) a reduced demand for oil because of high prices or regulations to reduce environmental damage. For instance, unlike the six projections illustrated in Appendix Table 42 and Figure 25, which show a fairly short interval of time during which the oil production peaks and begins to decline, it is possible to forecast that after reaching a maximum, oil production will level off for a few decades before starting to decline, which is a very distinct possibility.

As mentioned before, even the highest estimate of the ultimate recovery of oil (4000 billion bbl) is considered to be within the range of possibilities for the discovery of new oil fields. The discovery of very large wet-gas accumulations is also very probable; and the conversion of natural gas to synthetic liquids (GTL), a new and hopefully significant process, will probably become an important source of petroleum liquids in the 21st century. Reserve growth of known fields and additional amounts of oil to be recovered by enhanced oil recovery methods because of important technolog-ical developments will also contribute to increase the ultimate recovery of petroleum liquids.

The Persian Gulf region will remain the main source of oil supply and the key factor in the world's energy equation. It will also remain as an unstable region with considerable potential for unpleasant surprises. The possibility cannot be excluded that the countries in this rich oil province may, at some time in the future, decide to use their oil as a weapon to enforce their political agendas by the deliberate curtailment of production and the resulting decrease in world oil supply. It cannot be assumed either that the demand and supply of oil will be fully integrated geographically, so that oil-producing countries can always find a market for their oil, or that oil-consuming countries can always find an oil supply that satisfies their needs.

Because of all these unknown factors, the volumes and rates of oil production during the 21st century will not correspond to the smooth production curves shown in Appendix Table 42 and Figure 25. As in the past, embargoes, wars, revolutions, natural disasters, political upheavals, and economic vicissitudes and unrest, as well as technological advances impossible to predict at present, will account for periodic temporary disruptions and irregular ups and downs in the total rates of the world's oil production. The two oil shocks in the 1970s and the 2002 oil workers' strike in Venezuela are good examples. The world has probably experienced neither its last political surprise from key producing regions nor its last disturbance to the oil trade. Such potentially traumatic events cannot be easily modeled. It is hoped, however, that the oil production trends in the 21st century will fall within the overall projected trends, and that the estimates shown in Appendix Table 42 and Figure 25 represent plausible future orders of magnitude of the availability of petroleum liquids as a source of energy.

TAR SANDS: BITUMEN AND EXTRA-HEAVY-OIL DEPOSITS

ABSTRACT

Bitumen and extra-heavy-oil deposits, "tar sands," are known from many regions of the world. The volume of bitumen and extra-heavy oil in place has been estimated to reach 3800 billion bbl, of which 730 billion bbl is considered to be eventually recoverable at a profit. About 75% of these world totals is in two vast accumulations: one in western Canada and the other in eastern Venezuela. Not-well-known but probably large tar-sand deposits are also present in the former Soviet Union. Development of tar-sand deposits did not start until 1967 in Canada and 1998 in Venezuela. Production is expected to increase during the coming decades, as the technology to produce and upgrade the bitumen and extra-heavy oil continues to advance and their economic value improves. Tar sands in Canada, Venezuela, and elsewhere are expected to make an important contribution to oil production during the 21st century.

INTRODUCTION: DEFINITIONS

"Tar sands" (also called "bituminous sands" and "oil sands") have, in general, been defined as reservoirs containing oil too viscous to flow into a well in sufficient quantities for economic production or, in other words, oil essentially immobile in the reservoir. Several more specific definitions have been proposed, but none has been generally accepted. A United Nations Institute for Training and Research (UNITAR) Working Group on Definitions (Danyluk et al., 1984; Martinez, 1984) stated that, in determining the resource volumes of heavy oil and bitumen, viscosity of the oil should be used first to differentiate between crude oils, on one hand, and bitumens on the other. Density (specific gravity) should be considered next. Tar sands were therefore said to be characterized as containing bitumen, liquids, or semisolids with viscosities greater than 10,000 cp at original reservoir temperature, generally corresponding to a specific gravity of less than $10°$ API at $60°F$ ($16°C$). Carrigy (1983, p. 18) was critical of this definition and preferred to define tar sands as "reservoirs that contain low-gravity oil ($\sim 10°$ API or less) and need a large thermal input to reduce the oil viscosity to a level that will allow it to be produced through a well at economic rates." Other authors have defined tar sands as reservoirs containing oil with a specific gravity of less than $10°$ API (bitumen) and immobile in the reservoir.

General agreement exists regarding the idea that heavy oils have a specific gravity of $10–20°$ API [$10–22.3°$(?)] and a viscosity of less than 10,000 cp, whereas the extra-heavy oils are those having a specific gravity of less than $10°$ API but a viscosity low enough (less than 10,000 cp) to allow them to flow through a well without the need of thermal stimulation. This capability to flow through a well also distinguishes extra-heavy oils from bitumen, which is defined as heavier than $10°$ API and with a viscosity greater than 10,000 cp at reservoir temperature, which therefore makes it immobile in the reservoir.

The general term "tar sands" will be used in the following pages to include both bitumen and extra-heavy-oil accumulations.

The estimates of the oil or bitumen in place or of the amounts of recoverable oil or bitumen from tar-sand deposits, as reported in the published sources of information, do not always follow the above definitions of heavy oil, extra-heavy oil, and bitumen. Viscosity measurements, although apparently the most important factor for successful economic recovery, are not always available, and other information needs to be used to decide if a certain deposit contains extra-heavy oil or bitumen. If, for instance, a field is producing oil heavier than $10°$ API without the application of thermal stimulation, it is assumed to contain extra-heavy oil and not bitumen. In addition, in most tar-sand accumulations, the boundary between bitumen and extra-heavy oil or between extra-heavy and heavy oil is gradational, and placing the boundary at a certain value of the viscosity (10,000 cp) or at a certain specific gravity ($10°$API) is strictly arbitrary and difficult to do. The reported estimates of the volumes of bitumen or extra-heavy oil in place or of the amounts expected to be recovered economically cited in this study therefore include not only bitumen and extra-heavy oil but also undetermined, and undeterminable, amounts of heavy crude oil somewhat lighter than $10°$ API. This, however, is not believed to materially distort the discussions, estimates, and conclusions of the study.

Tar-sand deposits are known from many parts of the world, but present knowledge indicates that two giant deposits, one in western Canada and the other in eastern Venezuela, may contain as much as 80–85% of the approximately 3800 billion bbl of total bitumen and extra-heavy oil in place in the presently known tar-sand deposits. Following the definitions discussed above, the tar sands of western Canada are considered bitumen deposits, whereas the Orinoco Oil Belt of eastern Venezuela falls in the category of heavy- and extra-heavy-oil accumulation, because although the specific gravity of the oil ranges from 4 to $17°$ API and averages less than $10°$ API, the reservoir temperatures are high enough to lower the viscosities to less than 10,000 cp, making the oil mobile under reservoir conditions and allowing some of the oil in place to be recovered by primary production methods, i.e., without the need for thermal stimulation (Masters et al., 1983).

The only other major tar-sand deposits may be located in the former Soviet Union, but they are poorly understood, because available information concerning their size and characteristics is still ambiguous and inconsistent.

Besides the giant tar-sand deposits of Canada and Venezuela, and perhaps those of the FSU, smaller bitumen and extra-heavy-oil deposits have been reported and, in some cases, developed in Albania, Angola, China, Colombia, Iran, Italy, Jordan, Madagascar (Malagasy Republic), Nigeria, Peru, Romania, Trinidad, the United States, and Zaire (Phizackerley and Scott, 1967; Walters, 1974; Chilingarian and Yen, 1978; Meyer and Dietzman, 1981; Rühl, 1982; Schumacher, 1982; Meyer et al., 1984a, b; Roadifer, 1986a, b, 1987; Meyer, 1987; Meyer and Schenk, 1988). None of them appears at this time to be of major importance, certainly not comparable to the giant tar-sand deposits of Canada and Venezuela.

The low commercial value assigned until recently to tar-sand deposits may be responsible for other large bitumen or extra-heavy-oil accumulations having remained undiscovered, unreported, or undeveloped. All important tar-sand deposits may therefore not have been discovered yet.

This situation will no doubt change as concern for the potential exhaustion of conventional lighter oil deposits increases; as technology for extracting and upgrading the bitumen contained in the tar sands to an economical feedstock for refineries becomes available; and as new uses for the bitumen are developed, as, for example, the elaboration in Venezuela of the Orimulsion, a direct combustion product, an emulsion composed of 70% of extra-heavy oil suspended in 30% of water with the help of a chemical surfactant to stabilize the emulsion (Pacheco and Alonso, 1995).

Although problems of upgrading, transportation, and use need to be solved, the economics of producing bitumens and extra-heavy oils are improving, and attention is progressively being focused in the last few years on delineating and assessing the known bitumen and extra-heavy- and heavy-oil deposits and in searching for new ones. Many of the tar-sand deposits crop out and were identified early in the exploration for oil throughout the world. Most of those that do not crop out are at shallow depths along the rims of sedimentary basins and were discovered when drilling for deeper, lighter oil prospects. Not many of those that crop out have remained unknown, but some of those in the shallow subsurface may still be waiting to be discovered.

THE TAR SANDS OF WESTERN CANADA

The tar sands (or more commonly called "oil sands") of western Canada are found in four separate main deposits (Athabasca, Wabasca, Peace River, and Cold Lake) along the gentle eastern flank of the asymmetric western Canada basin in the province of Alberta. These deposits, as mentioned above, predominantly contain bitumen that ranges in gravity from 6 to 12° API and is too viscous to flow into production wells without the application of some sort of thermal stimulation. Most of the bitumen is found at shallow depths (above 600 m [2000 ft]) saturating Lower Cretaceous sandstones. The Athabasca deposit in part crops out or is close enough to the surface (less than 75 m [250 ft] of overburden) to make it possible to develop this sector of the deposit, which contains about 10–12% of the total oil sands bitumen, by open-pit mining methods.

Paleozoic carbonates (of Devonian, Mississippian, and Permian age) underlying unconformably the Lower Cretaceous oil sands are also found saturated with bitumen over an extensive area.

Estimates of the volume of bitumen in place in the western Canada oil sands have increased through the years, as more wells were drilled and more detailed studies were undertaken: most estimates have ranged from 750–950 billion bbl in the 1960s and early 1970s to 1700 billion bbl since the late 1980s to the present, excluding the poorly known bitumen deposits in the Paleozoic carbonates (Table 10; Figure 26).

To a certain degree, these estimates, as in the case of the estimates of the ultimate recovery of conventional lighter oil, lack independence, with many of them being based on data and interpretations of others, particularly on the estimates of those authors or organizations considered most authoritative. At present, the most commonly quoted estimate of the total amount of bitumen in place in the Lower Cretaceous oil sands is that of the Alberta Energy and Utilities Board (previously named the Alberta Oil and Gas Conservation Board and the Alberta Resources Conservation Board): 1700 billion bbl. The estimates of the bitumen in place in the Paleozoic carbonates are controversial; they range from 188 billion bbl (Strom, 1984) to 1560 billion bbl (Meyer et al., 1984a). The most recent estimates are about 800 billion bbl. Of the four main oil sand deposits, the Athabasca is the largest, with an estimated volume of bitumen in place that is now considered to amount to 950 billion bbl.

Estimates of the amounts of bitumen that can ultimately be recovered economically from the known accumulations in the Cretaceous oil sands have ranged from 280 to 400 billion bbl. Most recent estimates concur in predicting a recovery of about 300 billion bbl (Table 10; Figure 26), of which about 60 billion bbl will be recovered by open-pit mining techniques. It is interesting that although the estimates of the volume of bitumen in place have increased through the years, the estimates of the recoverable bitumen have remained essentially constant at about 300 billion bbl. This may reflect the realization, after a few years of bitumen production and upgrading, that the percentage of bitumen in place that can be profitably recovered is not as high as originally hoped for; although it was estimated in the 1970s and 1980s that as much as 25–30% of the bitumen in place could be profitably recovered, estimates in the 1990s range from only 12 to 17%, higher in the open-pit operations but lower in subsurface in-situ projects.

Open-pit mining started in 1967 upon completion of the first upgrading plant by Suncor Energy. Production of synthetic crude oil (SCO, liquid oil derived

Table 10. Estimates of bitumen in place and recoverable crude oil from tar sands in Canada.

	Year of Publication	Author and/or Organization	Bitumen in Place (billion bbl)		Recoverable (billion bbl)
			Total	Athabasca	
1	1963	AO&GCB* (later AERCB)	956	626	330(300?)
2	1967	Phizackerley and Scott (BP)	710.8	626	
3	1968	Vigrass (Imperial Oil, consultant)	750	575	
4	1972	AERCB** (vide Rühl, 1982)	895	626	330
5	1974	Govier (AERCB)	895	626	330
6	1974	Jardine (Imperial Oil)	895	625	
7	1977	Bowman and Carrigy (AOSTRA)[†]	950	626	
8	1978	Outtrim and Evans (AERCB)	1350 (+1350 in carbonates)	869	690
9	1978	Spragins (Syncrude Canada)	895		
10	1979	AENR[††]	969	721	283
11	1979	AERCB	980	720	
12	1980	Carrigy (AOSTRA)	981 (+1350 in carbonates)	720	
13	1981	Janisch (Petro-Canada)	1350 (+1350 in carbonates)	867	
14	1981	Strom and Dunbar (AERCB)	980	720	
15	1981	Meyer and Dietzman (U.S. Geological Survey)	1350 (+1560 in carbonates)	869	109–159 (50)
16	1981	Wennekers (Petro-Canada)	1351 (+1351 in carbonates)	869	
17	1982	Wennekers (Kassa Oil & Gas)	1000 + (+312 in carbonates)		
18	1982	Rühl	981 (+1258 in carbonates)	723	283
19	1982	Schumacher	965–1120 (+1350 in carbonates)	869	
20	1984	Strom (AERCB)	1170 (+189 in carbonates)		
21	1984	Meyer et al. (U.S. Geological Survey, U.S. Department of Interior, Energy Information Administration, U.S. Department of Energy)	1248 (+315 in carbonates)	962	48.76 (reserves)
22	1984	Lennox (Gulf Canada Resources, after AERCB, 1981, Rpt. 82-18)	1162	863	
23	1985	Seifert and Lennox (Gulf Canada Resources)	1289 (+318 in carbonates)	934	47.7 (reserves)
24	1986	Carrigy (AOSTRA)	±1000 (+300 in carbonates)		200–300
25	1986a, b; 1987	Roadifer (Mobil)	1350[‡]	869	
26	1987	AERCB	1233 (+1245 in carbonates)	937	308
27	1988	Meyer and Schenk (U.S. Geological Survey)	1673 (+982 in carbonates)	1336	
28	1989	World Energy Council–Survey of Energy Resources	1711		401
29	1989	Houlihan and Evans (AERCB)	1680	1315	
30	1989	Meyer and Duford (U.S. Geological Survey and BP)	2077 (+439 in carbonates)	951	311
31	1989	AOSTRA	1700 (+816 in carbonates)	950	
32	1990	Meyer and DeWitt (U.S. Geological Survey)	2517 (including carbonates)		
33	1992	AERCB	2517 (including carbonates)		308
34	1994	National Energy Board (Canada)	2516 (including carbonates)		308
35	1999	Newell (Syncrude Canada)	1700–2500		300
36	2000	National Energy Board (Canada)	2516 (including carbonates)		308

*Alberta Oil and Gas Conservation Board, 1963, A description and reserve estimate of the oil sands of Alberta, Oct. 1963, 60 p. (vide Govier, 1974; Outtrim and Evans, 1977; and Rühl, 1982).
**Alberta Energy Resources Conservation Board.
[†]Alberta Oil Sands Technology and Research Authority.
[††]Alberta Department of Energy and Natural Resources (vide Rühl, 1982).
[‡]Heavy oil and tar.

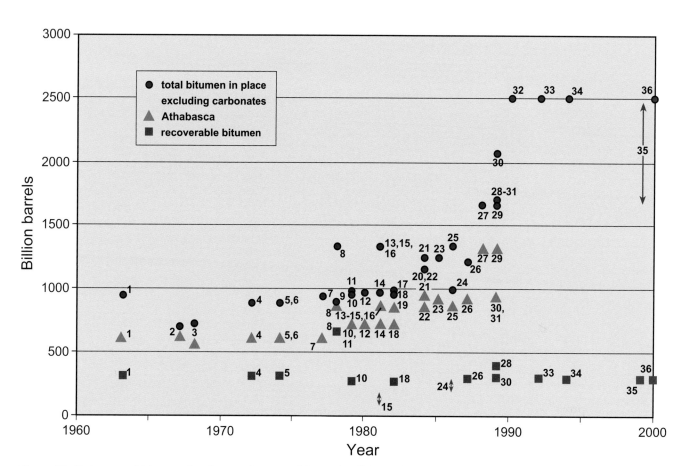

Figure 26. Estimates of bitumen in place and recoverable crude oil from the oil sands of Canada.

from the upgrading of crude bitumen) increased from 454,000 bbl in 1967 to 20.4 million bbl in 1978 (Appendix Table 43; Figure 27). When a second upgrading plant (of Syncrude Canada) started operations in 1978, production from open-pit mining operations increased more rapidly, reaching 102.7 million bbl of SCO in 1995, 117 million bbl in 2000, and 161 million bbl in 2002. A third open-pit mine, the Athabasca Oil Sands Project (AOSP), started operations in the spring of 2003 and is expected to reach a daily production of 155,000 bbl of SCO by the fall of the year. With the opening of the new mine and expected greater production from both Suncor Energy and Syncrude Canada, production of SCO from mining operations should exceed 230 million bbl in 2004.

Most major oil companies operating in Canada are involved in a variety of pilot and experimental programs for subsurface (in-situ) bitumen recovery principally using thermal methods involving the injection of heated fluids such as steam. Alberta's Energy and Utilities Board (AEUB) lists 37 pilot and experimental in-situ commercial projects in oil sands deposits at the end of 2002. Production of bitumen (shown as crude bitumen in Appendix Table 43; Figure 27) from these

projects started in 1976 and has increased from 2.7 million bbl that year to 54.3 million bbl in 1995, 105.5 million bbl in 2000, and 110.5 million bbl in 2002 (Appendix Table 43; Figure 27). This rapid increase has been largely caused by the development of the Steam-Assisted Gravity Drainage (SACD) production technique; two parallel, superimposed horizontal wells are drilled near the base of an oil sand, and steam is injected into the sand through the upper well. The steam heats the bitumen, which flows down by gravity into the lower well and is pumped to the surface. Recovery is expected to reach 50–60% of the bitumen in place, and production costs is expected to be slightly less than those for open-pit mining.

The combined annual production of crude bitumen and SCO from the Canadian oil sands reached 222 million bbl in 2000 and 271 million bbl in 2002.

As demand for liquid fuels in Canada increases and as the domestic production of conventional oil eventually starts to decrease, production of SCO and crude bitumen from the Canadian oil sands will continue to increase. It has been estimated that production from both open-pit mining operations and in-situ projects may reach 800 million bbl in 2010 if all of

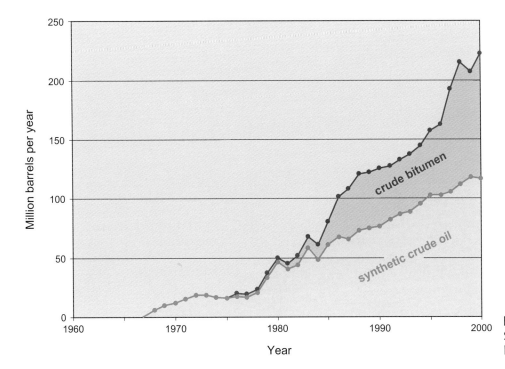

Figure 27. Canadian oil sands: Synthetic crude oil and crude bitumen production.

the currently proposed projects go into production. Bitumen and SCO production will make an important contribution to the oil production in Canada during the 21st century; it may provide more than half of the total Canadian oil production by the end of the first decade of the century.

THE ORINOCO OIL BELT OF EASTERN VENEZUELA

The heavy- and extra-heavy-oil deposits in eastern Venezuela, like the Canadian oil sands, are found in the gentle flank of an asymmetric sedimentary basin, the Eastern Venezuela Basin (Valera, 1981). However, unlike those in Canada, these deposits occur as a continuous belt 700 km (435 mi) long and 50–100 km (30–60 mi) wide just north of the Orinoco River. In addition, unlike the bitumen deposits of Canada, the Orinoco Oil Belt of Venezuela contains mostly extra-heavy oil, which, despite its low specific gravity (ranging from 4 to 17° API, but mostly 8–10° API) is movable at reservoir conditions because of the high reservoir temperature. The oil therefore flows into production wells without the need of thermal stimulation, although thermal methods, principally steam injection, have proved effective in increasing the flow rates.

The extra-heavy oil of the Orinoco Oil Belt occurs mostly in Tertiary sandstones, with lesser volumes in the underlying Cretaceous sandstones. The Venezuelan extra-heavy-oil accumulation does not crop out, occurring at depths ranging from 180 to 1100 m (600

to 3600 ft.). Toward the north (basinward), it grades into progressively lighter oil accumulations, some of which have been in production from numerous oil fields since the 1930s.

The first exploratory wells in the Orinoco Oil Belt were drilled in the 1930s by several private oil companies, but because of the very low specific gravity and high viscosity of the oil encountered, interest in the area faded until the late 1970s and early 1980s, when Petroleos de Venezuela S.A. (PDVSA, the Venezuelan national oil company) undertook a thorough exploration and evaluation program of the extra-heavy-oil accumulation, involving extensive geophysical surveys and the drilling of 662 wells.

Estimates of the volume of extra-heavy oil in place in the Orinoco Oil Belt have not changed as much through time as those for the Canadian oil sands (Table 11; Figure 28).

The first published estimates of the extra-heavy oil in place (Galavis and Velarde, 1967, 1972) amounted to about 700 billion bbl. Except for a few much higher estimates (as high as 2000–3000 billion bbl), estimates of about 700 billion bbl were generally accepted during the 1970s and early 1980s. In 1984, PDVSA announced the results of their detailed evaluation program of the Orinoco Oil Belt that indicated a volume of extra-heavy oil in place of 1200 billion bbl. Additional evaluation work since then apparently has not changed this estimate. Assuming an overall recovery of about 22%, PDVSA has estimated that 267–270 billion bbl of extra-heavy oil may be eventually produced from the Orinoco Oil Belt.

Table 11. Estimates of extra-heavy oil in place and recoverable oil for the Orinoco oil belt in Venezuela.

	Year of Publication	Author and/or Organization	Oil in Place (billion bbl)	Recoverable (billion bbl)
1	1967	Galavis and Velarde (MM&H)*	692	69.2 (10%)
2	1972	Galavis and Velarde (MM&H)	692.4	69.2 (10%)
			most interesting area = 582	58.2
3	1975	Swabb (Exxon)	600–3000	70
4	1975	Holmgren et al. (Mobil)	1050	
5	1977	Bowman and Carrigy (AOSTRA)**	700	
6	1979	Borregales (PDVSA)†	>1000	
7	1980	Carrigy (AOSTRA, after Gutierrez, 1981)	2000–3000	
8	1981	Meyer and Dietzman (U.S. Geological Survey)	700–3000	500
9	1981	Gutierrez (PDVSA)	2000	
10	1981	Wennekers (Petro-Canada)	700–3000	
11	1982	Schumacher	700–2100	
12	1982	Alcántara and Castillo (Lagoven)	>1000	
13	1982	Rühl	1100 (but as high as 2200–2700)	165–200 (15–20%) 330–540
14	1984	Zamora and Zambrano (Lagoven)	700–1000	
15	1984	Meyer et al. (U.S. Geological Survey, U.S. DOI,†† EIA (DOE)‡	700–1000	143
16	1984	Fiorillo (PDVSA)	1000	
17	1984	PDVSA	1200	267 (22%)
18	1985	Seifert and Lennox (Gulf Canada Resources)	1170	reserves = 57
19	1986	Meyer and Schenk (U.S. Geological Survey)	1200	264
20	1987	Martinez (consultant)	1182 (13° API or heavier)	
21	1987	PDVSA	1200	271 (22%)
22	1987	Roadifer (Mobil)	1200	
23	1987–1988	Fiorillo (PDVSA)	1200	267
24	1989–1990?	AOSTRA**	1524	
25	1990	Masters et al. (U.S. Geological Survey)	1200	267
26	1992	PDVSA Newsletter	1200	264 (22%)
27	1993	PDVSA Newsletter	1200	270 (22%)
28	1994	PDVSA Newsletter	1200	270
29	1996	Croft and Stauffer (Pantera Petroleum)	>1200 (6–13° API)	
30	1998	PDVSA (*Oil & Gas Journal,* October 19, 1998, p. 49)	1300–1800 (includes heavy oil)	270
31	2000	Moritis (*Oil & Gas Journal*)	1200	240–270
32	2001	PDVSA (*Oil & Gas Journal,* October 29, 2001, p. 48)	>1200	100–300
33	2001	Espinosa et al. (PDVSA-Intevep)	>1200	

*Ministerio de Minas e Hidrocarburos.
**Alberta Oil Sands Technology and Research Authority.
†Petroleos de Venezuela S.A. (according to *Oil & Gas Journal*, v. 78, no. 5 [February 4, 1980], p. 72–76; at UNITAR Conference on long-term energy resources (November–December 1979, Montreal).
††U.S. Department of Interior.
‡Energy Information Administration, U.S. Department of Energy.

These volumes of oil in place and of recoverable oil are still generally accepted as well grounded by PDVSA and other assessors (Table 11; Figure 28).

Commercial production of the extra-heavy oil from the Orinoco Oil Belt did not start until the late 1980s, when it was first used to prepare Orimulsion, an emul-sion composed of extra-heavy oil, water, and a sur-factant. The first shipment of this boiler fuel occurred in 1988. Since then, production of 7.5–9.5° API extra-heavy crude from the eastern part of the Orinoco Oil Belt (the Cerro Negro area) for the preparation of Ori-mulsion has gradually increased to 4.9 million bbl in

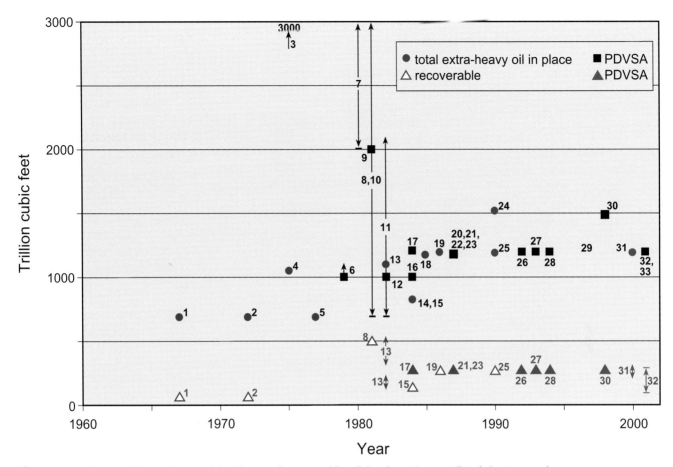

Figure 28. Estimates of extra-heavy oil in place and recoverable oil for the Orinoco Oil Belt in Venezuela.

1991, about 10 million bbl in 1993, 17.77 million bbl in 1995, and about 21 million bbl in 1997. If Orimulsion receives the acceptance as a fuel that PDVSA expects, production of this fuel from the extra-heavy oil of the Orinoco Oil Belt should increase in the future.

In the 1990s, PDVSA entered into four 35-year joint-venture projects (called "strategic associations") with several private oil companies for the commercial development of the Orinoco Oil Belt (Treviño, 1998). Better understanding of the reservoirs, the availability of 3-D seismic surveys and horizontal drilling, and the use of advanced production techniques have made the profitable production of the extra-heavy oil possible.

The extra-heavy oil, with a specific gravity of 8–10° API, is blended at the wellhead with a diluent (lighter crude oils, condensates, or 47° API naphtha) to produce a 16–17° API blend that is shipped north through 200–315-km (125–195-mi) pipelines to industrial complexes on the Caribbean Sea coast, where it is upgraded to 19–32° API synthetic crude and exported by tanker. The diluent is recovered and returned by pipeline to the production areas for reuse.

Production started in August 1998 at a rate of about 30,000 bbl/day and has increased progressively to about 240,000 bbl/day in 2000 and 470,000 bbl/day in 2003. It is expected that the production will reach about 630,000–650,000 bbl/day (230–237 million bbl/yr), the presently planned capacity of the upgraders, some time during the 2010s and remain at that level during the rest of the life of the joint-venture contracts. The 630,000–650,000-bbl/day level of extra-heavy-oil production will yield about 575,000 bbl/day of upgraded oil (210 million bbl/yr).

Over the 35-year lifetime of the projects, it is expected that about 7.5–8 billion bbl of extra-heavy oil will be recovered from the Orinoco Oil Belt, a small percentage of the extra-heavy oil that has been estimated that can be ultimately recovered. Higher recoveries could be obtained if the upgrading capacity of the industrial complexes is increased.

During the remaining years of the 21st century, oil production from the Orinoco Oil Belt can continue at the presently planned or higher rates. It will make, in any case, an important contribution to Venezuela's oil production during the 21st century.

THE BITUMEN AND EXTRA-HEAVY- AND HEAVY-OIL DEPOSITS OF THE FORMER SOVIET UNION

A problem with the published estimates of the bitumen and extra-heavy- and heavy-oil deposits of the former Soviet Union is that most of them do not specify whether they include just bitumen or bitumen and extra-heavy oil, or bitumen, extra-heavy oil, and heavy oil of unspecified gravity.

The most detailed review of the occurrence of bitumen and extra-heavy and heavy oil in the FSU is that of Meyerhoff and Meyer (1987). They reported the occurrence of 782 billion bbl of bitumen in place, of which 647 billion bbl (83%) is in the Siberian Platform and 109 billion bbl (14%) in the Volga-Urals basin. They estimated the occurrence of an additional 919 billion bbl of heavy and extra-heavy oil in place in the FSU (729 billion bbl [79%] in the Siberian Platform and 77 billion bbl [8%] in the Volga-Urals basin), of which 109 billion bbl may be recoverable. Meyer et al. (1984a) had previously estimated the volume of natural bitumen in place in the FSU at 1153 billion bbl and the recoverable reserves at 112 billion bbl. They also included estimates of the amounts of the combined heavy- and extra-heavy-oil accumulations, but they did not provide separate estimates for heavy and extra-heavy crudes. The Meyer and Schenk (1988) estimate is close to the one of Meyerhoff and Meyer: 792 billion bbl. The estimates of Meyerhoff and Meyer, Meyer and Schenk, and Meyer et al. are based on reviews of publications by FSU geologists. The inconsistent use of the terms "reserves" and "resources" make some of the estimates by FSU authors difficult to use.

Other published estimates of the bitumen, extra-heavy oil, and heavy oil in tar-sand deposits of the FSU are those of Rühl (1982) (732 billion bbl), Kruyer (1984) (600 billion bbl), and the Alberta Oil Sands Technology and Research Authority (1989) (1325 billion bbl). None of these estimates, however, specifies whether these figures are for oil in place or supplies the source of the estimates. Other estimates of the size of the bitumen and extra-heavy- and heavy-oil deposits in the FSU indicate that they may be large but not yet well defined. It appears, however, that unlike the Canadian and Venezuelan tar sands, extra-heavy oil and natural bitumen in the FSU occur in numerous separate accumulations distributed over a very large area. More precise and reliable estimates of the size and location of the extra-heavy oil and natural bitumen deposits of the FSU and their production statistics must await the availability of more recent publications not known or available to me at this time.

Production of heavy and extra-heavy oil in the FSU is reported by Meyerhoff and Meyer (1987) to have reached 31,295 bbl in 1983 [3.9 million bbl in 1985(?)]. More recent production reports have not been found.

ESTIMATES OF WORLDWIDE VOLUME OF BITUMEN AND EXTRA-HEAVY OIL IN PLACE AND AMOUNT TO BE EVENTUALLY RECOVERED

Careful estimates of the volume of bitumen and extra-heavy-oil accumulations have been carried out only in Canada and Venezuela. Little reliable information is yet available about the bitumen and extra-heavy-oil accumulations of the FSU, although enough is known about them to be able to state that they are probably large but smaller than those of Canada and Venezuela. No large bitumen and extra-heavy-oil deposits are known at present from anywhere else in the world, and although the possibility of finding other giant tar sands in little-explored basins cannot be entirely dismissed, such an eventuality does not appear likely.

Because of this uncertainty concerning bitumen and extra-heavy-oil deposits other than those in Canada and Venezuela, estimates made over the years of the in-place volume of the world's deposits of bitumen and extra-heavy oil and of the total amount of bitumen and extra-heavy oil to be eventually recovered from them have differed widely. The estimates of bitumen and extra-heavy oil in place that appear to be most authoritative and credible range between 2000 and 4000 billion bbl, but some are much higher. Carrigy (1980, p. 300), for instance, stated that "based on the scant information we have, it is estimated that bituminous sand deposits contain from 5 to 10×10^{12} (trillion) barrels of crude bitumen..." Wennekers (1981) estimated that the bitumen and heavy oil in place of Canada, Venezuela, the FSU, and the United States may reach 6000 billion bbl. He did not specify, however, how much of this amount is heavy oil or what gravity oil he included under this category. Meyer et al. (1984a) state that worldwide in-place heavy and extra-heavy oil and bitumen occurrences may exceed 5000 billion bbl. Similarly high estimates were favored by Meyer and Schenk (1988) and Meyer and Duford (1989). Other estimates of the worldwide volume of bitumen and extra-heavy oil in place (Demaison, 1977; Grathwohl, 1982; Schumacher, 1982; Roadifer, 1987) range from 2100 to 2800 billion bbl.

As previously discussed, at present, the most generally accepted figure for bitumen in place in the Canadian oil sands is 1700 billion bbl (excluding the more controversial accumulations in Paleozoic carbonate reservoirs) and 1200 billion bbl of extra-heavy oil in place in the Orinoco Oil Belt in Venezuela, for a total of 2900 billion bbl. To assign a figure to the bitumen and extra-heavy-oil deposits of the FSU is, of course, very risky. I have chosen 800 billion bbl, giving a total of 3700 billion bbl for the bitumen and extra-heavy oil in place in Canada, Venezuela, and the FSU. No substantial volumes of bitumen and extra-heavy oil in place are believed to occur in tar sands deposits in other parts of the world. A probably optimistic figure of 100 billion bbl has been assigned to these deposits for a total worldwide volume of 3800 billion bbl of bitumen and extra-heavy oil in place.

The estimates of the bitumen and extra-heavy oil to be eventually recovered worldwide also vary widely. Most range between 600 and 1100 billion bbl, but W. H. Dorset of Chevron estimated (*Oil & Gas Journal*, April 29, 1991, p. 75–76) that 3000 billion bbl of extra-heavy oil and bitumen may be recovered in the world at a producing cost of $40–60 per barrel.

For the purpose of this study, I will use an estimate of 700–750 billion bbl for the amount of bitumen and extra-heavy oil to be ultimately recovered from the known and potentially discovered deposits: 300 billion bbl from Canada, 270 billion bbl from Venezuela, and about 150 billion bbl from the FSU, which is approximately 18–20% recovery from the 3800 billion bbl of estimated worldwide bitumen and extra-heavy oil in place.

It would not be unreasonable, however, to predict that the ultimate amount of oil to be recovered from bitumen and extra-heavy-oil accumulations will be larger than is now estimated. The estimates of the ultimate recovery of oil have consistently increased during the last few decades, and there is no reason to believe that the same will not be true in the case of bitumen and extra-heavy-oil accumulations as future exploration identifies new ones and increases the size of those already known and as new and more efficient recovery technology is developed and used.

NATURAL GAS

ABSTRACT

Natural gas has become the fuel of the future, and it is expected to overtake and pass oil as a source of energy early in the 21st century. However, this was not always true. Until the last few decades of the 20th century, the high cost and difficulty of transporting and distributing natural gas, particularly from regions of the world far from centers of gas consumption, restrained interest in exploring for and producing natural gas. However, the realization that natural gas is abundant and the cleanest burning and most efficient of the fossil fuels (the fuel of choice in the generation of electricity) has created great interest in natural gas in the last four decades. Production of gas has increased fivefold, and its reserves have grown more than sevenfold since 1960. The FSU and the Persian Gulf region have 71% of the world's gas reserves. Significant volumes of gas are "stranded" in large but undeveloped fields, with transportation still remaining as the key factor in the production and use of natural gas. The very large estimates of the ultimate recovery of gas from conventional plus unconventional reservoirs (25,000–30,000 tcf) warrant that natural gas will be a major, if not the main, source of energy in the 21st century, particularly when the gas-to-liquids process becomes extensively used worldwide.

INTRODUCTION

Natural gas, a clean-burning, abundant fuel that is capable of highly efficient electric power generation, is now generally acclaimed as the fuel of the future, the fuel of choice, and it might very well be. It was not always so. Although gas is said to have been used in China as far back as 200 B.C. (Meyerhoff, 1980), until as recently as 1960, natural gas was regarded as a nuisance, a less-valued by-product of crude oil exploration and production to be gotten rid of as cheaply as possible, commonly by venting it to the atmosphere or by flaring it. Exploratory drilling for natural gas was not seriously considered, and finding gas when searching for oil was commonly considered a calamity.

The main reasons for this lack of interest in discovering and producing natural gas were the low price of gas and the difficulty and high cost of transporting and distributing it. Natural gas can only be transported through pipelines or in liquefied form in special, very expensive cryogenic tankers. At one time, it was estimated that to move comparable amounts of energy through a pipeline, the cost of moving gas was 4–5 times higher than that for oil and 20 times higher by tanker. There was little incentive, therefore, to discover and produce natural gas if considerable distance separated the natural-gas deposits from the centers of consumption, and such was certainly the case for what we now know are some of the largest accumulations of natural gas in the world, in the FSU and in the Persian Gulf region. During the 1950s, only in the United States was natural gas produced, transported, and consumed in substantial amounts. In 1950, the United States accounted for 90% of the world's natural-gas production and was still producing 75% of the world's total in 1960 (Masters, 1993).

The picture has changed dramatically in the last two or three decades as demand for natural gas has greatly increased: (1) markets for natural gas have developed in Europe, Japan, and elsewhere; (2) more gas was needed to be reinjected in oil fields in an advanced state of depletion; (3) the technologies of gas liquefaction and pipeline construction have improved, lowering the cost of long-distance natural gas transportation; and (4) environmental consciousness (emphasis on the protection of the environment) has become an important issue in the energy policy agenda, particularly in the developed world. Natural gas has indeed become the fuel of choice in most energy markets worldwide, particularly in the power-generation sector, where gas has been able to compete successfully with coal and has become the chosen fuel for the majority of new electric power-generating plants using combined-cycle gas turbines. Natural gas has been found to be clean, cost-competitive, and plentiful.

This increase in the demand for natural gas has made possible in the last decades the construction of long, large-diameter pipelines from important but remote gas fields to industrial centers of gas consumption: from Algeria, across the Mediterranean Sea to Spain and Italy; in the northwest shelf of Australia; and from the prolific gas fields of the FSU to Europe. Several others are under construction or in the planning stage. In addition, as a result of the growth in the demand for gas, exploration for gas has intensified, and the capacity of the cryogenic-tanker fleet to transport liquefied natural gas (LNG) has considerably expanded.

Still, as late as the late 1980s, 88% of all natural gas produced in the world was consumed in the country in which it was produced; it remained a domestic fuel. Even at present, transportation, instead of discovery and production, remains the key factor in the development of the natural-gas industry. Large volumes of potential natural-gas production still remain stranded for lack of commercial transportation to centers of consumption.

Forecasts for the increasing production and use of natural gas are generally glowing, but the challenges to reach these objectives are still formidable. In addition to transportation, most important perhaps are the large capital requirements and the need for technological developments that would result in cost reductions to ensure competitiveness with alternative energy sources. The development of close international cooperation between all components of the world natural-gas industry to be able to make possible the construction of large-diameter pipelines that need to cross international borders will also be critical.

Natural gas is known to occur under many different conditions. Gas accumulations have been assigned to two general groups depending on the feasibility of the recovery of their contained natural gas: Conventional natural-gas accumulations are those in reservoirs with a permeability high enough to allow gas to flow at rates that make its production profitable at current development and marketing costs. They are generally supported by water. Unconventional accumulations, on the other hand, are those from which gas is not profitably recovered for one reason or another at a certain time. Schmoker (2002, p. 1993) describes unconventional gas accumulations as "continuous accumulations... that exist more or less independently of the water column," and discusses several methods for their assessment.

Some of the sources of natural gas labeled unconventional in the past have produced gas profitably in the United States for at least a decade or two and will undoubtedly become major sources of energy in the decades to come. Not only in the United States but in many other countries of the world, such unconventional sources of natural gas are known to occur, and although they are not given serious recognition now, they will certainly be developed sometime in the future. Foremost in this category of former unconventional sources of natural gas in the United States are gas in coal beds (the coalbed methane), gas in low-permeability sandstones (tight sands), basin-centered gas accumulations, and gas in organic black shales (also known as fractured shales). They now contribute 26% of the total United States gas production (Law and Curtis, 2002).

Coalbed methane, which not too long ago was considered an unconventional type of gas accumulation (and the cause of deadly accidents in underground coal mines), is now being produced profitably in several basins in the United States (about 7% of the total United States natural-gas production) and in Australia and is being investigated in other parts of the world. Natural gas from tight-sand reservoirs is also becoming a substantial part of the total natural gas produced in the United States. Lesser production has been obtained for many years from organic black shales and from basin-centered reservoirs. Production of the natural gas in methane hydrates and in solution

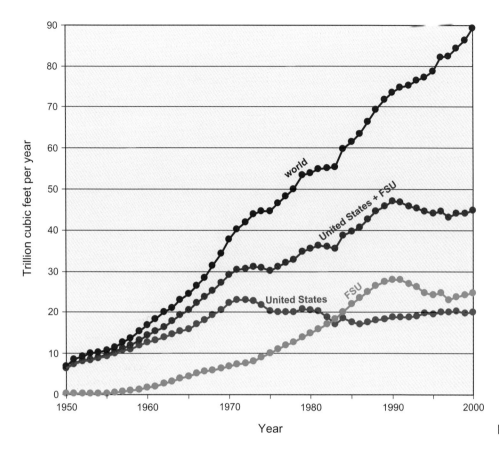

Figure 29. Natural-gas production.

in geopressured formation waters will probably not be economically viable for many years, if ever. These last two occurrences of natural gas may be better called speculative.

In the following discussion, conventional natural-gas deposits will be discussed first, followed by sections on coalbed methane, tight-sand gas, basin-centered gas accumulations, gas in organic black shales, and the speculative natural-gas accumulations.

Statistics on natural-gas production, reserves, etc., are generally given in cubic feet (or billions [bcf], or trillions of cubic feet [tcf]) or in cubic meters and multiples. When necessary to convert from cubic feet of gas to energy-equivalent barrels of oil to compare their contribution as sources of energy, the conversion factor commonly used is 6000 ft^3 of gas as equivalent to 1 bbl of oil.

CONVENTIONAL NATURAL GAS

Production

As mentioned above, until about 1960, the United States was the only country in the world producing, transporting, and consuming significant amounts of

natural gas. However, production of natural gas in the FSU, minor in the 1950s, rapidly increased in the 1960s and 1970s and caught up with and passed United States production in 1983. Despite a drastic reduction of its natural-gas production after the breakup of the FSU in 1991, the FSU has remained the world's largest producer of natural gas, followed closely by the United States. The United States and the FSU, which, until 1970, had together produced more than 75% of the natural gas in the world, progressively lost their prominent position as gas producers. By 1990, the combined gas production of the United States and the FSU amounted to 64% of the world's production, and by 2000, it amounted to only 50%.

Canada and the UK, having 7 and 4.5%, respectively, of the world's 2000 gas production, follow the FSU and the United States as important gas producers. Six other countries, Algeria, Indonesia, Iran, the Netherlands, Norway, and Saudi Arabia, account for another 15.5% of the world's gas production. These 10 countries produced 77% of the world's gas in 2000.

Yearly natural-gas production for the world and for the United States and the FSU from 1950 to 2001 is shown in Appendix Table 44 and Figure 29. Cumulative natural-gas production for the world during the same time span is shown in Appendix Table 44 and Figure 30.

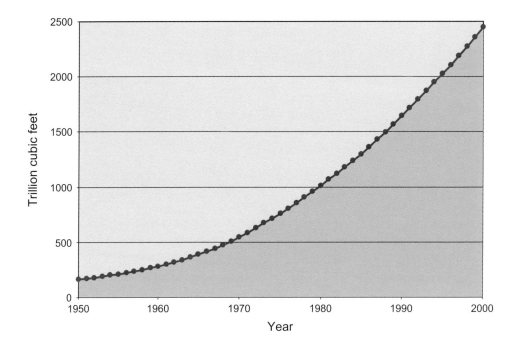

Figure 30. World cumulative natural-gas production.

It has been predicted that because oil production has increased slowly since the mid-1970s (see Appendix Table 36; Figure 19), whereas gas production has increased at a very rapid rate since the early 1960s (it increased more than five times from 1960 to 2000), natural gas is expected to overtake and pass crude oil as a source of energy early in the 21st century. In 1960, natural-gas production was equivalent to about 37% of the oil production; it increased to about 38% in 1970, 41% in 1980, 55% in 1990, and 60% in 2000.

The volumes of natural gas shown in Appendix Table 44 and Figures 29 and 30 do not distinguish associated gas (natural gas that occurs in association with oil in a reservoir, either as free gas or in solution) from nonassociated gas and do not include vented, flared, or reinjected gas. Until the 1950s, most of the gas recovered as a by-product of oil production was vented to the atmosphere or flared. Lesser amounts were reinjected to maintain reservoir pressure and enhance oil production. Although less gas is now flared as the demand for gas has increased, particularly in the United States and other developed countries, there are still considerable amounts of gas vented and flared in many oil-producing regions of the world, largely because of lack of market or lack of a collection and distribution system for natural gas. Gas flares in Russia, the Niger Delta, the Algerian Sahara, and the Persian Gulf region still provide the source of some of the brightest light spots on Earth at night.

How much gas has been vented, flared, or reinjected since oil production started almost 150 years ago is not known, and most of the few estimates that have been made generally cover only the most recent decades. Klett and Gautier (1993), using maximum and minimum values for the gas-oil ratios of known oil fields, have calculated that, from the time oil was first produced until 1990, a total of between 154 and 257 tcf of associated and dissolved natural gas were vented or flared in the world. This represents between 9 and 15% of the cumulative natural-gas production to 1990, a relatively small percentage. These figures, however, may be grossly underestimated. Bois (1982) has estimated that half the associated gas ever produced was either flared or reinjected, an average of about 7 tcf/yr, a figure considerably higher than the estimate of Klett and Gautier (1993).

Of the total amount of natural gas vented or flared, the major part occurred from the 1940s to the late 1970s. Since then, the volumes of gas vented or flared have decreased because of the increased profitability of gas production and the increase in the amount of gas reinjected into oil reservoirs.

It has been estimated that 4.35 tcf of gas was vented or flared in 1991 (about 5.8% of the world gas production), to which the OPEC members contributed about two-thirds (2.9 tcf). During the late 1990s, the amount of vented or flared gas has been estimated to range between 3.5 and 3.8 tcf/yr, representing about 4.3–4.6% of the total world natural-gas production, confirming the reduction in the volumes of natural gas vented or flared in more recent years.

Because some of the gas recovered during the production of oil has been vented, flared, reinjected for pressure maintenance into oil reservoirs, or used for

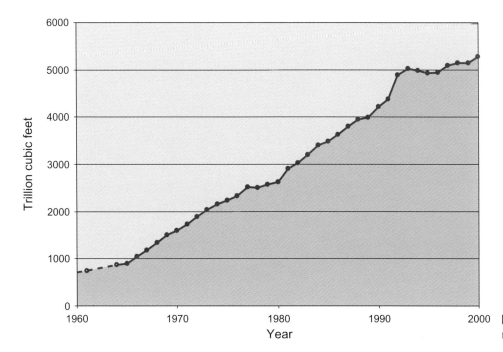

Figure 31. World natural-gas reserves.

plant and station fuel, some sources of gas-production statistics distinguish between total (gross) production and marketed production. As more of the produced associated gas is marketed, the difference between these two volumes has steadily decreased. In 1968, the marketed world production was 79.5% of the gross production; by 1978, it had increased to 83.4%; and by 1991, marketed production represented 94.5% of the world's gross production of natural gas. It has remained at this level during the 1990s.

The volumes of natural-gas production shown in Appendix Table 44 and Figures 29 and 30 are believed to represent marketed production, although most sources of these statistics do not specify whether they correspond to marketed or gross natural-gas production.

Reserves

For several reasons, the estimates of natural-gas reserves are not as influenced by political expediency and economic ulterior motives as those of oil. Unlike the case of oil, the amount of their gas reserves is not an issue with the OPEC members because they have no quotas for their gas production. However, the possibility that these and other countries may have made public natural-gas "political reserves" at one time or another cannot be entirely discarded.

However, how different countries use the terminology of reserve estimation is as much of a problem with natural gas as it is with oil. A problem also exists regarding the reserves estimates for discovered natural-gas fields that, because gas was not economically viable until recently, are underdeveloped or remain undeveloped and unproductive—stranded for lack of a market. The reserves of these fields may be greatly underestimated. However, also as in the case of oil, the order of magnitude of the discrepancies, particularly concerning the total world natural-gas reserves, is well within the degree of accuracy needed for this study. The ultimate resource of natural gas is known to be very large and will certainly become much larger in the future.

Appendix Table 45 and Figure 31 show the estimates of world natural-gas reserves since 1966 as reported by the *Oil & Gas Journal*. Other sources consulted report comparable figures. The 1961 estimate is from Weeks (1962, 1963), and those for 1964 and 1965 are from *World Oil*. No other estimates for natural-gas reserves were found for the years before 1966. All sources indicate a steady increase in the estimates of the world's natural-gas reserves since the 1960s; they have doubled in the last 20 years and more than tripled in the last 30 years.

Two regions in the world, the FSU and the Persian Gulf region, dominate the present estimates of the world's natural-gas reserves; combined, they contain about 71% of the total known natural-gas accumulations in the world (35.5% in the FSU and 35.5% in the Persian Gulf region). The share of Russia and Iran represents 45%. At least 16 supergiant gas fields are known in the FSU (including the Urengoy and Yamburg fields), and probably the largest known gas field

Table 12. Natural-gas reserves (as of December 31, 2002) (in billion cubic feet).*

Total world	5,501,424
Former Soviet Union	1,952,600
Russia	1,680,000
Persian Gulf region	1,952,670
Iran	812,300
Qatar	508,540
Saudi Arabia	224,200
Abu Dhabi	196,100
Iraq	109,800
United States	183,460
Algeria	159,700
Venezuela	148,000
Nigeria	124,000
Indonesia	92,500
Australia	90,000
Norway	77,300
Malaysia	75,000
The Netherlands	62,000
Canada	60,118

*Source: *Oil & Gas Journal*, December 23, 2002.

in the world, the North field, has been discovered in Qatar; it is now estimated to contain 500 tcf of gas but may be larger. Other supergiant gas fields have been identified in the Persian Gulf region (the South Pars field in Iranian waters is also among the largest in the world) but have not been developed because of the lack of large-scale industries in the region, its small population, and the long distance to centers of natural-gas consumption.

Table 12 shows the 16 countries containing the largest reserves of natural gas.

Natural-gas reserves, like gas production, are expected to catch up and pass oil reserves early in the 21st century, because they have been growing at a fast, steady rate, whereas oil reserves have not been reported to have increased appreciably in the last 10 years. In 1960, natural-gas reserves amounted to about 40% of the oil reserves. This percentage has increased steadily since then: 43% in 1970, 68% in 1980, 70% in 1990, and 85% in 2000. As mentioned above, natural gas will inevitably become an increasing source of supply of the world's energy demand during the 21st century.

Reserves-to-production Ratios

The computation of the world's natural-gas reserves-to-production ratios (R/PR), the time it will take to deplete the current reserves at the present rate of production, is recorded in Appendix Table 46 and shown in Figure 32.

While the oil R/PRs have remained fairly consistent at about 30–45 since the late 1950s, the R/PRs of natural gas have increased from about 40 in the early 1960s to 60–65 in the 1990s, reflecting the sustained discovery and development of new gas fields

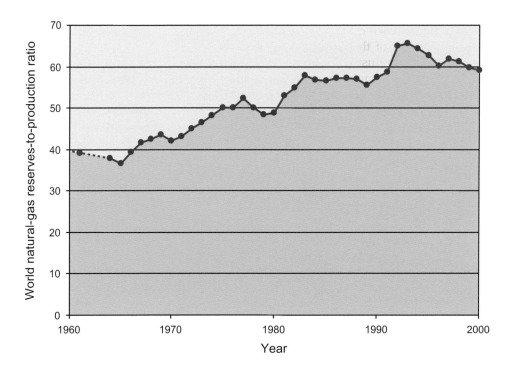

Figure 32. World natural-gas reserves-to-production ratios.

in the last 35 years and the consequent increase in the volume of natural-gas reserves. The gas R/PRs decreased in the mid-1990s, but they are expected to resume their increase as new gas fields are discovered, as presently known but undeveloped fields are put in production, and as a result, the addition of new reserves exceeds the rate of gas production.

Estimation of the Ultimate Recovery of Conventional Natural Gas

As in the case of estimates of the ultimate recovery of oil, the estimation of the ultimate volume of natural gas that will be recovered from all gas accumulations in the world (what has also been called the resource base) is the most critical factor in predicting future availability of natural gas as a source of energy. In addition, as in the case of oil, it includes (1) the gas already produced, (2) the reserves recoverable from presently known fields, (3) the volumes of natural gas yet to be discovered, and (4) the gas in known accumulations, uneconomic or undeveloped at present, but that may become profitable in the future.

Because gas has only been a desirable source of energy in the last four decades, the first estimates of the ultimate amount of natural gas to be produced were not made until the late 1950s. Understandably, the earlier ones were commonly very tentative. Those reported in readily accessible publications and which I have been able to document are listed in chronologic order in Table 13 and shown in Figure 33. The author and/or organization and the amount of the estimate are indicated in each case.

As in the case of the estimates of the ultimate recovery of oil, we must realize that the different estimates of the ultimate recovery of natural gas may have been made using somewhat different methods, perhaps for different purposes, and with variable degrees of knowledge of the various gas provinces of the world. Many of the estimates listed in Table 13 are not original; they are those of others considered particularly authoritative by the author or represent the average of previously published estimates.

In addition, as in the case of oil, the estimates generally do not specify the methods used, the assumptions made concerning future economic conditions, the future price of gas, or possible advances in the technology to explore for, discover, and produce natural gas, nor is it always clear whether the estimates are for gross or marketed production and whether they include the volumes of gas vented, flared, or reinjected. Many estimates do not state whether they include volumes of cumulative production. Others state that they include only current proved reserves plus potential discoveries and not cumulative production, apparently because natural-gas production statistics are not as reliable as those for oil. None of the estimates reviewed mentions reserve growth, just as important a factor in the estimation of the ultimate recovery of natural gas as it is in the case of oil. (An exception is the U.S. Geological Survey 2000 estimate that includes 3660 tcf of reserve growth.) No mention is made either if the estimates include the volumes of gas expected to be recovered by the application of enhanced recovery methods. This factor, however, is not as important in the case of natural gas as it is in the case of oil, because a very high percentage of the gas in gas reservoirs is recovered by primary recovery.

Nevertheless, the general trends of the order of magnitude of the estimates provide sufficient information to serve the objectives of this study.

As can be seen in Table 13 and Figure 33, the estimates of the volume of natural gas to be ultimately recovered from gas fields now known or expected to be discovered in the future have steadily increased since the first estimates were made in the late 1950s, from 5000 to about 14,000 or 16,000 tcf. Some higher estimates have been made; for instance, Fisher (1994a, p. 106) states that "if indeed the impact of technology in the exploratorily mature United States over the past decade were extrapolated globally, the remaining gas resource would double to about 20,000 tcf." I agree with him. Cedigaz (Appert, 1998) has estimated the ultimate volume of natural gas to be recovered in the world at between 14,124 and 17,655 tcf.

The U.S. Geological Survey 2000 estimate of ultimate recovery of natural gas is 15,401 tcf, including 3660 tcf of reserve growth and 5196 tcf of natural gas yet to be discovered. Reserve growth of natural-gas fields, also frequently overlooked or underestimated as in the case of oil, should be a crucial component of any estimate of the ultimate recovery of natural gas (see Schmoker and Klett, 2000).

Considering the world's cumulative natural-gas production of about 2500 tcf at the end of 2000 and the estimated reserves of about 5500 tcf, it would be necessary to add in the future about 7000 tcf of gas to the volume already discovered to reach an estimated ultimate recovery of 15,000 tcf. To reach the high estimates of 17,000 and 20,000 tcf, it would be necessary to add as much as 9000–12,000 tcf, i.e., 115–150% of the volume of natural gas already discovered. Will this be possible?

A critical ingredient in estimating the ultimate world recovery of natural gas is the part of such estimate corresponding to the FSU, about which not enough is known at the present time. Estimates of the ultimate

Table 13. Estimates of the ultimate recovery of natural gas (estimates are shown in Figure 33 by listed number).

	Year of Publication	Author and/or Organization	tcf
1	1956	U.S. Department of Interior*	>5000
2	1958	Weeks (Standard Oil Co., New Jersey)	>5000–6000
3	1959	Weeks (Standard Oil Co., New Jersey)	6000
4	1961	Weeks (Weeks Petroleum Corp.)	6000
5	1962	Weeks (Weeks Petroleum Corp.)	12,000
6	1962	Hubbert (Shell)	7500
7	1965	Hendricks (U.S. Geological Survey)	15,280
8	1967	Ryman (Standard Oil Co., New Jersey)**	12,000
9	1967	Royal Dutch Shell	10,200
10	1968	Weeks (Weeks Petroleum Corp.)	6900
11	1968	World Energy Council–Survey of Energy Resources	6000
12	1969	Hubbert (U.S. Geological Survey)	8000–12,000
13	1970	Weeks (Lewis G. Weeks Associates)	6900
14	1971	Weeks (Lewis G. Weeks Associates)	7200
15	1973	U.S. Federal Power Commission	16,370
16	1973	Hubbert (U.S. Geological Survey)	12,000
17	1975	Linden and Parent (Institute of Gas Technology)	9740
18	1975	Moody and Geiger (Mobil)	8164
19	1975	National Academy of Sciences	7821
20	1976	Adams and Kirkby (BP)	<6000
21	1976	Barthel et al. (Bundersanstalt für Geowissenschafteu und Rohstoffe)	8360
22	1976	Grossling (U.S. Geological Survey)	11,200–28,000
23	1977	Kalisch and Wander (American Gas Association)	10,510
24	1977	Parent and Linden (Institute of Gas Technology)	9150–9550
25	1977	Whiting (Texas A&M University)	7000
26	1978	Despaires (Institut Francais du Petrole)	7800
27	1978	McCormick et al. (American Gas Association)	11,430
28	1978	World Energy Council–Survey of Energy Resources	9980
29	1980	World Energy Council–Survey of Energy Resources	10,340
30	1980	Meyerhoff (Meyerhoff and Cox Inc.)	7670
31	1980	Parent (Institute of Gas Technology)	7900–9200
32	1980	Roorda (Shell)	10,200
33	1982	Bois (Institut Francais du Petrole)	10,343
34	1982	Parent (Institute of Gas Technology)	9000–10,560
35	1983	Riva (Library of Congress)	11,328
36	1983	Parent (Institute of Gas Technology)	9000–10,000
37	1984	Toens and Van der Merwe (Nuclear Development Corp., South Africa)	10,167
38	1987	Masters et al. (U.S. Geological Survey)	9280
39	1987	Pecqueur (Elf Aquitaine)	±7000–10,500
40	1989a, b	Bookout (Shell)	10,200
41	1990	Masters et al. (U.S. Geological Survey)	10,769
42	1991	Masters et al. (U.S. Geological Survey)	10,782
43	1992	Masters et al. (U.S. Geological Survey)	10,512
44	1993	Masters (U.S. Geological Survey)	10,517
45	1994	Masters et al. (U.S. Geological Survey)	11,568
46	1994	Global Gas Resources Workshop C.L. Ruthven, ed.	14,490
47	1994a, b	Fisher (University of Texas)	>20,000
48	1995	Enron	14,024
49	1997	Enron	15,457
50	1997	Masters et al. (U.S. Geological Survey)	11,568
51	1997	Edwards (University of Colorado)	11,625
52	1998	Krylov et al. (various Russian institutions)	14,124
53	1998	Appert (Cedigaz)	14,124–17,655
54	2000	U.S. Geological Survey 2000	15,401
55	2001	Edwards (University of Colorado)	13,141

*According to Adams and Kirkby (1976).
**According to Hubbert (1969).

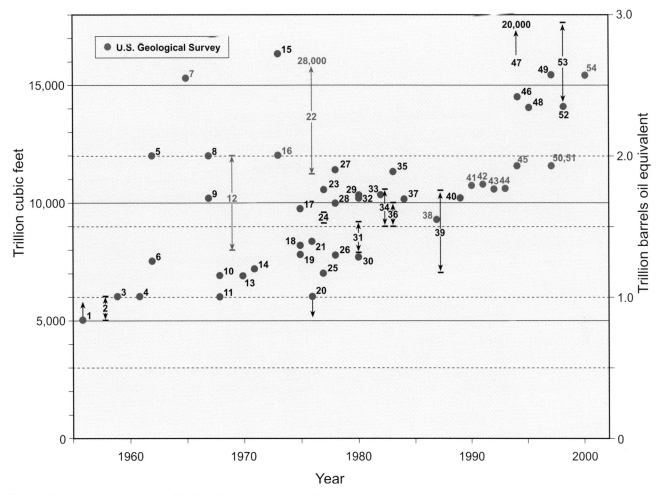

Figure 33. Estimates of the worldwide ultimate recovery of natural gas.

volume of natural gas to be recovered in the FSU have varied widely over the years. Most recent estimates range from about 7475 tcf for Russia only (Nemchenko, 1994) and 8850 tcf for the entire FSU (Staroselsky, 1994) to as much as 12,350 tcf (Ellis, 1994). Grace (1994, 1995) believes that the estimates of Nemchenko and Staroselsky and other official FSU figures are overly optimistic about the volumes of gas yet to be discovered in the FSU. He also points out (1995, p. 73) that "the Soviet concept of 'reserves' stretches the western concept beyond its normal limits." He proposes to reduce the estimates of Nemchenko and Staroselsky by 20% to about 5800 tcf for Russia and 6000 tcf for the FSU. Enron (1997) estimated the natural-gas potential resources of the FSU as of the end of 1995 (which apparently does not include the cumulative gas production to that date of about 527 tcf) as 5181 tcf.

Irrespective of what the potential for discovery and development of new natural-gas accumulations in the FSU may be, it seems safe to predict that the ultimate recovery of conventional natural gas will reach

at least 14,000 tcf. Gas reserves, R/PRs, and estimates of the ultimate volume of natural-gas recovery have been steadily increasing in the last three decades and can be expected to continue to increase for yet a few more decades.

The only factor that has held up the exploration for and the discovery of new gas fields in several prospective regions of the world, and the development of some known but undeveloped large fields, is the location of these prospective regions and of some known large fields far from consumption centers. However, the practical, economic, and environmental attractiveness of natural gas can be expected to overcome this current impediment, particularly if the process of converting gas to liquids (GTL) becomes economically attractive, solving the current long-distance transportation problems.

New gas accumulations are being discovered in many parts of the world; additional discoveries will be made. Reserve growth will inevitably occur in the presently producing fields. Many sedimentary basins in the world have not yet been adequately explored for gas, some because they are known to be predominantly gas prone and are distant from centers of gas consumption, others because of insufficient deep drilling to depths where gas is more likely to be found than oil. While no major new oil province is likely to be discovered and not many new giant and supergiant oil fields have been discovered in the last two or three decades, rich new gas provinces containing major natural-gas fields undoubtedly will be found as the result of more intensive and widespread search for gas.

Estimates of ultimate recovery of conventional natural gas of 17,000–20,000 tcf do not seem unrealistic and will be used in this study to estimate natural-gas production in the 21st century.

COALBED METHANE

The occurrence of natural gas in coal beds has been recognized for hundreds of years, because it is commonly released from the coal and occasionally causes deadly explosions in underground coal mines. Coal, as will be discussed in a following section, is the most abundant energy source in the Earth, and it should therefore be expected that large volumes of gas, generally called coalbed methane (CBM), should be present in coal deposits throughout the world, providing an enormous potential energy resource (Bibler et al., 1998). Estimates of global coalbed methane resources go as high as 12,640 tcf.

Attempts to drill into coal beds to draw out their contained gas and, in this way, reduce mining hazards were first made in Europe in the late 19th century, but recognition of its potential as an energy source and the exploration for and commercial production of coalbed methane did not occur until the late 1970s and early 1980s in the United States (in the Black Warrior basin of Alabama and in the San Juan basin of New Mexico and Colorado). Significant exploration for and production of coalbed methane in these basins began in the mid-1980s, encouraged by a federal tax credit granted for the production of this abundant energy resource (Section 29 of the Crude Oil Windfall Profits Tax Act of 1980). It was predicted by some that interest in the production of coalbed methane would decrease or even die with the termination of the tax credit at year-end 1992, but this has not been the case, probably because natural gas is considered a desirable clean source of energy, and because the development of new production technology has made possible the profitable production of coalbed methane even without the tax-credit stimulus. From its origin as an experimental coal-mine degasification procedure, coalbed methane

is becoming a promising new worldwide source of energy. In the United States, it has now grown to provide 7% of the total natural-gas production. Coalbed methane no longer deserves to be called an unconventional gas source; a mining hazard for many decades, it has become an economically viable conventional natural gas source.

Coalbed methane is generated during the coalification process, and most of it (98%) is found adsorbed in the coal, which is a microporous solid with large internal surface areas in the matrix pore structure that can adsorb very large amounts of gas. Lesser amounts occur as free gas in fractures (called cleats) and large pores, and/or dissolved in the groundwater present in pores and fractures; coal serves as both the source and the reservoir rock for the coalbed methane.

Gas produced from coal beds is composed mainly of methane (in excess of 95%) and very minor amounts of heavier hydrocarbons (mostly ethane and propane), nitrogen, and carbon dioxide (CO_2).

The gas content of coal beds generally varies with the rank of the coal and its depth of burial; the higher the coal rank and the greater the depth, and the resulting pressure to which the coal has been subjected, the greater its potential gas content. For a given coal rank, the adsorbed gas content increases with increasing pressure and decreases with increasing temperature.

Coalbed methane has been commercially produced from coal beds 1–30 m (3–98 ft) thick at depths ranging from 50 to 2500 m (160 to 8200 ft). Thicker producing coal seams generally sustain higher gas flow rates. In those close to the surface (less than 150 m [500 ft]), the potential of the coal for retaining the methane is considerably reduced, and those at depths greater than 1000–1500 m (3300–4900 ft) are commonly afflicted with various production problems. At present, most economically profitable coalbed methane production is from depths of 150–1500 m (500 to 4900 ft).

Profitable production and ultimate recovery of the gas depend on the desorption characteristics of the coal, on its water saturation, and particularly on the development of effective fracture permeability (cleat formation, orientation, and spacing, which depend on regional tectonic stresses, bed thickness, and coal rank). Commonly, permeability in coalbed methane reservoirs is low (1–10 md), virtually all of which is caused by the occurrence of fracture systems. To attain gas flows high enough to make profitable the production of coalbed methane, it is therefore generally necessary to enhance the existing fracture systems of the coal beds by means of hydraulic-fracture treatments.

One of the large obstacles to the economic recovery of coalbed methane is the detrimental effect of the water contained in the potentially productive coal beds; it

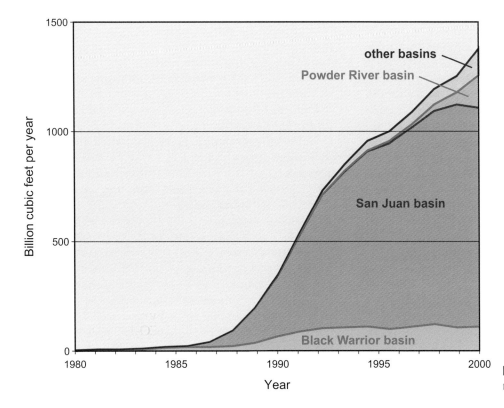

Figure 34. United States coalbed methane production.

inhibits desorption of the gas from the coal matrix and preferentially flows to the wellbore. Some wells may produce only water for months before gas begins to flow, and water production is reduced or ceases. The disposal of the water is commonly a serious problem. This and other problems, however, undoubtedly will be solved by new and better drilling, completion, and production technology as the demand for coalbed methane as a source of clean energy increases.

For some time, coalbed methane has been gathered and used in the United States, as well as in other countries, as a result of coal-mine degasification procedures to ensure safe mining operations and to avoid venting the gas into the atmosphere. Records of the volumes of coalbed methane obtained from this production are not readily available but are not believed to be significant. Similarly, only a minor percentage of the gas that is liberated during coal-mining operations is now being gathered and used.

Significant commercial production of coalbed methane is, at present, limited to the United States and Australia. Production in the United States started in the Black Warrior basin in Alabama in 1982, and soon after, it was joined by production in the San Juan basin in northern New Mexico and southern Colorado, from which the major part of United States production of coalbed methane now comes. The main increase in production in these two basins has occurred since 1986, stimulated by the federal tax credit and by important

improvements in exploration and production technology. During the late 1990s, production in the Powder River basin in Wyoming and Montana developed rapidly, and it became the second most productive basin in 2000. It is now the most active area of coalbed methane development in the United States. The large volumes of produced water, however, have caused considerable public concern and controversy. Production of coalbed methane in the Black Warrior, San Juan, and Powder River basins and the total annual and cumulative production in the United States are shown in Appendix Table 47 and Figure 34. Minor production (a total of about 122 bcf in 2000, less than 10% of the total United States coalbed methane production) has also been developed in the Appalachian, Raton, Uinta, and Piceance basins and is shown as "other basins" in Appendix Table 47 and Figure 34. Coalbed methane production has grown from minor in the early 1980s to 1% of the total United States natural-gas production in 1990, 4.8% in 1995, and 7% in 2001.

Production of coalbed methane in the United States, however, may peak in the early 21st century. Production in the Black Warrior basin, the birthplace of coalbed methane production, has started to decline, because no new coalbed-gas fields have been brought in for several years, and the San Juan basin production may have peaked around 1999 or 2000, although it will remain the main source of coalbed methane in the United States for many years. Possible maintenance or increase of coalbed methane production will have to rely on increases in the production of other basins: the Powder River, Appalachian, Raton, Uinta, Piceance, Green River, and other prospective basins.

Spurred by the economically viable production of coalbed methane in the United States, interest in coalbed methane is currently developing worldwide. Projects in the early stages of exploration or pilot testing, some moving toward commercial development, are currently occurring in many other countries: in the FSU, China, and Canada (which, together with the United States, contain nearly three quarters of the world's coal resources) as well as in India, Germany, Poland, the United Kingdom, Spain, France, Hungary, the Netherlands, the Czech Republic, New Zealand, South Africa, and Zimbabwe (Boyer et al., 1992; Kelafant et al., 1992; Schraufnagel, 1993; Davidson et al., 1995; Cairn Point Publishing, 1997). In these and other coal-rich countries, the governments have come to recognize the potential importance of coalbed methane in the development of their domestic energy supply and have started to evaluate the magnitude of their coalbed methane resources and to estimate the economic feasibility of producing this potentially important source of energy.

China, one of the countries with the largest coal resources in the world, much of it of high rank and "gassy," and in desperate need of a clean source of energy, has promoted during the last three decades the development of coalbed methane production, whether the operation is profitable or subsidized by the government. More than 100 wells have been drilled by Chinese and foreign operators (Enron, Arco, Amoco, Texaco, Phillips Petroleum, and BHP) in the coal basins of northeast China (Cairn Point Publishing, 1997; Stevens, 1999). China's resources of coalbed methane have been estimated to range from 1060 to 1240 tcf (Kuuskraa et al., 1992). Several papers presented by Chinese authors at the International Coalbed Methane Symposia at the University of Alabama (1995, 1997, 1999) report estimates of 30–35 trillion m^3 (1059–1236 tcf) for China's coalbed methane resources above 2000-m (6600-ft) depth. Sun et al. (1999, p. 565) state that "the total coalbed gas resources [of China] buried shallower than 2000 m [6600 ft] is about $326,366 \times 10^8$ m^3 [1152 tcf]." Kelafant et al. (1992, p. 81) report that "the Chinese government estimates *in-place* [my emphasis] coalbed gas to be 1060–1240 tcf." At present, although about 35 bcf of methane is being recovered from hundreds of coal mines in China, only a small amount is being produced experimentally from wells in Jin Chen, Shang Xi province.

Minor commercial production has also been developed in Australia, where two fields (projects) in the State of Queensland have begun delivering coalbed methane (called "coal-seam methane" in Australia) into a gas pipeline system since 1996 (Australian Gas Association, 1996; Cairn Point Publishing, 1997; Murray and Schwochow, 1997; Scott, 1999). In 2002, production reached 20 bcf. However, several large-scale projects in Queensland are well advanced, and major commercial production of coalbed methane in this Australian state will increase in the next few years. It is evident that there are substantial volumes of coalbed methane in Queensland, and they are close to major population centers.

Available estimates of the proved reserves (recoverable at a profit at present) of coalbed methane are limited to the United States, and even in this case, they are scarce and not always consistent. It is not always clear, for instance, if, by reserves, the authors refer to volumes of gas recoverable at present or to gas that is uneconomic now but recoverable profitably in the future. Appendix Table 48 shows the published estimates of proved reserves of coalbed methane in the United States. They progressively increased from 1421 bcf in 1988 to about 10,000 bcf during the mid-1990s and to 17,531 bcf in 2001 (9.5% of the natural gas reserves of the United States).

Most early estimates of coalbed methane in place in the United States are grouped around 400 tcf (Ayers and Kelso, 1989; Kuuskraa and Brandenburg, 1989; Gas Research Institute, 1993, 1995 [vide Curtis, 1996]; Kuuskraa and Boyer, 1993; Rice et al., 1993; Schraufnagel, 1993). Scott et al. (1994, their figure 2) raised this figure to 665 tcf [690(?) 683 in their figure]. Tyler et al. (1995) estimate the coalbed methane resources in place in the United States as 675 tcf. Murray (1996) mentions a range of 275–649 tcf.

The estimates of the percentage of this gas-in-place that ultimately will be recovered range widely among the various authors (from 25 to 70%), partly because of differences in methods of study and economic factors. When no reliable information is available, some authors suggest that, for the purpose of resource estimation, 50% of the gas in place should be considered recoverable over the economic life of a coalbed methane well. This would indicate a coalbed methane resource of 200 tcf for the United States. The Potential Gas Committee estimates of the coalbed methane total potential resource [recoverable gas(?)] in the United States range from 134.2 tcf in 1995 (*Oil & Gas Journal*, September 4, 1995, p. 109) to 146.3 tcf in 1996, 141.4 in 1998 (*Oil & Gas Journal*, April 12, 1999, p. 36), 155.2 tcf in 2000, and 168.8 in 2002 (*Oil & Gas Journal*, January 19, 2004). This last estimate includes 17.1 tcf of probable, 56.7 tcf of possible, and 95.0 tcf of speculative resources. The Potential Gas Committee's estimate of 168.8 tcf

Table 14. Estimates of world coalbed methane resources.

	tcf
Kuuskraa, 1980 (resource base)	2280
Kuuskraa et al., 1992	
major world coalbed methane resources, table 2, p. 51	2980–9260
worldwide coalbed gas resource, p. 52	4000–7000
Boyer, 1992 (resources)	3990–9490
Gayer and Harris, 1996 (resources)	8827
Murray, 1996 (methane resource, tcf in place)	2976–12,640
Murray and Schwochow, 1997 (potential coalbed methane)	12,000
Sun Maoyuand and Fan Zhiquiang, 2000 (240 tm^3)	8474
Ayers, 2002 (resources = gas in place)	2980–9260

has been mentioned by several other sources as "resources," "technically recoverable," and "recoverable" coalbed methane in the United States.

The immaturity of the coalbed methane industry, even in the United States, makes the estimation of the worldwide coalbed methane resources (the amount of coalbed gas to be ultimately recovered) still very tentative and commonly ambiguous. It is not clear in some cases, for instance, if the figures for "resource" refer to gas in place or to resources (recoverable gas). Early estimates of the worldwide resources (resource base, ultimate recovery) of coalbed methane are few and surprisingly similar (probably because they are made by the same authors). It is not clear, however, if the estimated volumes of coalbed methane refer to gas in place or to recoverable gas.

Table 14 shows the estimates.

Kuuskraa et al. (1992) assigned the bulk of the coalbed methane resources to five countries: Russia, China, the United States, Canada, and Australia, on account of their large coal resources (perhaps as much as 90% of the world's coal resources), as follows:

	tcf
Russia	600–4000
China	1060–1240
United States	400
Canada	200–2700
Australia	300–500

The range of these estimates of coalbed methane resources clearly reflects the scarcity of detailed information about this source of gas supply. This should be expected considering that the potential of coalbed methane as an important source of energy has been seriously considered in just the last two decades, and

that significant commercial production has only been developed so far in the United States. As coalbed methane reservoirs are better known, as they are in a more advanced stage of development, as the potentiality of coalbed methane production is given more careful consideration throughout the world, and as more substantial study is devoted to the occurrence of coalbed methane in the coal-rich countries, it should be expected that the estimates of the worldwide coalbed methane resources will grow considerably, in view of the enormous magnitude and wide distribution of the coal resources of the world.

Additional, more detailed information on coalbed-methane reservoirs may also make it possible to develop more accurate procedures to estimate the total volume of recoverable coalbed methane for the world and for particular countries. One such procedure could be to generate a value or range of values for the volume of recoverable coalbed methane per ton of coal, and then to apply this value or range of values to the estimates of the volumes of coal in the world or in particular countries suitable for the recovery of coalbed methane.

Values for the amount of the coalbed methane contained in coal beds (gas in place) that can be economically recovered are not easy to obtain. The values reported in the literature refer, in some cases, to gas content [gas in place(?)], in others, to gas yield, and in still others, to gas resource per ton of coal. A review of available information on more than 100 known coalbed methane reservoirs discloses that most values for the coalbed methane in place or for coalbed methane resources [gas in place(?), gas ultimately recoverable(?)] range from less than 10 to 900–1000 ft^3/t (0.3 to 25–28 m^3/t). The gas content in coal is variable, changing within a field or even within the same coal seam. Most, however, range between 200 and 600 ft^3/t (6 and 17 m^3/t). On the basis of these values and assuming that, on average, about 50% of the coalbed methane in place can be recovered, a range of 200–350 ft^3 (6–10 m^3) of recoverable gas per ton of coal appears to be a reasonable figure that can be used for estimating the potential amount of natural gas that may be produced from coal beds. The more recent estimates of world coal resources (ultimately recoverable coal) amount to about 11 trillion t. Adding another 9 trillion t of coal now too deep to be mined economically, an estimate of 20 trillion t is reached for the amount of coal expected to be present in the world as a potential source of coalbed methane. Using the range of 200–350 ft^3 (6–10 m^3) of gas recoverable/t of coal and the worldwide coal resource of 20 trillion t, the volume of the resources of coalbed methane would range between 4000 and 7000 tcf, an estimate within the range of those of other authors mentioned earlier. This range of coalbed methane resources would represent between 20 and 40% of the estimates of conventional natural gas resources of 17,000–20,000 tcf discussed earlier.

The above procedure for estimating the coalbed methane resources has, however, several weaknesses. Estimates of coal resources have been made primarily from the standpoint of mining the coal and not from the recovery of its contained gas; they are, therefore, not always suitable for the estimation of coalbed methane recovery. Coal-resource estimates for coal-mining purposes are also generally restricted to coal-bearing sections not much deeper than 900 m (3000 ft), whereas recovery of coalbed methane may be possible from greater depths where the coalbed methane yield per ton of coal may be highest. Finally, the use of a value for the coalbed methane yield of a coal reservoir requires not only the knowledge of the volume of the coal but also the characteristics of the coal: rank, thicknesses of the seams, etc.

The development of coalbed methane as an important source of energy in the 21st century will not be free of problems and constraints (see Kuuskraa and Boyer, 1993, for instance), but even the above-mentioned estimates of the world's coalbed methane resources, if considered realistic, indicate that coalbed methane can contribute an important percentage of the ultimate volume of natural gas to be produced during the 21st century.

NATURAL GAS IN LOW-PERMEABILITY SANDSTONE RESERVOIRS (TIGHT SANDS)

The term "tight sands" has been used to refer to sandstone (and, in limited cases, to carbonate) gas reservoirs having a permeability too low (0.1 md or less, exclusive of fracture permeability) for the contained gas to flow naturally at rates profitable under existing economic conditions and with proven drilling and production technologies. To be economically viable, therefore, their permeability must be artificially enhanced by means of hydraulic fracturing or other stimulation techniques to increase the gas-flow rate. The boundary between tight-sand reservoirs and conventional sandstone reservoirs is rather vague and varies with changes in the price of gas, distance from consuming centers, and the availability of drilling and production technology, mainly hydraulic-fracturing efficiency.

Tight-gas reservoirs should occur in almost all petroleum provinces and in a variety of rock types. Because their occurrence is related to the distribution of the permeability of the reservoir rock, the boundaries of tight-sand accumulations are not always controlled by structural or stratigraphic features. The gas in tight-sand reservoirs is mainly composed of methane (about 85%), with lesser amounts of ethane and higher molecular-weight hydrocarbons, and generally less than 2% of CO_2. The geologic controls and characteristics of tight-gas reservoirs, as well as the common drilling and completion practices used in their development, are reviewed by Dutton et al. (1993) and Law and Spencer (1993).

Although the occurrence of gas in low-permeability reservoirs has been known for a long time, only in the last 20 or 25 years have efforts been made to develop this potential source of energy. This interest has evolved as a result of a better understanding of the characteristic properties of the reservoir rocks and the improved drilling and completion technologies developed during the last few decades.

Despite the worldwide occurrence of tight-sand gas reservoirs, the economic development of this potentially important source of energy has been restricted mainly to the United States and Canada. However, although the distribution and magnitude of tight-sand accumulations elsewhere in the world is not well known, it can be said with assurance that large volumes of gas are present in these low-permeability reservoirs; much larger, most authors believe, than in coalbed methane reservoirs.

Natural-gas production from tight sands has been and continues to be the most important source of gas from unconventional gas accumulations in the United States. Production from 13 basins has gradually increased from about 1 tcf in the mid-1970s to 3 tcf in the late 1990s, representing 15% of the total natural-gas production in the United States.

Proved reserves for tight-sand gas in the United States have also increased from about 12 tcf in 1980 to 30 tcf in 1990 and 36 tcf in the late 1990s. Estimates of the volumes of gas in place, of the gas recoverable in the future (gas resource), and of the ultimate amount of tight-sand gas recoverable in the United States vary considerably for several reasons. Production of gas from tight sands is, to a great extent, an experimental practice, and the various estimates use different approaches and cover different basins or combinations of basins, from only 3 to as many as 113 (National Petroleum Council, 1980; MacDonald, 1990; Dutton et al., 1993; Gautier and Brown, 1993; Law and Spencer, 1993). Most recent estimates of gas in place in tight-sand reservoirs in the United States range from 600 to 925 tcf, and estimates for recoverable gas range from 200 to 550 tcf (generally assuming a recovery of 50% of the gas in place).

No statistics have been found for production or reserves of tight-sand gas of other countries in the world, although apparently, some production is being obtained in Canada.

The only estimate of world resources of tight-sand gas is that of Meyer (1977, p. 654), who states that "world resources of gas in tight sandstones might be in the range of 393 to 744 $\times$ 10^{12} ft^3."

As mentioned earlier, tight-sand accumulations should occur in all or nearly all petroleum provinces of the world. They will undoubtedly become increasingly important as the demand for natural gas intensifies, as the price of gas rises, and as the technology of drilling for and producing from tight-sand reservoirs improves, particularly the methods for enhancing the permeability of the reservoir rocks.

BASIN-CENTERED GAS ACCUMULATIONS

Law (2002, p. 1891, 1893), in his comprehensive recent discussion of basin-centered gas accumulations (BCGA), defines them as "regionally pervasive accumulations that are gas saturated, abnormally pressured, commonly lack a downdip water contact, and have low-permeability reservoirs." They are located in the deeper, central part of sedimentary basins. The natural gas is, for the most part, reservoired in sandstones and siltstones and less commonly in carbonates. Basin-centered gas accumulations may consist of a single, isolated reservoir a few feet thick or of single or multiple associated reservoirs totaling thousands of feet in thickness.

At present, natural-gas production from basin-centered gas accumulations is limited to the United States, where they make a modest contribution to the country's gas supply.

Exploration and production activity for basin-centered gas accumulations other than in the United States is currently very reduced. The presence of this type of gas accumulation elsewhere in the world is therefore poorly known, and references to basin-centered gas accumulations in the literature are limited. As a result, reliable data on natural-gas production, reserves, and resources in basin-centered gas accumulations are not available except for the United States. Even there, these data are commonly combined with those for natural gas from tight-sand reservoirs (Law, 2002).

The potential volume of natural gas in basin-centered gas accumulations, however, is believed to be very large. Basin-centered gas accumulations could be the most important economically exploitable category of the unconventional natural-gas accumulations.

As demand for natural gas increases during the 21st century, we can expect that exploration for and production of basin-centered gas accumulations will increase worldwide in the coming decades, and that these now-unconventional reservoirs will make a considerable contribution to the future production of natural gas in the world.

NATURAL GAS IN ORGANIC BLACK SHALES

Dark shales, rich in organic matter, called organic shales, black shales, gas shales, fractured shales, or, in the United States, eastern shales, have been a source of gas for more than 160 years in the Appalachian basin (DeWitt, 1986; Milici, 1993; Roen, 1993; Curtis, 2002). Milici (1993, p. 254) reports that "the first well drilled specifically for natural gas in the United States was completed in 1821 in shale beds.... in Fredonia, Chautauqua County, New York. It was sited above a natural-gas seep and produced gas for local uses, probably from fractures at a depth of 8 m." The Big Sandy gas field in eastern Kentucky has produced gas from Upper Devonian shales since 1921. Upper Devonian shales have also been a source of gas in the Lake Shore fields in northern Ohio and adjacent Pennsylvania for more than 100 years. More recently, gas production from the Devonian to Mississippian Antrim Shale has been actively developed in the Michigan basin (Milici, 1993; Reeves et al., 1996; Curtis, 2002). Appreciable gas production from organic black shales may soon also be developed in the Illinois and several other basins in the mid-continent and Rocky Mountains region (Milici, 1993; Reeves et al., 1996).

As in the case of the tight-sand and basin-centered gas accumulations, substantial gas production from organic black shales has been limited to the United States. Organic black shales, however, should occur in almost all petroleum provinces of the world, and gas production from such rocks in time should be developed in many other countries.

Organic black shales, like the gas-bearing coal beds, are the source, the reservoir, and, commonly, the seal of the gas. Moreover, like the coalbed, tight-sand, and basin-centered gas reservoirs, they have very low permeability. They require, therefore, either extensive natural-fracture systems or to be artificially fractured by a variety of techniques to increase their permeability to an economically viable level. Wells producing from organic black shales are generally long-term but low-volume producers. The estimation of the fraction of the gas in place recoverable from organic black shales is still controversial and difficult to compute. It is variable from region to region and from rock unit to rock unit, depending on the presence and degree of natural fracturing, the richness of the organic matter, the thickness of the shale, and other reservoir properties.

Production of gas from organic black shales in the United States has increased from about 70 bcf in 1979 to 135 bcf in 1985 and 280 bcf in 1995, principally from the Ohio Shale in the Appalachian basin and from the Antrim Shale in the Michigan basin, the most important sources of black-shale gas at this time. By 1999, production reached 380 bcf as production from the New Albany Shale in the Illinois basin, the Barnett Shale in the Fort Worth basin, and the Lewis Shale in the San Juan basin became increasingly important sources of gas. The Barnett Shale is now one of the most active gas plays in the United States. Shale-gas production now represents about 1.9% of the total gas production in the United States, the smallest in magnitude among the unconventional gas sources.

According to DeWitt (1986) and Milici (1993), more than 3 tcf of gas have been produced from the Devonian Ohio Shale in the Appalachian basin, mainly from the Big Sandy area of eastern Kentucky and adjacent West Virginia, where more than 10,000 wells had produced more than 2.5 tcf of gas by 1985. Milici (1993, p. 254) states that "it is estimated that about 20 tcf of gas can still be recovered from [the Devonian] formations [in the Appalachian basin] using currently available methods."

Estimates of the gas in place, resource, and recoverable gas from organic black shales vary widely. For the Appalachian basin, DeWitt (1986) estimates the amount of gas in place to range between 200 and 1860 tcf; Charpentier et al. (1993) estimate that the Devonian shales of this eastern basin contain between 577 tcf (95% probability) and 1131 tcf (5% probability), with a mean of 855 tcf of in-place natural gas. They state that the amount of recoverable gas "would be considerably lower." Gautier and Brown (1993) quote estimates by various organizations of the gas in place in organic black shales in the Appalachian basin, ranging from 225 to 248 tcf, of which 25–37 tcf may eventually be recovered with current technology and as much as 40–50 tcf with more advanced technology. Other estimates range as high as 2500 tcf of gas-in-place in the Appalachian basin.

For the Michigan basin, Gautier and Brown (1993) quote estimates of gas-in-place in black shale units ranging from 35 to 72 tcf, of which 11–21 tcf may be recovered with current technology and as much as 15–29 tcf with more advanced technology.

MacDonald (1990) estimates the gas "resource base" for the "Devonian and Mississippian shale deposits of the eastern United States" as 600 tcf. Production of gas from organic black shales in other basins has been limited, and their potential has not been as well studied as that of the Appalachian and Michigan basins. The volumes of gas-in-place and recoverable gas from black shales in the Illinois, Fort Worth, and Rocky Mountain basins may be considerable, but not enough is known about them yet to venture credible potential gas-production estimates.

Even if the estimates of gas-in-place in organic black shales in the sedimentary basins of the United States range up into the hundreds of trillions of cubic feet, the latest reported proved gas reserves in black-shale reservoirs (Kuuskraa and Stevens, 1995; Reeves et al., 1996) reach only about 3.9 tcf, a very small percentage of the total gas-in-place. This may indicate that the fraction of the gas-in-place in black shales recoverable by current production methods is very low, or perhaps that the limited experience of producing from these low-permeability reservoirs does not yet allow a realistic estimate of their potential as a future source of gas. Estimates of recoverable shale gas may increase with the advent of more advanced drilling, completion, and stimulation technologies.

The ultimate recovery of gas from organic black shales in the United States and elsewhere in the world still remains an unknown factor in estimating potential sources of energy in the 21st century.

GAS HYDRATES

Gas hydrates have been commonly mentioned as potentially vast sources of natural gas. The high cost and the technical, operational, and environmental problems of recovering gas from these reservoirs make them unlikely sources of gas for many years to come, certainly as long as cheaper, more readily produced sources of gas are available.

Gas hydrates (also called gas clathrates) are icelike, crystalline solids composed of natural-gas molecules, principally methane, trapped in rigid crystalline cages formed by frozen water molecules. They are known to be stable under conditions of high pressure and low temperature that have been recognized in polar regions at depths from 130 to 2000 m (425 to 6500 ft), where temperatures are low enough for the formation of permafrost, and in the uppermost part of deep-water sediments below the sea floor at depths of more than 100–1100 m (330–3600 ft) (Collett, 2001, 2002). Most methane in gas hydrates is of microbial origin, although some is believed to be the result of thermogenic processes (MacDonald, 1990; Kvenvolden, 1993a, b). Kvenvolden (1993a, b) reports that the total worldwide amount of methane in gas hydrates is certainly large, but that all of the estimates that have been made are speculative and uncertain. He adds that "a convergence of current ideas suggests that the amount of methane in gas hydrates worldwide is about 7×10^5 trillion cubic feet" (Kvenvolden, 1993b, p. 557). If this estimate is correct, he states that "the amount of methane carbon in gas hydrates is a factor of two larger than the carbon present in known fossil fuel deposits (coal, oil, and natural gas)" (Kvenvolden, 1993b, p. 559). He concludes, however, that "because of unsolved technological problems in producing methane from gas hydrates... wide-scale recovery of methane from these substances probably will not take place until sometime in the 21st century" (Kvenvolden, 1993a, p. 279). Other authors are less optimistic about the future availability of gas from gas hydrates, although it has been estimated (Collett, 1993) that 183 bcf of gas has been produced from gas hydrates in the Messoyakha field, in northern Siberia, since its discovery in 1968. The field produces from both hydrates and free-gas reservoirs.

More recently, Collett (2001, p. 85, 103; 2002, p. 1971) confirms that "current estimates of the amount of gas in the world's marine and permafrost gas hydrates accumulations are in rough accord at about 20,000 trillion m^3 [706,200 tcf]."

Collett (2002, p. 1971) agrees with Kvenvolden that "significant to potentially insurmountable technical issues must be resolved before gas hydrates can be considered a viable option for affordable supplies of natural gas." He adds that "we likely will not see significant worldwide gas production from hydrates for the next 30–50 years" (Collett, 2002, p. 1991). He may be too optimistic, considering the large volumes of conventional gas now available and to be added in the future.

GAS IN SOLUTION IN GEOPRESSURED BRINES

Gas in solution in underground brines is also found in many parts of the world. Although estimates of the volume of gas in these reservoirs are very high, technical, economic, and environmental problems will preclude for many years, if not forever, any significant development of this potential source of gas. The most daunting obstacle, perhaps, would be the disposal of the vast volumes of hot brine that would be produced in the process of recovering large volumes of the dissolved gas—on the average, a barrel of brine contains at most 30–35 ft^3 of natural gas under high temperature and pressure. To produce 1 mmcf of gas, it would, therefore, be necessary to dispose of about 30,000 bbl of hot salt water.

ESTIMATION OF NATURAL-GAS PRODUCTION DURING THE 21ST CENTURY

As discussed in previous sections, the ultimate volume of natural gas to be recovered from conventional accumulations has been estimated for the purpose of this study to range from 17,000 to 20,000 tcf. The volume of natural gas ultimately to be recovered from coal beds (coalbed methane) has very tentatively been estimated to range from 4000 to 7000 tcf. An estimate of the amount of gas to be recovered worldwide from tight-sand reservoirs is much more difficult to make; the only estimate found in the literature (Meyer, 1977) ranges from 393 to 744 tcf. Although important production of natural gas from coal beds, basin-centered accumulations, and tight sands is now limited to the United States, it is not unreasonable to predict that, when more and longer production histories result in a better knowledge of these types of reservoirs, the estimates of the volumes of natural gas to be ultimately recovered from them will be considerably larger than those made at present. An estimate of 25,000–31,000 tcf for the volume of natural gas to be ultimately recovered worldwide will therefore be used in this study.

This estimate, including both conventional and other sources of natural gas, is higher than those that have been made so far and reported in the literature and may, therefore, be considered overoptimistic. However, there are good reasons for believing that this estimate is realistic and for thinking that, unlike in the case of oil, large volumes of natural gas and important new gas provinces are yet to be discovered and developed in several regions of the world.

As discussed earlier in this section, the difficulty and high cost of transporting and distributing natural gas during the early years of the development of the petroleum industry were responsible for restraining the exploration for natural gas, particularly in regions of the world far from centers of gas consumption. In fact, the discovery of gas accumulations during the search for oil was not welcomed. Although this may still be true in some regions where the production of gas is not economically attractive because of the impracticability of transporting it to consuming markets, several factors have made natural gas a very desirable target for exploration in other regions of the world in the last two or three decades; new technology has made possible, for instance, the exploration of basins located in very deep water and in the deeper parts of sedimentary basins where gas is more likely to be found than oil. The building of long-distance, large-diameter pipelines; the rapid development of a liquified natural gas (LNG) industry; and the more efficient distribution of gas in regions where gas, a cleaner source of energy, is in high demand have made the exploration for and production of natural gas more economically rewarding.

However, the most important development in the process of natural gas becoming the prime source of energy during the 21st century will be the achievement of an economically profitable gas-to-liquids (GTL) scheme of natural-gas production, conversion, transportation, and marketing.

The gas-to-liquids process is the chemical conversion of natural gas to synthetic, clean (sulfur-free) liquids (gas oil, kerosene, diesel fuel, and light distillates) that can be used in refineries or sold as liquid fuels. It would not only make possible the development of currently stranded natural-gas accumulations but would also reduce gas flaring and venting. In liquid form, natural gas could be transported long distances in tankers and traded worldwide.

Since the late 1990s, several governments and a good number of international petroleum companies (Shell, Exxon, Mobil, Chevron, Texaco, and British Petroleum) have shown strong interest in developing better and cheaper ways to convert natural gas to liquids. The basic gas-to-liquids technology has been known since the 1920s, and considerable improvements have been made in the last 10 or 15 years, but financial and technological hurdles still exist that have prevented the widespread development of a gas-to-liquids industry.

Shell put in production in 1993 a pilot plant in Bintulu, Malaysia, and ExxonMobil has operated a demonstration plant in Baton Rouge, Louisiana, for several years. ConocoPhillips expects to have a gas-to-liquids plant in production in Ponca City, Oklahoma, by 2005, about the same time that ChevronTexaco looks forward to having another plant working in Escravos, Nigeria. Several other companies are considering having gas-to-liquids plants in operation in Qatar, Indonesia, Peru, and the United States. Prospects for an emerging gas-to-liquids industry in the 21st century are auspicious. Gas-to-liquids technology is so promising that its successful development will change not only the gas industry but the entire petroleum industry and the oil- and gas-producing countries.

As a result, during the last couple of decades, exploration for natural gas has intensified, and considerably more natural gas than oil has been discovered. Although world oil reserves and oil reserves-to-production ratios have shown little growth (see Appendix Tables 38, 41; Figures 22, 23), natural-gas reserves and gas reserves-to-production ratios (Appendix Tables 45, 46; Figures 31, 32) have increased steadily, indicating that unlike in the case of oil, more gas is still being discovered than is being produced.

It is therefore not unreasonable to predict that large volumes of conventional natural gas remain to be discovered in regions not yet carefully explored, as well as in regions known to contain major gas fields, some of which have not been fully developed because of their distance from gas-consuming centers, regions that have the potential for the discovery of additional sizable natural-gas reserves. The Persian Gulf region and the West Siberian basin of the FSU fall in this category. It is also to be expected that the currently estimated reserves of the gas fields in production at present will progressively increase (reserve growth), adding to the volume of gas that will eventually be produced in the world.

In addition, as mentioned above, natural gas contained in coal beds and in tight sands, although being recovered profitably so far only in the United States and only in the last two decades, may represent a very large resource to be developed throughout the world in the 21st century, as new technologies and higher prices for gas make its production technically possible and economically attractive.

Several possible estimates of natural-gas production during the 21st century are shown in Appendix Table 49 and Figure 35. The estimates were not based on any mathematical formulation; they were prepared by assuming different trends of gas production during the next 30 or 40 years and then completing the production schedules to fit various estimates of the ultimate amount of natural-gas recovery as discussed above.

It is assumed in three cases (Appendix Table 49; blue, green, and red in Figure 35) that natural-gas production will increase rapidly during the next few decades, as gas provides an increasingly large share of the energy supply throughout the world, particularly for the generation of electricity. Gas, it was predicted in these cases, will become the electric industry's most important source of fuel; most new power plants will be fueled with gas; the consumption of electricity will increase at a fast rate; and the electric industry will therefore become the gas producers' most important market. Gas may also be in high demand as an important source of liquid fuels when the development of the technology for the conversion of gas to liquids makes such a conversion economically attractive. The increasing worldwide concern about the environment, particularly about a controversial possible global warming, will make gas the fuel of choice of the early decades of the 21st century, and if a "hydrogen economy" develops during the new century, as will be discussed later, natural gas will be one of its crucial components.

In these first three cases, it was estimated that natural-gas production will increase from about 80 tcf/yr in the late 1990s to about 145 tcf/yr in 2030, an 80% increase. Production estimates were then projected to peak and start decreasing during the rest of the 21st century, from 2070 to 2095, depending on the volume of ultimate gas recovery they represent: 24,000 tcf for

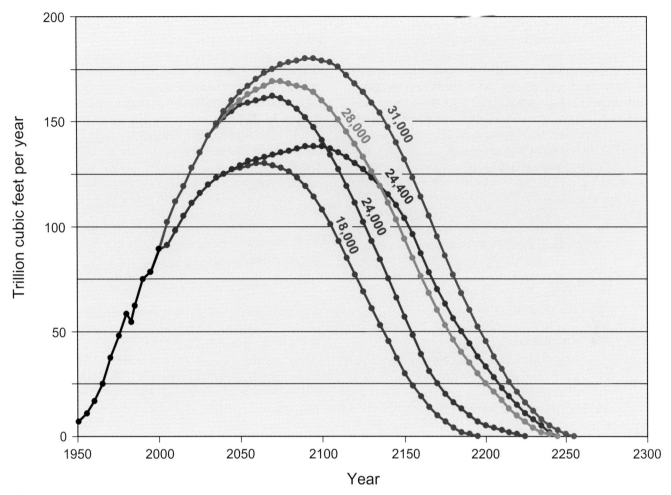

Figure 35. Natural-gas production projections in the 21st century.

the lower estimate (blue), 28,000 tcf for the middle estimate (green), and 31,000 tcf for the higher estimate (red).

A fourth estimate (Appendix Table 49; brown in Figure 35) assumes that natural-gas production will increase more slowly during the next few decades, reaching 120 tcf/yr in 2030 and peaking at 138 tcf/yr in 2090–2100. The ultimate recovery of gas for this estimate is 24,400 tcf.

Finally, a more pessimistic estimate (Appendix Table 49; purple in Figure 35) represents an ultimate recovery of natural gas of only 18,000 tcf, in line with some of the latest estimates listed in Table 13 and shown in Figure 34 (about 15,000 tcf for conventional gas and 3000 tcf for coalbed methane, tight-sand reservoirs, and other unconventional natural gas).

Many other estimates, of course, can be made by assuming different schedules of natural-gas production and different estimates of the total amount of natural gas to be ultimately recovered. They would, however, most likely fall somewhere between the most pessi-

mistic (purple) and most optimistic (red) estimates discussed above.

The estimated gas-production schedules will undoubtedly not resemble the smooth production curves shown in Figure 35. As in the case of oil, many industry uncertainties, unforeseen political and economic events, as well as possible technological innovations, also impossible to predict at present, will account for a much more uneven future schedule of natural-gas production. It is hoped, however, that the natural-gas production schedules shown in Appendix Table 49 and in Figure 35 represent a plausible order of magnitude of the role that natural gas will play as a source of energy in the 21st century.

Natural gas undoubtedly will be a major source of energy and will become a major factor in the world's economic development during the 21st century, particularly critical during its middle and late decades. How important a source of energy will depend on the development of new technologies to produce natural gas from coal beds and tight sands (and perhaps from other unconventional gas reservoirs). Most important will be the development of efficient and long-distance means of transporting the gas from producing to consuming regions, globalizing the construction of large-diameter pipelines, so they can securely cross international borders, as well as reducing the cost of transporting gas as LNG. Perhaps most critical, however, will be the successful development of an economically profitable gas-to-liquids industry, so gas can evolve from a local to a truly international source of energy.

The estimates of the ultimate gas resources, both conventional and unconventional, are now very large. They will undoubtedly increase in the future. Gas is becoming the preferred fuel for the generation of electricity, because the use of coal is increasingly found to be undesirable, and because nuclear power and hydropower suffer from strong environmentalist opposition. If these developments continue during the 21st century, natural gas will undoubtedly become the international fuel of the future, available to supply a major share of even the highest demand for energy.

COAL

ABSTRACT

Coal is, by far, the most abundant fossil fuel in the world. At the average rate of production of the last 10 years, the current 1 trillion metric tons (t) of coal proved reserves in the world should last 200–230 years. At present, eight countries (the United States, the FSU, China, India, Australia, Germany, South Africa, and Poland, all but two of them in the northern hemisphere) produce 80% of the world's coal. They contain 70% of the coal in place and about 90% of the coal reserves and resources of the world. The use of coal, however, has serious operational and environmental problems; it is dangerous to mine, dirty to transport and use, and when burned, it vents undesirable gases and particulates into the atmosphere. Most operational and environmental problems would be easier to overcome if coal was to be converted to liquid or gas to be used as transportation fuel and in the generation of electricity. Coal contribution to energy supply during the 21st century will depend not on its plentiful occurrence but on the development of technology for the clean use of coal.

INTRODUCTION

Although oil is, at present, the principal source of energy in the world (easier to produce, transport, store, and use) and gas is acclaimed as the "fuel of the future" because it is the cleanest burning of the main sources of energy, coal is, by far, the most abundant and lowest cost fossil fuel source of energy; coal reserves are 5 times as large as either oil or gas reserves, and coal resources are estimated to be about 17 times the oil resources and 20 times the gas resources.

Coal is also a versatile source of energy. It can be used for the generation of electricity and as a source of heat; it can be converted to liquid and gaseous fuels; and it is a source of chemical feedstock and metallurgical coke.

These advantages, however, are countered at present by major drawbacks, the most important of which is the serious environmental pollution resulting from the mining, transportation, and burning of coal.

Coal is a readily combustible rock containing more than 50% by weight and more than 70% by volume of carbonaceous material, including inherent moisture. Coal is formed by the alteration, compaction, and induration of variously altered plant remains accumulated in conditions similar to the conditions that favor the formation of modern-day peat deposits.

Coal is most commonly and usefully classified by rank, which essentially indicates the degree of coalification or stage of metamorphism that a coal reaches in its slow, progressive transformation from the oxygen-rich plant debris in peat deposits to lignite, subbituminous, and bituminous coals, and finally to the carbon-rich, oxygen-poor anthracite coals. As this transformation occurs, the moisture content and the volatile matter decrease, whereas the percentage of fixed carbon and the calorific value increase. The demarcation levels in this characterization of coal by rank, unfortunately, are not yet internationally uniform.

The terms "hard coal" and "brown coal" are commonly used, although not consistently. Brown coal sometimes includes both lignite and subbituminous coal, whereas at other times, it is applied only to subbituminous coal or only to lignite. Hard coal is used in some cases for bituminous coals and anthracite, but in other cases, it also includes subbituminous coal. In the following discussion of coal production, as well as in tables and figures, the term "lignite" will be used, whenever possible, for coal with a calorific value of less than 8300 Btu/t, and hard coal will be used for subbituminous and bituminous coals plus anthracite.

Peat, the precursor of coal, is generally considered separately from coal. It has been defined as an unconsolidated deposit of semicarbonized plant remains in a water-saturated environment, such as a bog or fen, and with persistently high water content (at least 75% and as much as 95%). Peat represents only an insignificant source of energy (less than 1% of the total recoverable reserves of solid fuels). As much as 70% of peat harvested worldwide is used for nonenergy purposes. Peat is, at present, used as a fuel principally in Russia, Ireland, Canada, and the Scandinavian countries. The FSU, Canada, and the United States have the largest known peat deposits. Peat, as a source of energy, will not be included in the following discussion of coal.

Coal deposits are known from all post-Silurian systems. Most prolific are the Permian (about 29% of the world's coal deposits) and the Carboniferous (20%), followed by the Tertiary (21%), the Jurassic (15%), and the Cretaceous (15%). Small deposits are known in the Devonian and the Triassic.

The use of coal has undergone profound changes in the last 70 or 80 years. Whereas in the early part of the 20th century, coal was used mainly as an industrial, transportation, and domestic fuel, the greater part (about 85%) of the coal mined today is used in the generation of electric power. As a percentage of the total commercial energy consumption, coal has declined from 90–95% at the beginning of the 20th century to about 25% today.

All figures for coal production, reserves, and resources will be given in metric tons. Converting other kinds of units to metric tons (t) caused some problems because some publications reported coal production, reserves, and resources figures in tons without specifying whether they were in short tons or metric tons. Even more disconcerting was to find out that the figures given in "tons" in certain publications were short tons in some cases and metric tons in others.

To account for the increasing calorific value of coals with increasing rank, the volumes of coal mined are commonly given in "tons of coal equivalent" (tce); on burning, a ton of lignite or subbituminous coal produces much less heat than a ton of bituminous coal or anthracite. Therefore, consideration of the rank of the coal under discussion is needed for the conversion of tons of coal equivalent to other units.

PRODUCTION

Coal has been used as a fuel for more than 2000 (perhaps 3000) years—the Chinese are reported to have burned coal in 100 B.C., and the Romans are known to have mined and used coal at about the same time. Records of the use of coal in several European countries more than 1000 years ago exist. Modest but continuous mining of coal began in England, Scotland, and several European countries during the 12th or 13th centuries. Demand for coal for fueling brick kilns, for the manufacture of glass and iron, and for the smelting and casting of brass increased during the 15th and 16th centuries. It has been estimated that by the middle of the 16th century, England's yearly coal production had reached 220,000 t, the highest in the world then. However, it was during the Industrial Revolution in the late 18th and early 19th centuries that a dramatic expansion in the mining and use of coal occurred. By 1900, about 90–95% of the world's commercial energy was derived from coal. Since then, coal's share of the energy consumption has declined steadily, having been gradually replaced as the preferred fuel by oil, which is cleaner and easier to produce, transport, and use. In 1925, coal still provided 83% of the world's consumed energy, but coal's contribution progressively decreased to 55% in 1950, 46% in 1960, 27% in 1980, and 25% in 2000 (see Appendix Table 8; Figure 12A, B). Nevertheless, the total volume of coal produced and consumed in the world has continued to increase during the 20th century (Appendix Table 50; Figure 36). Before 1925, three countries (the United States, the United Kingdom, and Germany) accounted for 80% of all coal (hard coal and lignite) production. Since 1925, the greatest increases in coal production were registered in the FSU, China, India, Australia, South Africa, and some Eastern European countries. (For a history of coal production, see Eaverson, 1939; Elliott and Yohe, 1981; Landis and Weaver, 1993).

Fairly reliable coal production statistics are available since 1865, when world coal production was 190 million t/yr. Production increased progressively to about 770 million t in 1900, about 1350 million t in 1935, and 1850 million t in 1950. Production rose rapidly after 1950 as demand for electricity increased briskly throughout the world, and coal became the most common fuel for power plants (see section on electricity). Coal production reached more than 4700 million t in 1990. The collapse of the former Soviet Union in 1991 and its effect on the countries of Eastern Europe resulted in a decrease in the coal production of these countries, particularly in Russia and the former East Germany, a decrease that was reflected in the total world production of coal.

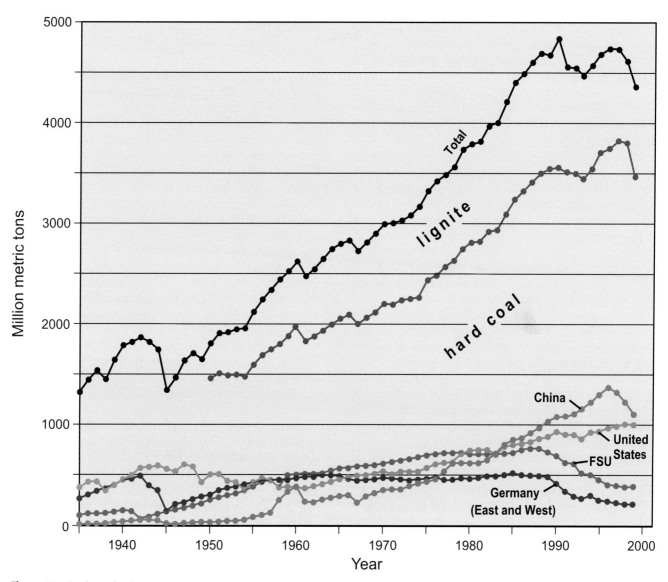

Figure 36. Coal production.

Appendix Table 50 shows coal production volumes (hard coal plus lignite) from 1950 to 1999 for the whole world and for the four major coal producers: the United States, the countries of the former Soviet Union, China, and Germany (including West and East Germany for the years when the country was divided). During these last 50 years of the 20th century, the combined coal production of the four countries as a percentage of the total world production increased from 65% in the early 1950s to 67–68% in the 1960s, 1970s, and 1980s but decreased to 62–63% in the late 1990s because of the reduction of brown coal (lignite) production in the former East Germany after its reunification with West Germany. These four countries and six more (Poland, Australia, South Africa, India, the former Czechoslovakia, and the United Kingdom) now contribute 85%

of the world's total coal production. Figure 36 shows coal production from 1935 to 1999.

Coal production in the United States increased consistently from 508 million t in 1950 to about 1000 million t in the late 1990s. The production of the FSU also rose from about 300 million t in 1950 to 800 million t in 1988 but decreased rapidly to about 400 million t during the late 1990s.

China's coal production increased more than 30-fold between 1950 and 1996, from less than 50 to 1400 million t/yr. China became the top world producer of coal in 1985. Production, however, decreased after 1996 to a little more than 1000 million tons in 1999, about the same as that of the United States.

The coal production of the combined West and East Germany changed little from 1950 to 1990; after

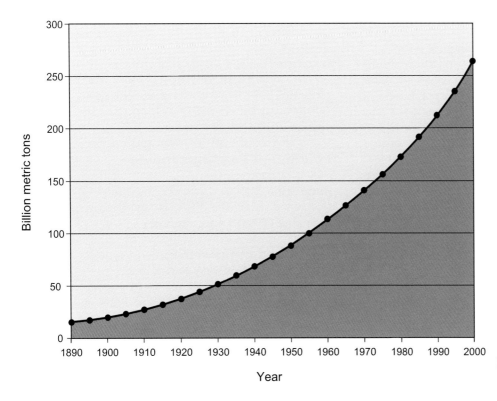

Figure 37. World cumulative coal production.

increasing from 343 to 500 million t/yr from 1950 to 1962–1963, it remained at between 460 and 500 million t/yr until 1990. During those years, the percent of brown coal production in Germany increased from 62 to between 77 and 83% of the total coal production. Production decreased rapidly during the 1990s, reaching about 200 million t in 1999, mainly because of the drastic reduction in the production of brown coal in the former East Germany after the reunification of Germany, and because of the decrease in hard coal production in the Ruhr and Saar districts, where coal reserves were depleted because of the fact that the mining of potential new resources was limited by depth and competition from imports became stronger.

Appendix Table 51 and Figure 36 show the production of hard coal from 1935 to 1999 for the whole world and for the five largest hard coal producers: the United States, the countries of the former Soviet Union, China, India, and Germany. These five countries accounted for between 70 and 77% of the world's total hard coal production during most of the second half of the 20th century.

Lignite (or brown coal) production from 1950 to 1999 for the whole world, the United States, the FSU, and Germany is shown in Appendix Table 52 and Figure 36. The share of these three countries of the world's lignite production decreased steadily from 79% in 1950 to 38% in 1999. The most important lignite production is from Europe, between 64 and 74% of the total world production. Besides Germany, sig-

nificant volumes of lignite are mined in Bulgaria, the Czech Republic, Greece, Hungary, Poland, Romania, and the former Yugoslavia. Other important lignite producers are Australia, Canada, India, North Korea, and Thailand. No lignite mining has been reported from China, a major hard coal producer.

Cumulative world production of coal is shown in Table 50 and Figure 37. Amounts previous to 1935 are based on various estimates by the U.S. Bureau of Mines, Hubbert (1962), and the World Energy (Conference) Council. The amount of coal mined so far represents 25% of the current coal reserves and a very small percentage (about 2.25%) of the estimated coal resources.

RESERVES AND RESOURCES

Terminology

Despite the commendable efforts made in the last 40 years by the World Energy Council (WEC [previously known as World Power Conference and World Energy Conference]) to arrive at a uniform international terminology for expressing the volumes of coal reserves and resources, considerable confusion still reigns concerning the use and meaning of terms used in the estimation of these volumes. A standard, clear, and consistent terminology, recognized and used worldwide, does not yet exist. For no other potential source of energy do we find a more careless use of the terms "reserves" and "resources" than for coal.

Local terminologies are based on historical usage (evolved under different social, legal, economic, and technical systems) and on commercial practices and are subject to the vagaries of political expediency. The multitude of terms used by different countries, different organizations, and different authors ("reasonably assured reserves," "exploitable reserves," "demonstrated reserve base," "speculative geological resources," "indicated resources," "probably discoverable resources," "estimated hypothetical resources," and "prospective resources," to mention just a few) makes the interpretation of the figures for coal reserves and resources as reported by different sources extremely difficult. As recognized by most authors and organizations, the different compilations of coal reserves and resources on a worldwide basis include data that are not necessarily comparable and represent a considerable degree of inaccuracy.

Even after the World Energy Council established precisely worded definitions of "reserves" and "resources," it is obvious, in many cases, that different countries and different authors have not interpreted them consistently. Considerable judgment is therefore required when comparing the figures for reserves and resources obtained from different sources.

Because of this lack of agreement about a standard terminology for reporting coal reserves and resources and because of the confusing multiplicity of terms used for this purpose by different countries, organizations, and authors, preference has been given in this study to the figures reported by the World Energy Council in its successive issues of the *Survey of Energy Resources*.

The *Survey of Energy Resources* has been issued at various intervals starting in 1962. The data for these surveys have been compiled from a wide range of reliable sources. Each of the member countries of the World Energy Council is requested to complete a carefully prepared questionnaire regarding the coal reserves and resources of their country. The data obtained from these sources are supplemented with data from earlier surveys, from various governmental and nongovernmental organizations, from specialist journals, from the publications of major international energy companies and other organizations, and from consultants active in the field of coal mining, transportation, and use.

Unfortunately, not all countries responding to the World Energy Council questionnaires followed the required specifications, so the figures contained in the issues of the *Survey of Energy Resources*, although the best available, are not always entirely comparable. For instance, the request for information for the preparation of the *Survey of Energy Resources* specifies

the following standards in coal-seam thickness and depth of burial: 0.3 m (1 ft) or more in thickness for both hard coal and lignite and a maximum overburden of 1200 m (4000 ft) for hard coal and 500 m (1600 ft) for lignite or brown coal. These specifications, although mostly accepted and used in determining the potential economic viability of coal deposits (and, consequently, the estimates of their reserves and resources), are not followed by all countries. However, the tables in the *Survey of Energy Resources* generally include the minimum seam thickness and the maximum depth of the deposits used by the different countries in estimating their coal reserves and resources.

The soundness and authority of the data for coal reserves and resources contained in the issues of the World Energy Council's *Survey of Energy Resources* are confirmed by the acceptance they have received from the many other organizations (national and international, governmental, industrial, academic, and research) that have used them in their work.

The terminology adopted by the World Energy Council in the preparation of their *Survey of Energy Resources* can be summarized as follows:

- *Proved reserves in place*: the total identified and carefully measured amount of coal estimated to be in place in known deposits as revealed by outcrops or by drilling or mining and by detailed sampling to establish its rank and quality; these deposits have been assessed to be exploitable under current or expected local economic conditions and using currently available mining technology.
- *Proved recoverable reserves (economically recoverable reserves)*: the fraction of the proved reserves in place, within specified limits of thickness of seam and overburden, that can be recovered economically under current or expected local economic conditions, using currently available mining technology; the percentage of the coal in place that can be recovered at a profit ranges from as low as 50 to almost 100% in open-pit mines.
- *Estimated additional resources*: the estimated amount of coal in addition to the proved reserves in place that are of foreseeable economic interest; it includes the coal deposits not yet discovered but estimated to exist in unexplored extensions of known deposits or in undiscovered deposits in known coal-bearing areas based on the presence of geological conditions favorable for the occurrence of coal.
- *Total resources*: the sum of the reserves in place and the estimated additional resources; the total volume of coal available in the Earth that may be successfully exploited and used by humans in the foreseeable future.

This classification and terminology of reserves and resources may be viewed as overly simplistic, principally because it does not recognize the degree of geological assurance in estimating the coal reserves — the use of proved, probable, and possible or equivalent categories. It provides, however, the basic information needed in this study to estimate the possible sources of energy in the 21st century: the order of magnitude of the volume of coal in the world; how much of it is now known (the proved reserves in place); how much of it is recoverable economically at this time (the recoverable reserves); and how much additional coal has been estimated that will eventually be discovered and developed. In other words, how much coal will be available during the 21st century as a source of energy. The accuracy of this information should be expected to be higher in the estimation of reserves than in the case of resources.

Coal statistics from sources other than the World Energy Council, although commonly difficult to interpret, generally fall within the general order of magnitude of the World Energy Council figures and were useful in this study.

Reserves

Over the years, numerous different organizations and authors have reported estimates of the coal reserves for individual countries, for certain regions, and for the world as a whole. As discussed above, these different estimates have not been reached on the basis of uniform procedures, definitions, and specifications (seam thickness and depth) and are, therefore, not altogether comparable.

The selection of coal-reserve estimates shown in Table 15, Appendix Table 53, and Figure 38 has favored those reported by the World Energy Council in its periodic issues of the *Survey of Energy Resources*. Those selected are based on a clear and consistent definition of reserves, are considered most reliable, and have been universally accepted and used. For instance, as shown in Table 15, the coal-reserve estimates included in British Petroleum's excellent annual issues of *Statistical Review of World Energy* are taken from World Energy Council's *Survey of Energy Resources*. The coal-reserve estimates of other accredited organizations, such as the Energy Information Administration of the U.S. Department of Energy, correspond very closely to those of the World Energy Council.

Following the World Energy Council, Table 15 and Figure 38 include numbers for proved recoverable reserves and, in some cases, for proved reserves in place. Reserve estimates by organizations and authors that do not follow the World Energy Council definitions

and terminology have been interpreted, to the best of my ability, to correspond to one or the other of these two categories of reserves. The table and figure include only coal-reserve estimates reported since 1960, because they represent investigations based on more recent and complete information (earlier estimates are discussed by Fettweis, 1979).

The major part of the world's coal reserves is in the northern hemisphere, north of latitude 30°N. In the southern hemisphere, only Australia and South Africa have sizable coal reserves. South America is particularly devoid of coal deposits.

Most estimates of the world's proved recoverable coal reserves made in the last 25 years range from 800 to 1100 billion t (Table 15). The last estimates by the World Energy Council in the 1998 and 2001 issues of the *Survey of Energy Resources* total 984 billion t. Of this amount, 60% is in three countries: the United States (25%), the FSU (23.4%), and China (11.6%). Eight countries, the above three plus India (8.6%), Australia (8.3%), Germany (6.7%), South Africa (5.0%), and Poland (2.3%), account for 91% of the world's proved recoverable coal reserves.

Appendix Table 53 shows the proved recoverable coal reserves of the United States, the FSU, and China and their percentage of the total world coal reserves. The most recent estimates of the proved recoverable reserves of the United States range from 240 to 250 billion t, and those of the FSU range from 230 to 240 billion t. The recoverable coal reserves of China have been estimated for the last 10 years to range from 114.5 to 118.5 billion t.

The world's proved recoverable reserves of coal are five times as large as the proved reserves of either oil or natural gas.

Most estimates of the world's proved coal reserves in place made during the last 20 years range from 1300 to 2400 billion t (Table 15). The latest estimate by the World Energy Council in the 1995 *Survey of Energy Resources* amounts to 2400 billion t. Of this volume, 42% is in three countries: the United States (18%), the FSU (12%), and China (12%). Eight countries, the above three plus India, Germany, South Africa, Australia, and Poland, account for 70% of the world's proved coal reserves in place.

The estimates of the world's proved recoverable coal reserves and proved coal reserves in place have changed surprisingly little in the last 40 years or so, despite a total coal production of about 155 billion t during those years. In the 1960s and 1970s, the proved recoverable reserves ranged between 600 and 800 billion t. The estimates increased to about 1000 billion t by the late 1980s and have remained at that level since then (Table 15; Figure 38). The estimates of the proved

Table 15. World coal reserves (hard coal plus lignite).

		Proved Recoverable Reserves (Billion Metric Tons)	Reserves in Place (Billion Metric Tons)
1	1962, World Power Conference, SER*	800 hc** = 600 bc† = 200	
2	1968, World Power Conference, SER	730 hc = 460 bc = 270	
3	1974, World Energy Conference, SER	591	1420
4	1976, U.S. Department of Interior, Energy Perspective 2	607	
5	1976, World Energy Conference, SER	713	1125
6	1978, World Energy Conference, Coal Resources (Peters and Schilling)	795 hc = 615 bc = 180	
7	1979, Fettweis	1076	2152
8	1980, World Coal Study	829	
9	1980, Tatsch	651	1566
10	1980, World Energy Conference, SER	882 hc = 488 bc = 394	1320
11	1980, Bestougeff (after 6)	795	
12	1981, Feys	695	1297
13	1982, Bender (U.S. Geological Survey) (after 6)	795	
14	1982, U.S. Department of Energy-Energy Information Administration, *International Energy Outlook*	796 hc = 616 bc = 180	
15	1983, World Energy Conference, SER	946	1520
16	1984, Matveev et al.	1239††	4298‡
17	1986, World Energy Conference, SER	827	2094
18	1988, BP *Statistical Review of World Energy*	1926 hc = 580 bc = 446	
19	1989, World Energy Conference SER	1596 hc = 1074 bc = 521	2446
20	1990, BP *Statistical Review of World Energy*	1083 hc = 640 bc = 443	
21	1991, BP *Statistical Review of World Energy*	1079 hc = 637 bc = 442	
22	1992, World Energy Council, SER	1039 hc = 521 bc = 518	1956
23	1992, BP *Statistical Review of World Energy* (after 22)	1040	
24	1992, U.S. Department of Energy-Energy Information Administration, *International Energy Outlook*	1059	
25	1993, BP *Statistical Review of World Energy* (after 22)	1039	
26	1994, U.S. Department of Energy-Energy Information Administration, *International Energy Outlook* (after 22)	1039	

Table 15. World coal reserves (hard coal plus lignite) (cont.).

		Proved Recoverable Reserves (Billion Metric Tons)	Reserves in Place (Billion Metric Tons)
27	1994, BP *Statistical Review of World Energy* (after 22)	1039	
28	1995, World Energy Council, SER	1032 hc = 519 bc = 513	2400
29	1995, U.S. Department of Energy-Energy Information Administration, *International Energy Outlook*	1039	
30	1995, BP *Statistical Review of World Energy* (after the World Energy Council)	1044	
31	1996, U.S. Department of Energy-Energy Information Administration, *International Energy Outlook*	1039	
32	1996, BP *Statistical Review of World Energy* (after 29)	1032	
33	1997, U.S. Department of Energy-Energy Information Administration, *International Energy Outlook*	1037	
34	1997, BP *Statistical Review of World Energy* (after 29)	1032	
35	1998, World Energy Council, SER	984 hc = 509 bc = 475	
36	1998, U.S. Department of Energy-Energy Information Administration, *International Energy Outlook*	1036	
37	1998, BP *Statistical Review of World Energy* (after 29)	1032	
38	1999, BP Amoco *Statistical Review of World Energy* (after the World Energy Council)	984	
39	2000, U.S. Department of Energy-Energy Information Administration, *International Energy Outlook*	987	
40	2000, BP *Statistical Review of World Energy* (after the World Energy Council)	984	
41	2001, World Energy Council, SER	984 hc = 519 bc = 465	
42	2001, BP *Statistical Review of World Energy* (after the World Energy Council)	984	
43	2002, BP *Statistical Review of World Energy* (after the World Energy Council)	984	
44	2002, U.S. Department of Energy-Energy Information Administration, *International Energy Outlook*	988	
45	2003, U.S. Department of Energy-Energy Information Administration, *International Energy Outlook*	982	
46	2003, BP *Statistical Review of World Energy* (after the World Energy Council)	984	

*SER = *Survey of Energy Resources.*
**hc = hard coal.
[†]bc = brown coal [lignite(?)].
[††]Established reserves.
[‡]Total reserves.

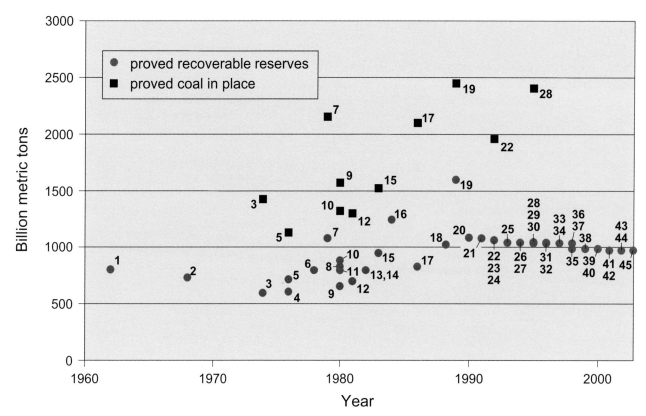

Figure 38. Estimates of world coal reserves.

coal reserves in place have ranged somewhat more, generally increasing from about 1300–1500 billion t in the 1970s and early 1980s to 2000–2400 billion t in the late 1980s and 1990s (Table 15).

Eight countries (the United States, the FSU, China, India, South Africa, Australia, Poland, and Germany) have 95% of the world's 519 billion t of proved recoverable reserves of anthracite and bituminous coal (hard coal), and five countries (the United States, the FSU, China, Australia, and Germany) have 86% of the world's 465 billion t of proved recoverable reserves of subbituminous coal and lignite (brown coal).

Reserves-to-production Ratios

At the average rate of production of the last 10 or 15 years, the approximately 1 trillion t of proved recoverable reserves of coal now estimated to be reasonably well known in the world should last for 200–230 years (Appendix Table 54; Figure 39). The estimates represent a 40–50% recovery of the coal in place. If higher percentages of recovery can be attained in the future because of the advent of new mining technology, the time it will take to deplete the known coal deposits will be longer, in the order of 300–400 years. However, if, as expected, the production of coal increases in the future, the depletion of the world's coal deposits will

take less time. Still, even assuming considerable future increases in production, coal will be a plentiful potential source of energy during the 21st century. Its use may only be limited by its detrimental effect on the environment.

Resources

The first scientifically based and consistent attempt to estimate the coal resources of the world was made in connection with the 12th International Geological Congress in Toronto in 1913 (Dowling, 1913). The total resources were estimated to be 6402 billion t to a depth of 4000 ft (1200 m) and 7397 billion t to 6000 ft (1800 m), of which 671 and 714 billion t would be recoverable, respectively. In the process of preparing this estimate of the world's coal resources, considerable effort was made to ensure compatibility of national and regional estimates. It received considerable international recognition.

Most estimates made between the years 1913 and 1962 ranged between 5000 and 7000 billion t. Appendix Table 55 and Figure 40 include only the estimates of the world's coal resources reported since 1960, because, as in the case of the estimates of the coal reserves, they represent investigations based on more recent and complete information. The estimates of the World Energy Council, considered as the most

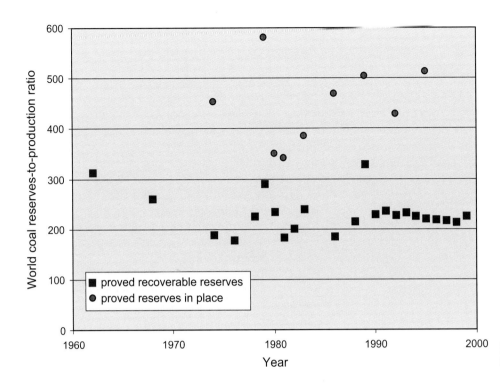

Figure 39. World coal reserves-to-production ratios.

reliable, are shown in red in Figure 40. Other estimates, although not always labeled as resources, are believed to correspond to the resources as defined by the World Energy Council.

With the exception of the estimates made by Averitt (1969, 1973, 1975, 1981), which amount to about 15,000 billion t, most estimates of the world's coal resources range between a little less than 10,000 and 12,000 billion t. Like the reserve estimates, they have not varied much over the last 30 or 40 years. The most recent resource estimates by the World Energy (Conference) Council (1989, 1992, 1995) range between 10,555 and

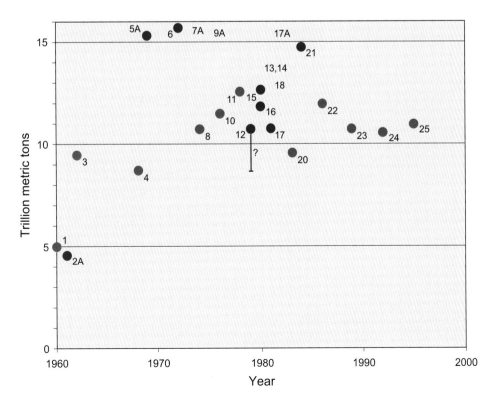

Figure 40. Estimates of world coal resources.

11,000 billion t. This amount represents about 4.5 times the proved coal reserves in place and 10–11 times the proved recoverable reserves. More recent estimates by the World Energy Council, as reported in the 1998 and 2001 issues of the *Survey of Energy Resources*, do not include the totality of the coal-producing countries; they do not, for instance, include China, one of the countries with important coal deposits. Those estimates, therefore, are not included in Appendix Table 55 and Figure 40.

Of the total volumes of the world's coal resources, about two-thirds is hard coal, and one-third is lignite (or brown coal).

As in the case of the production volumes and the estimates of coal reserves, the estimates of the world's resources of coal are dominated by those of the United States, the FSU, and China (see Appendix Table 55). However, the percentage of the total world coal resources assigned to these three countries has decreased somewhat during the last 20 years or so, because more information has become available about the coal deposits of other countries. Until the mid-1980s, the combined coal resources of the United States, the FSU, and China were estimated to represent about 80–90% of the world's resources, but the latest estimates assign these three countries only about 70–75% of the world's coal resources: 14–15% in the United States, 50% in the FSU, and 9–10% in China. Six additional countries, Australia (7.5%), India and Germany (2.8% each), Poland (2%), the United Kingdom (1.7%), and South Africa (1.1%), account for another 18%. According to these estimates, 87–92% of the world's coal resources are in just nine countries, with all but two of them in the northern hemisphere.

The estimated world coal resources of about 11,000 billion t represents 17 times the estimated resources of oil and 20 times the resources of natural gas (Figure 40). Even assuming a 50% recovery of the estimated 11,000 billion t of coal resources and a considerable increase of production in the future, an abundant supply of coal as a source of energy is assured for several hundred years.

FORECASTS OF COAL PRODUCTION DURING THE 21ST CENTURY

All evidence, as discussed above, indicates that coal is an extremely abundant solid fuel and that, possible disadvantages aside, it could supply the world with a source of energy for several hundred years, even if its production rate was to increase substantially in the 21st century.

Although the current estimates of coal reserves and resources are judged by some to be too low, most authors

believe that the distribution of coal deposits in the Earth is reasonably well known, and that it is unlikely that entirely new major deposits of coal will be discovered at depths that would make them economically exploitable by current or foreseeable mining methods. Most agree, however, that there is a good probability that numerous small deposits will be found, and that the presently estimated reserves of known coal deposits will increase as they are further developed (reserve growth). Increases may also be expected in countries that at present have coal deposits sufficient to supply their demand for many years and have therefore no incentive to explore further to identify new coal deposits. The availability of coal as a fuel in the 21st century is therefore expected to increase and not to decrease.

The extent to which the vast volumes of coal that are known or estimated to occur throughout the world will be produced and used as a source of energy during the 21st century will depend not on its availability but on the demand for it, more specifically, on the purpose for which the coal will be used.

The forms of energy for which there is greater demand at present are liquid fuels (for various means of transportation), gaseous fuels (for domestic and industrial use and, increasingly, for the generation of electricity), and electricity. As mentioned before, coal is more and more being used for the generation of electricity (and a small percentage for metallurgical applications), less and less as a domestic, industrial, and transportation fuel, and because of its serious drawbacks as a source of environmental pollution. Coal has been replaced as the main fuel for the generation of electricity by gas, nuclear and hydroelectric power, and, to a small extent, by "non-hydro" renewable sources of electricity.

Oil and gas are, at present, the sources of energy for which there is the most demand in the world. As long as they are available, the demand for coal will increase only to a limited degree.

However, coal will be available in sufficiently ample volumes to replace oil and gas as sources of energy if oil and natural gas become scarce and inadequate to satisfy the demand for liquid and gaseous fuels during the 21st century and beyond.

It will be necessary, however, to make the mining, transportation, and use of coal possible in an environmentally less detrimental manner. This can be accomplished to a great degree by converting coal into a liquid or gaseous form. Processes to accomplish this conversion have been known for many decades, but they are expensive at present, and as a result, the liquid or gaseous products derived from coal are not economically competitive with oil and natural gas.

The environmental drawbacks to the use of coal can also be overcome in part by the development of new, more efficient, and more environmentally compatible technologies for the mining and transportation of coal by building more efficient coal-burning electric-generation plants at the coal mines and by devising advanced, more efficient methods of transportation of the electricity produced in these plants.

Important operational obstacles to the use of coal as a replacement for oil and natural gas as a source of energy during the 21st century also need to be vanquished. At present, 90% of the world's production of coal is consumed in the countries where coal deposits are located and mined; only 10% is exported. Demand for additional large volumes of coal for international trade would place considerable strain on the availability of facilities for the mining, transportation, distribution, and use of coal, although there are vast resources of coal to be developed. An adequate infrastructure for such an expanded development is not available today; new mines will need to be opened, and satisfactory and sufficient means of land and sea transportation and trans-shipment will need to become available. This will involve the building of a large fleet of coal-carrying trains and ships; the possible laying of coal-slurry pipelines; and the improvement, enlargement, or building of loading and unloading port facilities in many countries for the export and import of coal, particularly in those countries where coal has not been traded in the international market.

To develop such an expanded infrastructure will require substantial investments in both the exporting and importing countries, and the fact that the great majority of the world's coal deposits are in three countries, most of them distant from seaboard, will contribute to the enormous cost of the needed financial commitments. The magnitude of such commitments may be increased by environmental regulations requiring the use of control techniques during the mining, transportation, and combustion of coal. To overcome the mining and transportation obstacles, however, will be easier than to find the means of sequestering the CO_2 produced by the combustion of coal.

Thus, all the environmental, economic, and operational problems connected with the contribution of coal to the supply of energy in the 21st century would be easier to overcome if coal was converted to liquid or gaseous form as early as possible in its journey from the mine to the consumer.

In the 21st century, the world will require liquid fuels and electricity. Coal, in solid but preferably in liquid or gaseous form, can be a plentiful source of both. New or improved technology should be able to overcome coal's economic and environmental disadvantages to a considerable extent. If the past is any guide, a review of the great technological advances during the 20th century provides confidence that such new or improved technology will be developed in the coming decades as energy demand grows, as oil and gas are increasingly unable to supply that demand, and as pressure to develop cleaner and more efficient ways to use coal builds up. To develop such new or improved technology probably will be one of the most important missions of governments and the energy industry. Along with the development of an efficient and economic technology to convert natural gas to liquids, to convert coal to liquid or gaseous form will be essential to assure an adequate source of energy during the 21st century.

Will an environmentally 100% clean energy source, which will be able to supply most of the world's energy demand, be developed sometime in the 21st century? It seems unlikely. Even if nuclear, hydroelectric, and various other renewable energy resources considerably increase their share of the world's energy consumption in the 21st century, they can only supply electric power. Demand for liquid fuels will remain strong for a long time yet; oil—and natural gas and coal, if they can be made available in liquid form at a competitive price—will remain the main suppliers of this demand.

OIL SHALES

ABSTRACT

Oil-shale deposits are known from many countries in the world, and their potential as a source of oil has been described as "vast." Nevertheless, significant production of shale oil has not yet been attained without some kind of economic subsidy. In addition to its unprofitability, the production of shale oil presents some serious environmental problems: the damage caused by surface mining, the large volumes of fresh water needed to retort the oil shale to produce shale oil, and the disposal of huge volumes of spent shale that would result from a commercial operation. The production of shale oil is now uneconomic and will remain so, as long as other sources of energy (oil, natural gas, and coal) are readily available at more attractive prices. This will likely be the case for a long time.

INTRODUCTION

An oil shale is defined as a fine-grained sedimentary rock that contains a high proportion of endogenous organic matter (kerogen) mostly insoluble in ordinary petroleum solvents, from which substantial amounts of synthetic oil and/or gas can be extracted by heating it to a sufficiently high temperature, a process called "retorting." Oil shales have a low calorific value and high ash and mineral content.

The potential of an oil shale as a source of energy depends on the economic recoverability of oil (and gas) from it. The lower limit of the oil yield for an oil shale to be considered potentially economic now ranges between 10 and 15 gal of shale oil/t, but the development of new mining and processing technology for the oil shale or a substantial increase of the price of oil may make oil shales with lower oil yields economically attractive in the future. At present, only the oil shales with the higher oil yields are used to obtain shale oil, about 25% of the total world oil shales mined. The remaining 75% is mainly used as a solid fuel in the generation of electricity and heat (69%) and for the production of synthetic domestic gas, cement, and specialty chemical products (6%). Only the recovery of synthetic oil will be considered in the following discussion of oil shales.

Oil shales are known from many countries throughout the world. They range in age from Proterozoic to Tertiary, but not all are potential commercial sources of energy. Russell (1990) discusses the occurrence of oil shales in 51 countries (see also Duncan and Swanson, 1965; Congress of the United States, Office of Technology Assessment, 1980). Dyni (2003, personal communication) lists 36 countries containing oil-shale deposits in a table with their corresponding estimates of in-place shale-oil resources and discusses the 14 countries with the largest of these deposits. The most important oil-shale deposits are those of the Eocene Green River Formation in the states of Colorado, Utah, and Wyoming in the United States and those of the Permian Irati Shale of southern Brazil. Other significant oil-shale deposits are known in Estonia, China, Russia, Australia, Canada, Morocco, Israel, and Jordan. More than 40 other countries are reported to have oil-shale deposits, but few of them are, at present, considered as possible commercial sources of energy.

In most countries of the world, the oil-shale deposits are insufficiently studied, and information concerning their size or the quality of the shale is, in many cases, lacking or, worse, greatly exaggerated.

The production of oil from oil shales is not new; a synthetic crude oil (SCO) was first manufactured in Scotland in 1694 by retorting oil shale. Significant production began from these Scottish oil shales in the 1840s and peaked in 1913 with a cumulative production of about 1.5 million bbl of oil/yr. Production ended in the 1960s because it was not possible to compete with low-cost imported oil.

In 1815, commercial oil-shale retorting was started in New Brunswick, Canada, and between 1850 and 1860, more than 50 commercial plants were constructed in the United States to retort oil from shale imported from Canada. "Colonel" Drake's discovery of oil near Titusville, Pennsylvania, in 1859 closed down the oil-shale industry in the United States and Canada, but minor production continued in several other countries. Oil shale has been used in Estonia since 1916, principally as solid fuel in the generation of electricity; retorting of the shale to obtain oil has been only a small part of the oil-shale industry in Estonia; at most, only about one-third of the oil shale mined was retorted to obtain shale oil.

Since the beginning of the oil-shale industry, more than 1 billion t of oil shale are believed to have been mined in 19 countries, 80% of it in China and Estonia.

Two general processes of recovery of shale oil have been attempted: (1) mining the oil shales, either underground or in open pits, crushing, and above-ground retorting (the process so far used in all oil-shale projects); and (2) in-situ processing, involving drilling into the oil-shale unit, fracturing it to increase its permeability, igniting the shale, and recovering the oil thus generated through other wells. This second process is still in an experimental stage but holds promise, because it reduces the harm to the environment.

PRODUCTION

Early production of shale oil occurred mainly in Scotland, Estonia, France, Canada, and China. Duncan and Swanson (1965) estimated that about 400 million bbl of shale oil had been produced up to 1961, most of it in these five countries. Reliable statistics about volumes of early shale-oil production, however, are understandably difficult to obtain. An additional 280–300 million bbl are estimated to have been produced since 1961 in Estonia, China, and Brazil (Figure 41), for a total of 680–700 million bbl.

Rühl (1982, p. 129) states: "Without having statistics, it can only be estimated that probably no more than approximately 100 million tons [733 million barrels at 7.33 barrels per ton] of shale oil has been recovered cumulatively all over the world during the past 150 years."

Currently, sizable production of shale oil is occurring in Brazil and, to a lesser extent, in Estonia and China (Appendix Tables 56, 57; Figure 41). Production

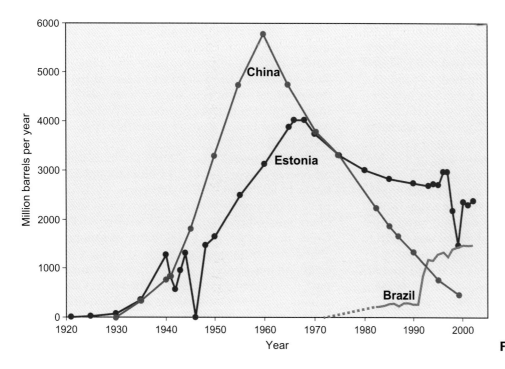

Figure 41. Shale-oil production.

of shale oil has been undertaken in Australia in the early 2000s. Use of oil shales as a source of energy is in the planning stage in other countries.

Brazil

Efforts to produce oil from the Permian Irati Shale started in southern Brazil in the 1880s but met with little success until Petrobras, the national oil company, undertook the study of the Irati oil-shale deposits in the 1950s (Padula, 1969). Production of shale oil started in 1972 on an experimental basis, reached 208,141 bbl by 1982, and has grown progressively since then, reaching about 1.46 million bbl in 2001, as shown in Appendix Table 56 and Figure 41. The Irati oil shales yield about 20 gal of shale oil/t.

Cumulative production of shale oil from the Irati oil shale reached about 15 million bbl by the end of 2001.

Estonia

The kukersite oil shales of the Baltic region are present mainly in Estonia but extend eastward into the St. Petersburg area of Russia. They are Ordovician in age and have a high shale-oil yield (24–48 gal/t), which makes them one of the richest oil shales in the world. They are not, however, well suited as feedstock for the production of liquid fuels and have, therefore, been predominantly used as solid fuel in the generation of electricity, in industrial boilers, in railroad locomotives, and in the manufacture of a variety of petro-

chemicals not easily obtainable from petroleum, coal, and other oil shales.

The mining and use of oil shale in Estonia started in 1916, both in underground mines and in open pits. Mining peaked in 1980 at a little less than 30 million t/yr and has since declined to about 12 million t in 1995, remaining at this level during the last few years (Kattai and Lokk, 1998).

Of this volume of mined oil shale, only a part was processed for the production of shale oil and petrochemical products, a part that decreased from 36–37% in the late 1950s to about 11–12% in the 1980s and 1990s (Yefimov et al., 1994).

As shown in Appendix Table 57 and Figure 41, production of shale oil in Estonia increased progressively from 115 t (843 bbl) in 1921 to a peak of 550,000 t (about 4 million bbl) in 1968. It decreased to 366,000 t (2.68 million bbl) in 1993, increased to 405,000 t (2.712 million bbl) in 1996 and 1997, and dropped sharply to 200,000 t (1.466 million bbl) in 1999. Higher crude oil prices encouraged an increase in the production of shale oil to 322,000 t (2.36 million bbl) in 2000. It remained at this level during 2001 and 2002.

The estimated cumulative volume of shale oil produced in Estonia since 1921 amounts to about 160 million bbl.

China

Oil-shale deposits are widespread in many regions of China, of which only a few have been so far developed. Most important are those in the Fushun, Maoming, and Huadian areas. They range in age from Silurian to Tertiary, with the principal ones being Tertiary. Some oil shales are associated with coal deposits, and in some cases, they have been burned as solid fuel together with the coal. Most of the oil shales, however, have been used to produce shale oil by retorting.

Mining of oil shales, mostly in open pits, started in China in 1929. Production of oil increased gradually since then to a peak of 780,000 t (about 5.7 million bbl) per year in 1959–1960, but the availability of cheaper oil after the discovery of the Daqing oil field in 1959 and the initiation of significant oil production in 1962 reduced the demand for shale oil, and production began to decline. Production of shale oil was down to about 300,000 t (2.2 million bbl) per year in the early 1980s and only about 60,000 t (440,000 bbl) in 1999 (Figure 41). Production has remained at the 60,000–80,000-t/yr level (440,000–600,000 bbl) since then. On average, Chinese oil shales yield about 18 gal of oil/t of shale.

It has been estimated that about 1 billion t of oil shale has been mined during the 75 years of development of the oil-shale industry in China. Cumulative production of shale oil during these 75 years amounts to about 25 million t (183 million bbl).

United States

Oil shales (or black shales high in organic matter) occur in many parts of the United States. They range in age from Ordovician to Tertiary. Most important, better studied, and having the best potential for economic development are the oil shales of the Eocene Green River Formation in the states of Colorado, Utah, and Wyoming. They are recognized as the largest known oil-shale deposit in the world. The total shale-oil resource of the Green River Formation is estimated at 1.5 trillion bbl. The richest and best explored deposits occur in Colorado's Piceance Creek basin. Those in Utah and Wyoming are of somewhat lower quality.

The potential of the Green River oil shales as an important source of oil was first recognized in the early 1900s, but as described by Russell (1990, p. 82): "Each time it appeared that oil shale was on the verge of commercial development, discoveries of new oil fields postponed actual development. This occurred in the 1920s, the 1940s following World War II, and again in the 1979–85 period. Each time development appeared imminent, new sources of oil were found or world prices declined. The economics of oil-shale production were always 'just around the corner.' This condition exists today and will continue to do so as long as the world supply and prices of oil are below that needed to produce shale oil at a profit." The situation has not improved since the publication of Russell's book. (For a detailed review of the ups and downs of the attempts to develop an oil-shale industry in the United States, see Fearon and Wolman's *Always a Bridesmaid? Forty Years of Estimating Costs and Prospects*, 1982).

Over the years, several attempts to produce shale oil operated sporadically in Colorado and Utah, particularly in the late 1970s and early 1980s when the price of oil reached all-time highs; but by late 1985, all operations other than Unocal's project on Parachute Creek in the Piceance Creek basin of Colorado were abandoned or suspended because of high operating costs and low crude oil prices. They were, for the most part, experimental projects designed to test alternative mining and retorting methods and to acquire technological data. None of them achieved economic production or became a commercial success.

Unocal, after having produced 50,000–55,000 bbl of shale oil in 1957–1958 before suspending operations in 1958, resumed production in 1986. More than 600,000 bbl was produced in 1987, and production reached 1 million bbl in 1988 and 1989, but the operation was shut down in 1990 in view of unfavorable economic results.

No shale oil is now being produced in the United States.

Cumulative production of shale oil in the United States has been estimated to have reached somewhat more than 500,000 bbl since 1919, 350,000–400,000 bbl of it in large-scale pilot operations since 1964 (Donnell, 1980).

Marine black shales (potential oil shales) occur in many other regions of the United States. They are lower in oil content, with most of them yielding only 1–15 gal/t of oil. The Devonian–Mississippian black shales of the northern mid-continent (also called "eastern shales"), extending over 14 states east and west of the Mississippi River, have the best potential. The richest parts yield 10–15 gal of oil/t by retorting, although they may have better yields by other methods of oil recovery. It is difficult to consider even the richest Devonian–Mississippian black shales as an economic source of oil in the future, even in the distant future. Work that has been carried out for the recovery of oil from these black shales is only of research nature: resource characterization and evaluation of the mining and processing methods. Natural-gas production, however, has been obtained from the eastern shales of the United States mid-continent since the mid-19th century (see section on natural gas in organic black shales).

RESERVES, RESOURCES, AND ULTIMATE RECOVERABLE SHALE OIL

Because most oil-shale deposits have not been extensively studied and developed, the estimates of their reserves, resources, and ultimate recoverable oil are tentative and not always reliable. As discussed above, shale oil has been produced over an extended period of time from just a few oil-shale deposits; most have not gone past the feasibility or data acquisition stages.

In addition, the terminology used to refer to the volumes of shale oil known or estimated to be recoverable from oil-shale deposits or to the likelihood that this oil may eventually be produced, is extremely confusing and burdened by a complete lack of standardization. The terms "reserves," "resources," and "oil in place" are used with a variety of meanings and are not always recognizable. Terms like "reserves in place," "indicated reserves," "inferred reserves," "proved oil in place," and many others are used without any clear definition of their meaning.

Criteria that can universally be used to estimate shale-oil reserves and resources have not been developed. Reserves and resources have been assessed in many different ways. Maximum depth to the oil-shale deposits, minimum thickness of the oil-shale beds, and minimum yield (gallons of oil per ton of shale) vary from country to country and from one author to another. These important factors are not specified in many cases. This makes it difficult or impossible to compare or compile the estimates of shale-oil reserves and resources in different countries or those given by different authors.

With this background in mind, the following are some estimates of the volumes of shale oil that are available today or will be available in the future as a possible source of energy.

Reserves

The generally accepted meaning of the term "reserves" comprises the volumes of solid, liquid, or gaseous materials that are known at present to exist and that can be extracted at a profit with existing technology and under present economic conditions. By this definition, no true oil-shale reserves are now known in the world, with the possible exception of those in the Irati Shale of Brazil, which have been estimated by Petrobras as 2.88 billion bbl at the end of 1999.

Other estimates reported as reserves or recoverable reserves are shown in Table 16.

Table 16. Shale-oil reserves.

	Billion Barrels
World (excluding the lowest and highest estimates)	550–2000
United States (mainly from the Green River Formation)	80–280
Estonia	
Proved reserves*	22
Proved reserves**	21.8
Probable reserves*	21.5
Probable reserves**	15.1
China: proved reserves	7.33
Brazil	
Measured reserves	2.88
Estimated reserves	59.1

*Kattai and Lokk, 1998.
**Kattai (1999, personal communication).

Resources

The term "resources" is defined as the estimated volumes of solid, liquid, or gaseous materials known or not yet discovered that can be economically extracted at present or in the future—in other words, reserves plus all other deposits of the particular material, including those that may eventually become economically recoverable. Resources should be differentiated from the volumes of a given material in place, which include the volumes of the material that, for several different reasons, may never be recovered at a profit.

Estimates of worldwide shale-oil resources vary widely. The estimates of one group of authors (World Energy Council *Survey of Energy Resources*, 1980; Rühl, 1982; Knutson et al., 1990, among others) range between 2480 and 6230 billion bbl for shales, yielding more than 10 gal/ton. Other much higher and most likely unrealistic estimates have been made by other authors: Duncan and Swanson (1965) and Culbertson and Pitman (1973) believe that the world's shale-oil resources may be as high as 342 and 345 trillion bbl, respectively. These two high estimates are for oil shales yielding more than 10 gal of oil/t and include 328 and 333 trillion, respectively, of undiscovered and speculative resources. Russell (1990, p. 1) states that "worldwide resources are very large, and a rough estimate indicates that oil shales occurring in at least 50 countries may contain in excess of 2,000 trillion (10×10^{12}) barrels of oil in place." At a recovery of 50%, this would mean that, according to Russell, 1000 trillion bbl of shale oil could eventually be recovered from worldwide oil-shale deposits.

Dyni (2003, personal communication) states, "Total resources of a selected group of oil-shale deposits in

33 countries is estimated at 411 billion tons of in-place shale oil, which is equivalent to 2.9 trillion U.S. barrels of shale oil." He believes, however, that "this figure is very conservative."

Possibly as much as three-fourths of the shale-oil resources of the world are in the United States and Brazil (50% in the United States and 20–25% in Brazil). Lesser oil-shale deposits are known from China, Estonia, and Russia.

The estimates of most authors for the shale-oil resources of the United States range between 1380 and 2900 billion bbl, 90% or more of them in the Green River Formation. Unrealistically high estimates, 28,000 and 27,518 billion bbl, have been proposed by Duncan and Swanson (1965) and Culbertson and Pitman (1973). As in the case of the worldwide estimates of these authors, those for the United States include 22,550 billion bbl of undiscovered shale oil in the first case and 23,600 billion bbl of speculative shale oil in the second. Dyni (2003, personal communication) states that the Green River oil shale "contains an estimated 215 billion tons of in-place shale oil (ca. 1.5 trillion U.S. barrels)."

Most recent estimates of shale-oil resources in Brazil (also called "oil in place" by some authors) range around 800 billion bbl. The Irati Shale is generally recognized as the second largest oil-shale deposit in the world, after the Green River Formation.

Shale-oil resource estimates for China range between 190 and 300 billion bbl, and those for Estonia range to about 40–45 billion bbl.

Large resources of shale oil have been estimated to occur in Russia, principally in the Olenek basin of Arctic Siberia (for which a shale-oil resource of 250 billion bbl has been estimated) and in the Volga basin. The shale-oil resources of the Volga basin (about 33 billion bbl) are considered uneconomic, because the high sulfur content of the shale yields an undesirable synthetic oil.

SHALE OIL AS A SOURCE OF ENERGY IN THE 21ST CENTURY

The potential of the world's oil-shale deposits as a source of oil has been described as "vast," "huge," "tantalizingly enormous," "tremendous," and "immense" (whatever these terms may mean) by several authors. At this point, however, only in Brazil, China, and Estonia is shale oil being produced, but only in small volumes and probably under government economic subsidy. Production of shale oil is now uneconomic.

In the immediate future, shale oil may not be able to enter the energy market, certainly not while other sources of energy (oil, natural gas, and coal) are readily available at reasonably low prices. The present technology to develop oil-shale deposits is complex and expensive, and it is not likely that much effort will be devoted to improve this technology as long as the profitable production of shale oil remains a distant possibility.

In addition to its unprofitability, a detrimental aspect of oil-shale development is the significant potential damage it may cause to the environment, mainly the harmful aftermath of mining (particularly surface mining), the disposal of the huge volumes of spent shale resulting from a commercial operation, and the possible contamination of streams. In addition, the retorting of the oil shales requires large volumes of fresh water—several barrels of water to produce 1 bbl of oil. Government policies aimed at the protection of the environment would, therefore, affect negatively the competitiveness of shale oil as a source of energy.

In the more distant future, the development of the world's oil-shale deposits and the profitable production of shale oil will depend on many factors: world crude oil and gas prices; the development of more effective and cheaper oil-shale production and processing technologies, for example, in-situ recovery processes; government policies; taxation; environmental considerations; and the impact of competing fuels.

Although the present estimates of the volumes of shale oil in place in the world are very large (perhaps trillions of barrels), they are based, for the most part, on limited and very rudimentary information. For most oil-shale deposits in the world, information on quality (oil chemistry and yield), thickness, and geographic extent, all crucial elements in their evaluation, is inadequate for the detailed appraisal of their potential. Most oil-shale deposits have not been systematically mapped in detail in the surface and seldom in the subsurface.

It is not known at present if future studies, to be undertaken when the prospects of the profitable development of oil-shale deposits are brighter than they are now, will confirm the present very large reserves and resources estimates, reduce them, or increase them. However, such studies, followed by attempts for substantial commercial production of shale oil, may not be undertaken until it appears that other, now-abundant and cheaper sources of energy—mainly oil, natural gas, and coal—begin to show signs of depletion with possible consequent increases in price.

If the estimates of the availability of oil, natural gas, and coal as sources of reasonably priced energy in the 21st century made in this study are correct and, particularly, if new technology for the conversion of gas and coal to liquid fuels develops as expected, oil, natural gas, and coal will be more than sufficient to satisfy the estimated demands for fossil fuels in the 21st century, and shale oil can be expected to make only a modest contribution to the world energy supply.

NUCLEAR POWER

ABSTRACT

Nuclear power plants are large-scale sources of electricity that do not contribute polluting products to the environment. However, after supporting the expansion of a nuclear industry during the 1970s and 1980s, in the last 20 years, particularly after the Chernobyl accident, governments and the public have become increasingly concerned about the potential problems of nuclear plants: the disposal of high-level radioactive waste, the possibility of disastrous accidents, the generation of plutonium (the fuel of nuclear weapons), and the fear of anything "nuclear." As a result, the number of nuclear plants in operation, under construction, or being connected to the electric grid recently has declined over most of the world. This, added to the progressive retirement of older plants, will inevitably lead to the decline of the share of the world's electricity generation held by nuclear power during the first half of the 21st century. A nuclear "renaissance" during the second half of this century will depend on governments, the power industry, and the public recognizing that (1) the problems of nuclear power can be solved, and that (2) nuclear power is needed if the damage to the environment caused by the burning of fossil fuels must be lessened.

INTRODUCTION

Energy obtained directly or indirectly from the nuclei of atoms has been called nuclear power. The large nuclei of high-atomic-weight elements, particularly uranium and plutonium, can be split apart into smaller particles, a process called fission. The nuclei of low-atomic-weight elements, hydrogen or isotopes of hydrogen, on the other hand, can be combined to form larger nuclei in a process called fusion. In both cases, energy is released. Both are considered as sources of nuclear energy, but only fission is technologically feasible at present and is the type of nuclear reaction used to produce steam in all operating plants for the generation of electricity.

Nuclear fusion is likely to remain an elusive ideal for many decades and may never become a feasible source of energy. A self-sustaining fusion reaction has not yet been achieved in the laboratory, and at this time, the development of a commercial fusion reactor during the 21st century appears improbable. Enthusiasm and funding for research on fusion has diminished appreciably in the last few years. Nuclear fusion is therefore not considered further in the following discussion of nuclear power.

At the time when the world is becoming increasingly concerned about the discharge of carbon dioxide (CO_2) and other harmful gases and particulates into the atmosphere, and about the presumed resulting global warming (for which oil-, gas-, and mostly coal-burning electricity-generating plants are blamed), nuclear electric plants have the indisputable advantage, along with hydroelectric plants, of being the only potential large-scale suppliers of electricity that do not contribute polluting combustion products to the environment. Nuclear power plants generate electricity without combustion. In addition, the fuel is so concentrated that only small amounts are needed per unit of energy delivered.

Other advantages of nuclear power put forward by its supporters are that it is cheaper and safer than other means of generating electricity (both of which are questioned by its detractors) and, potentially, an almost inexhaustible source of energy. This is particularly the case if the fast breeder reactors, which can generate (breed) new fuel (i.e., convert uranium 238 into plutonium) in quantities as large as, or even larger than, the amount consumed, can be accepted and operated safely and economically.

Rhodes and Beller (2000) capably defend the desirability of nuclear power. Concerning the disputed cost advantage, they state (p. 47): "Larger nuclear power plants require larger capital investments than comparable coal or gas plants only because nuclear utilities are required to build and maintain costly systems to keep their radioactivity from the environment. If fossil-fuel plants were similarly required to sequester the pollutants they generate, they would cost significantly more than nuclear power plants do."

Defending the safety of nuclear power, they remind us of the recurring and deadly coal-mine accidents, of the oil- and gas-plant fires and pipeline explosions, not to speak of the deaths caused by air pollution. By comparison, they state (p. 48): "nuclear accidents have been few and minimal.... As for the Chernobyl explosion [the only major nuclear accident resulting in many fatalities], it resulted from human error in operating a fundamentally faulty reactor design that could not have been licensed in the West."

As another great advantage of nuclear power, they mention (p. 47): "its ability to wrest enormous energy from a small volume of fuel."

Supporters of nuclear power believe that it will not only be a preferable source of electricity but will also be a necessary, even essential, one if the inevitable future increases in demand for electricity throughout the world are to be met.

However, there are serious reasons for concern about nuclear power plants. Most worrisome problems are the disposal of the high-level radioactive waste (HLW), the product of the operation of nuclear reactors; the radioactivity of the plutonium generated as a result of the fission reaction, particularly in the case of the breeder reactors; and the possibility that the availability of the high-grade plutonium may increase the possibility of the manufacture of nuclear weapons.

Because of these concerns, intensely scrutinized and widely publicized in the press, television, and journals and stridently magnified by vocal environmental organizations, there is now a lack of public confidence in nuclear power. Mistaken public perception of the reality of nuclear power has resulted in a strong opposition to the increase in electric generation in nuclear plants.

Beck (1999, p. 113) summarizes the situation by stating: "The worldwide future of nuclear energy is a highly disputed subject; one side is certain that nuclear energy will have to expand in the next century to meet energy demand, whereas the other side is equally certain that this energy form is too dangerous and uneconomical to be of long-term use."

He adds (p. 114): "Both sides believe so strongly in the logic of their case that they see the opposition as either illogical or deliberately untruthful and, therefore, not worth talking to... Both parties try to convince the public that their position is correct, and it has to be said that in most democratic countries the antinuclear lobbies seem to have been more convincing. Although this has convinced only a few governments to withdraw from the production of nuclear energy, it has made politicians reluctant to be seen to support nuclear power, so that decisions that are needed, such as the destination of nuclear waste, are not made; thus, the industry is drifting."

Beck believes that despite the drawbacks of the use of nuclear power in the generation of electricity, "there is a strong case for keeping the option for nuclear expansion open. Yet, there has to be doubt whether today's technology is adequate for such expansion" (Beck, 1999, p. 113).

THE RISE AND FALL OF THE NUCLEAR POWER INDUSTRY

As described by Beck (1994, 1999), the development of the use of nuclear power for electricity generation was a result of the U.S. Manhattan Project during World War II. It was first enveloped in great secrecy as part of a defense program, but in the 1950s, it became a promising potential source of electricity as part of President Eisenhower's 1953 Atoms for Peace Program.

The industry started off small. In 1957, the U.S. Government beached a submarine reactor at Shippingport, Pennsylvania, and converted it into a power plant with an output of 72 MW hr, the first commercial nuclear reactor in the world. It closed in 1982. The first entirely new nuclear power plant began operating in 1960 in Rowe, Massachusetts.

Nuclear power technology advanced markedly during the 1960s, bringing in during the late 1960s and the 1970s the building of nuclear electric plants and the development of a nuclear power industry in many countries of the world, principally in the United States, France, the United Kingdom, Germany, Japan, and the FSU.

In the United States, the oil crises of the 1970s and the resulting misgivings concerning the possible future unavailability of fossil fuels encouraged the expansion of the nuclear power industry. The fast growth of nuclear-generated electricity seemed inevitable. However, the coming on line of an increasing number of operating nuclear power plants also brought to the attention of the industry and the public the serious problems of nuclear electric power: Many of the new nuclear plants were found to be less profitable and to have more problems than expected, raising fears about their safety. The disposal of the spent fuel became a difficult issue made harder by the United States Government's decision to prohibit the reprocessing of the spent fuel and to stop research and development of fast breeder reactors. The increasing activity of a vocal antinuclear movement, raising fears of a proliferation of the availability of bomb materials, sabotage, nuclear theft, and terrorism, became perhaps the most serious impediment to the consistent development of nuclear-power. Their arguments were widely promoted, encouraged, and publicized by the media, making the public react irrationally against nuclear power. A series of minor mishaps and the Three Mile Island nuclear reactor accident on March 1979 added to the United States public's concern, supported the fearful predictions of the antinuclear movement, and rocked the nuclear power industry. As a result, no new nuclear plants have been ordered in the United States since 1978, and many previous orders were canceled.

Other countries, particularly in Western Europe, the FSU, and Japan, were not discouraged by the nuclear industry problems in the United States and aggressively pursued the development of nuclear generation of electricity during the late 1970s and early 1980s; they built new plants, developed reprocessing facilities, and planned for the operation of fast breeder reactors.

By the mid-1980s, nuclear power seemed to be the promising and important participant in the supply of energy in the 21st century that had been heralded in the 1960s and early 1970s. Then, on April 26, 1986, came the serious accident at the Chernobyl nuclear plant in the Ukraine, the first accident in a nuclear plant resulting in deaths of plant workers and inhabitants of the neighboring country. It killed 32, sickened hundreds, impelled the evacuation of 100,000, and spread clouds of radioactive material over large areas of the FSU and Europe. Estimates of the number of people possibly affected over the coming years range from 500,000 to as many as 4 million. And it added indignation and irascibility to the antinuclear activist movements all over the world. Even in countries such as France and Japan, which had accepted nuclear power as a viable source of electricity, the public began to doubt the desirability to continue to increase the development of the nuclear power industry.

Increasing apprehension about the safety of nuclear power has been evident in the last 20 years:

- In 1982, Mexico canceled plans to purchase up to 20 reactors by the end of the century.
- In Western Europe, the overall trend in the 1990s has been away from the building of new nuclear plants. Spain, in 1983, and Switzerland, in 1990, declared a moratorium on the building of new nuclear plants, and public and political opposition is limiting or completely stopping new nuclear plant construction in several other countries: Italy and Austria abandoned plans to enlarge their nuclear power capabilities. Belgium and the Netherlands seem to be considering phasing out their nuclear power programs. In June 1997, the Swedish Parliament, responding to a previous referendum, decided to shut down all nuclear power plants by 2010. The contribution of nuclear power to the generation of electricity in Western Europe is therefore projected to decrease because of the lack of new orders and the retirement of old plants. This trend, however, can change with changes in government, whether liberal or conservative, pronuclear or antinuclear, and in the political parties that are in control (see Energy Information Administration's *International Energy Outlook* 2003, p. 104–105).
- Fast breeder reactors previously operating in France, the United Kingdom, and the FSU have

been shut down, and the building of additional ones has been canceled (in the United States, the Clinch River reactor).
- Several minor accidents in Japanese nuclear plants in the late 1990s, one of which killed a worker, exposed scores of people to radiation, and forced partial evacuation of a nearby town, have increased public resistance to nuclear power in Japan.
- On June 15, 2000, the German Government and the nuclear power industry agreed on a plan to progressively phase out the country's use of nuclear energy during the next 20 years. Almost all of the country's nuclear plants may be shut down by 2020.

In addition, with increasing deregulation and privatization of electricity generation in the developed countries, the economic competitiveness of nuclear power, some of it owned or subsidized by the government, is progressively losing ground. Generation of electricity from new nuclear power plants is not competitive in open markets today because of their high startup cost and longer building time, and most developing countries do not have the economic capacity for the construction of nuclear power plants without financial and technical assistance.

This increasing opposition to nuclear power and the generally unfavorable economics of the construction and operation of nuclear plants is clearly reflected in the number of units in operation (Appendix Table 58; Figure 42), in the number of construction starts of new plants (Appendix Table 59; Figure 43), and in the number of new plants connected to the grid (Appendix Table 59; Figure 44).

The number of units in operation in the world increased rapidly from 81 in 1970 to 408 in 1987, but only 33 more units have been added in the last 16 years. In the United States, the number of units in operation grew to 111 in 1990 but has decreased during the 1990s to 104, because some of the older plants have been decommissioned, and no new ones have been placed on line. In France, the number of nuclear units reached 57 in 1993 but has since increased to only 59. In Japan, the number of nuclear units has also leveled off at 51–54 since 1994 (Appendix Table 59; Figure 42).

Information on the number of worldwide construction starts of new plants and of the number of new plants connected to the grid show even more dramatically the decline of interest in nuclear power. The number of construction starts peaked in 1968–1975 at 35 and has dropped sharply since then to 6 or less new starts during the 1990s and the early 21st century (Appendix Table 59; Figure 43). As a result, the number of new plants connected to the grid peaked in the mid-1980s

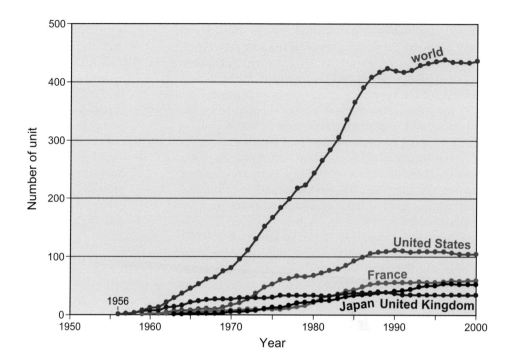

Figure 42. Nuclear units in operation.

at 33, decreased sharply during the late 1980s, and averaged 5 during the 1990s and early 2000s (Appendix Table 58; Figure 44). Most of the nuclear power units that were under construction in 2003 were in India (7), China (4), Ukraine (4), Japan (3), and Russia (3) and none in North America or Western Europe.

If these declining trends in new construction starts and in new nuclear plants connected to the grid continue during the early decades of the 21st century, the share of the world's electricity supply provided by nuclear power will inevitably decline sharply during the first half of the century.

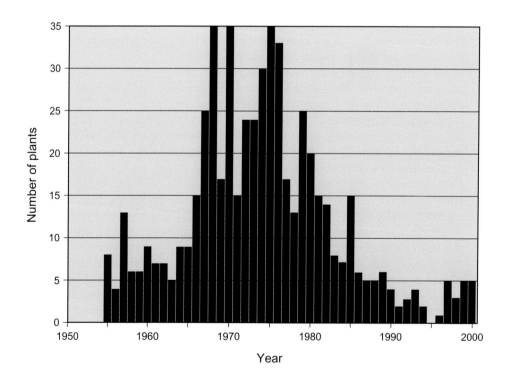

Figure 43. Nuclear plants that started construction.

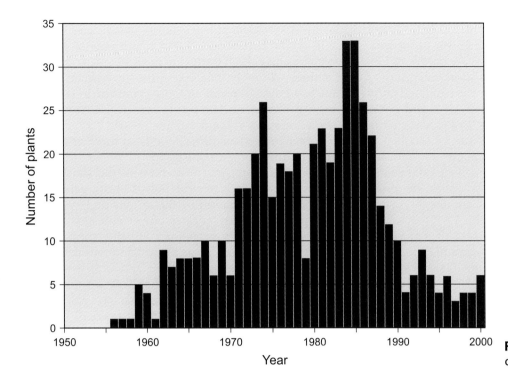

Figure 44. Nuclear plants connected to the grid.

NUCLEAR ELECTRICITY GENERATION

World nuclear electricity generation grew rapidly in the 1970s and accelerated in the 1980s, but the growth slowed down in the 1990s (Appendix Table 58; Figure 45). Similar patterns can be recognized in the generation of nuclear electricity in the United States, France, Japan, and other principal consumers of nuclear power. In the United States, the percentage of the electricity generated by nuclear plants, which grew rapidly from 4.5% in 1973 to 11% in 1980 and to 20% in 1990, has remained at the 19–20% level during the 1990s. In Canada, this percentage peaked in the mid-1990s at about 19% and declined to 16% by the end of the decade.

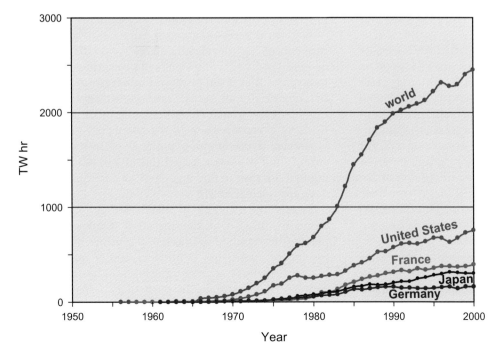

Figure 45. Nuclear electricity generation.

In the combined countries with nuclear power plants, the contribution of nuclear power to the total electricity generated increased from 3% in 1970 to 12% in 1980 and to 25% in 1985, but also in this group of countries, the nuclear-power share has remained fairly constant at about 25% during the late 1980s and 1990s. As in the case of the United States, this leveling off of the contribution of nuclear power can be ascribed to increases in the retirement of old plants and the lack of new orders, as well as concerns about the profitability of the nuclear plants, safety issues, the problems of radioactive waste disposal, and the increasing public opposition to nuclear power.

Since the mid-1970s, the United States, France, Japan, Germany, and the FSU have accounted for more than 70% of the world's nuclear-generated electricity (Appendix Table 60). At the end of 2002, nuclear power provided more than 20% of the total electricity supply of 19 countries. Those in which nuclear power contributes the highest percentage at present are Lithuania (80%), France (78%), Belgium (57%), Slovakia (55%), Bulgaria (47%), Sweden (46%), Ukraine (46%), Slovenia (41%), Armenia (40%), and Switzerland (39%) (International Atomic Energy Agency, 2003).

The contribution of nuclear power to the world's total supply of energy has grown from less than 1% in the mid-1960s and early 1970s to about 4% in the mid-1980s and has remained fairly constant at about 6.5% during the late 1990s and early 2000s (Appendix Table 8; Figure 12A, B).

THE ROLE OF NUCLEAR POWER IN THE 21ST CENTURY

At present, the short-range (20–30-year) future of nuclear power as a source of electricity does not look good—some have said it looks bleak. The information presented above on the number of nuclear units in operation, on plant construction starts, and on new plants connected to the grid certainly justifies such an opinion.

The projections of several organizations (World Energy Council, Energy Information Administration of the U.S. Department of Energy, International Institute for Applied Systems Analysis) range from little or no growth to a decrease in the contribution of nuclear power to the generation of electricity during the first 20 or 30 years of the 21st century.

The Energy Information Administration, for example, in its *International Energy Outlook* (2003, p. 101), presents a reference case in which "the nuclear share of the world's total electricity supply is projected to fall from 19% in 2001 to 12% by 2025. The reference case assumes that the currently prevailing trend away from nuclear power in the industrialized countries will not be reversed, and that retirements of existing plants as they reach the end of their designed operating lifetimes will not be balanced by the construction of new nuclear power capacity in those countries. In contrast, rapid growth in nuclear power capacity is projected for some countries in the developing world."

Other projections differ little from those of the Energy Information Administration.

These unpromising projections of the short-range future of the role of nuclear power in the generation of electricity are caused by the serious problems, real and perceived, that nuclear power still has, as discussed above: safety, economic competitiveness, radioactive waste disposal, fear of proliferation of nuclear weapons, sabotage, nuclear theft, and terrorism.

Despite the fact that more than 9000 reactor years of commercial operations had accumulated worldwide by 2000, and that only one serious accident has occurred during the operating life of the industry (in the Chernobyl plant, generally recognized as of fundamentally faulty design), the public still fears the release of radiation that may maim generations to come and the possibility of explosions in nuclear plants.

The addition of new safety equipment has added greatly to the cost of building new and upgrading existing nuclear plants. Nuclear plants are, at present, considered too costly to be able to compete on purely economic grounds with electricity-generating plants fueled by natural gas or coal, and the need of large upfront capital puts them out of reach of most developing countries. Nuclear power, it has been said, cannot simply be cheaper than coal-generated power; given its present opposition, it has to be substantially cheaper.

A general agreement exists among experts that the problem of high-level radioactive waste disposal can be readily solved using standard procedures available today, and that it can be stored underground in deep, dry, chemically unreactive, and impermeable geologic units that have been stable for millions of years, particularly if reprocessing of the spent fuel is allowed. However, overwhelming political opposition and a strong "not-in-my-backyard (NIMBY)" attitude have made it difficult to find a geologically adequate repository for nuclear waste.

Finally, the increased availability of radioactive materials resulting from the operation of nuclear power plants (the enriched uranium and plutonium, if reprocessing of spent fuel is allowed and if fast-breeder reactors are built and brought on line) has raised the concern of the public. In particular, accessibility to plutonium, the fuel of nuclear weapons, has made the public worry about sabotage and the possible theft of this radioactive product and its use by rogue countries or terrorist organizations for building nuclear weapons. Even if nuclear power, along with hydropower, is a potential, environmentally safe, large-scale source of electricity, anything "nuclear" is now dreaded and opposed over most of the world. A significant fraction of the public believes that nuclear power is unsafe, unreliable, uneconomic, and unnecessary.

The long-range future of nuclear power and its role in the generation of electricity in the mid- and late-21st century may depend primarily on the way governments, the public, and the politicians view the nuclear industry.

To have a "second nuclear era," a "nuclear renaissance," if nuclear power is to make an important contribution to the generation of electricity during the mid- or late 21st century, governments and the nuclear industry will have to dispel the current strong opposition to the use of nuclear power in the generation of electricity, to reach a greater public consensus in favor of nuclear power. To do this, the public needs to be assured that the nuclear plants are safe, that the spent fuel can be safely stored, that there is no danger of radioactive materials falling in the hands of terrorist organizations, and that nuclear plants can compete economically with other sources of electricity. This is not an easy task. Both the public and the electricity industry need to regain confidence in nuclear power. Convincing the public of the safety or of the low risk of nuclear power plants, for example, will require a very long time free of destructive and deadly accidents. Trust is quickly lost and very slowly, if ever, regained.

Scientists and technicians knowledgeable about nuclear power believe that these challenging objectives can be met, that the current problems are resolvable, that new advanced technology will be developed, and that improving economic conditions will make nuclear power fully acceptable. They believe that smaller, technologically advanced modular nuclear reactors, easier and cheaper to build, simpler to maintain, and quicker to go on line than the enormous reactors that dominate the industry today, may be an important part of solving the current problems, both operational and financial.

Standardization of the nuclear reactors and modification of policies, regulations, procedures, and organizations will also contribute to improve the practicability of nuclear power.

The disposal of high-level radioactive waste presents a considerable challenge, particularly if the number of nuclear plants increases considerably during the 21st century. However, as mentioned above, it can be solved with the right strategy and leadership and if freed from emotional resistance and not-in-my-backyard tactics. The large volume of radioactive waste of a greatly expanded nuclear power industry, mentioned by some detractors as an insurmountable problem, can effectively be handled by reprocessing and the placing in operation of fast breeder reactors by the mid-21st century.

The decommissioning of old plants, considered by some as dangerous and expensive, will also require attention, but it is not expected to become a serious obstacle in the development of an expanded use of nuclear power in the generation of electricity.

Too much attention has been paid in the last 10 or 15 years to the negative aspects of nuclear power, and too little attention has been paid to its advantages as a nonpolluting, potentially large-scale source of electricity. If the protection of the environment against the combustion gases and particulates develops into an important policy directive throughout the world during the 21st century, particularly in the developed countries, a renaissance of the nuclear industry is possible, perhaps even inevitable. Nuclear power is the only realistic alternative to natural gas and coal as fuels for the generation of electricity. Other sources are not viable for one reason or another. As will be discussed below, hydroelectric power will continue to account for no more than a 15–20% share of the world's electricity, and it does not seem, at this time, that other possible sources (geothermal, solar, wind, and biomass) will be able to contribute more than a small percentage of the electricity consumed in the 21st century. If the consumption of electricity throughout the world is to increase as now predicted, and environmental pollution is to be kept at a minimum, nuclear power will inevitably have to become a major factor in the generation of electricity during the second half of the 21st century. Otherwise, coal and natural gas will have to be the main fuels for the generation of electricity.

The nuclear power option will stage a comeback when governments, the industry, and, most important, the public realize that the option is needed and when it has been demonstrated that the problems that have stalled it during the late 20th and early 21st century are correctable and have been corrected. Not using nuclear power to generate electricity may well entail more and more serious problems than using it. The nuclear power industry, like all other industries and all human activities, cannot be expected to be 100% accident-free. Its risks need to be evaluated and balanced against the risks of rejecting it.

The wisest strategy for satisfying the demand of electricity in the 21st century will be the proper mix of fuels—natural gas, coal, hydroelectric power, and nuclear power—and a serious effort to conserve energy.

HYDROELECTRIC POWER

ABSTRACT

The flow of water provides one of the cleanest, most efficient ways of generating electricity. It is now the second largest source of electricity in the world. However, it has negative environmental and social impacts: extensive flooding upstream from large hydroelectric dams, with the consequent displacement of human population and the disruption of the environment. The development of hydroelectric power also has geographic constraints: sites favorable for building large dams are limited and unevenly distributed throughout the world. In addition, hydroelectric plants have questionable economic viability, particularly for developing countries, because they require large initial investments that are not always affordable. These unfavorable impacts on people, river basins, the environment, and aquatic life have, in the last 20–30 years, generated considerable opposition to the building of new dams, particularly in the developed countries. As a result, the pace of construction of new dams has slowed. Hydroelectric power has grown steadily during the 20th century, but its share of the world's total electricity generation has declined from 40% in the 1920s to about 18% in the 1990s. Although hydroelectric power output will continue to grow during the 21st century, it should not be expected to contribute more than 15–20% of the world's total power generation and no more than 5% of the world's total energy requirements.

INTRODUCTION

The energy supplied by flowing water has been used by humans since shortly before the start of the Christian era—the waterwheel, the geared water mill, and the Vitruvian mill received increased use as civilization developed in the Mediterranean region.

During the Middle Ages, the use of the geared water mill extended throughout Western Europe, where it was used to grind grain, to saw wood and marble, and to crush metallic ores. Water mills became the base for the beginning of industrial development. Flowing water as a source of energy extended to many other regions of the world in subsequent years.

However, it was not until the flow of water was used in the generation of electricity that water started to make a major contribution to the world's supply of energy. Hydroelectric power is now the second largest source of electricity in the world (18%) after coal, which supplies about 38%, but ahead of nuclear power, natural gas, and oil, which contribute 17, 16, and 10%, respectively, of the total generated electricity.

The world's first hydroelectric power plant was built at Godalming, in England, in 1881. Hydroelectric power developed rapidly during the 20th century, even in countries with other abundant sources of energy.

Hydroelectric power was at first believed to be an attractive source of energy, potentially widely distributed around the world, environmentally clean, inexhaustible, efficient, and economically competitive, especially attractive to countries with limited fossil fuel resources. It has been found that this was not always the case.

Although dams, not only those built for the generation of electricity but also those designed for flood control, irrigation, water supply, or a combination of these purposes, have delivered significant benefits, they have often been the cause of disastrous social upheavals and major negative environmental impacts, and their construction has serious geographic limitations.

• Large areas upstream from the dam have been flooded, in some cases covering rich and fertile agricultural land.
• The populations in these flooded areas have been displaced, causing them to lose not only their homes but their livelihoods and their sociocultural foundations as well. It has been estimated that between 40 and 80 million people may have been displaced between 1950 and 1990, 26–58 million of them in India and China. According to the World Commission on Dams (2000, p. 102), "the impacts of dam-building on people and livelihoods—both above and below dams—have been particularly devastating in Asia, Africa, and Latin America, where existing river systems supported local economies and the cultural way of life of a large population containing diverse communities."
• Dams alter and divert river flow affecting access to water, transforming landscapes, and resulting in significant irreversible impact on the environment both upstream and downstream from the

dams. They often have a negative impact on rivers, watersheds, and aquatic ecosystems.

- Evaporation of the water in the reservoirs upstream from the dams becomes a problem in some cases, particularly in arid regions where fresh water is a priceless commodity, needed for agriculture and human consumption.

- Hydroelectric plants are not necessarily inexhaustible sources of electricity. They have a limited operating life. The reservoirs behind the dams become silted up in many cases with the consequent long-term loss of water storage capacity. The sediment caught at the dam can sometimes cause the erosion of the turbines if it reaches the power intakes. The retention of sediment behind the dam can also be the cause of serious problems downstream, where the reduced influx of sediment often results in important changes of the river valley, sometimes all the way to its delta. The Aswan High Dam in Egypt and the Hoover Dam in southwestern United States are good examples of such problems. In Egypt, for thousands of years, farmers along the Nile Valley counted on the yearly floods to bring organically rich silt that fertilized their intensely cultivated plots. Now, the silt piles up behind the Aswan Dam.

- Finally, the development of large-scale hydroelectric power requires sites that allow the building of dams that provide a high head (height of water fall), located in a river with high water flow, with large water-storage capacity, and, preferably, near centers of electricity consumption. These dams are sometimes called "large dams" and are defined as having a height of 15 m (50 ft) or more from the foundation or, if only 5–15 m (16–50 ft) in height, having a reservoir volume of more than 3 million m^3 (177 million ft^3). Sites favorable for the building of large dams are limited in number and unevenly distributed throughout the world. They are not found, for instance, in arid regions with low rainfall and, therefore, with no permanent streams, or in regions of low relief. Although most of the world's sites for the building of hydroelectric power plants have been identified, only a relatively small proportion have been developed, mostly in the industrialized countries. At least seven European countries (Austria, France, Italy, Norway, Spain, Sweden, and Switzerland) generate substantial percentages of their electricity in hydroelectric plants.

- The building of large dams requires large initial capital investment that is not always rewarded with favorable economic profitability. The reported returns on the investments made in hydroelectric power plants have increasingly been questioned.

Small-scale hydroelectric power plants (small dams or "run-of-river" plants) are widely distributed throughout the world. They can serve local needs and are often the foundation of rural electrification, but they provide overall only a small percentage of the total world hydroelectricity. Many of them are located in mountainous regions with steep, fast-flowing rivers, in which large dams backing sizable reservoirs cannot be built.

Despite these serious problems, the building of hydroelectric dams increased rapidly during the middle years of the 20th century (1930s to 1970s) and peaked in the 1970s, when an average of two or three large dams were commissioned each day somewhere in the world. Their number increased from 5000 in 1949 to 45,000 in the late 1990s (World Commission on Dams, 2000). However, information on the unfavorable impacts of dams on people, river basins, the environment, and aquatic life, as well as their not-always-favorable economic performance, in the last 20–30 years generated considerable opposition to the building of dams. This opposition has become more widespread, vocal, and organized and has contributed to a decline in the pace of construction of new, large hydroelectric dams over most of the world in the past two decades. The exceptions are China, where the huge Three Gorges hydroelectric dam is expected to be commissioned between 2003 and 2009, India, and Brazil, where a considerable amount of hydroelectric capacity is under construction or planned. Also contributing to this decline in North America and Europe has been the fact that most technically attractive sites have already been developed. The average large dam today is about 35 years old.

Momentum for river restoration is also accelerating in many countries, especially in the United States, where the decommissioning rate for large dams has overtaken the rate of construction since 1998. Anti-dam strategies developed principally by environmental and human rights activist groups worldwide have been effective in reducing the pace of construction of dams, of all kinds and sizes, in most countries of the world since the 1970s. Although the global effects on the environment are much less damaging than those of other energy options, the local effects have been made to appear greater.

Nevertheless, hydroelectric power is still important in some countries, especially in those that are short of other sources of electricity but have large rivers originating in mountainous regions. (For a discerning although somewhat supportive review of the pros and cons of hydropower, see World Energy Council, 1995, 1998, and 2001 issues of the *Survey of Energy Resources*).

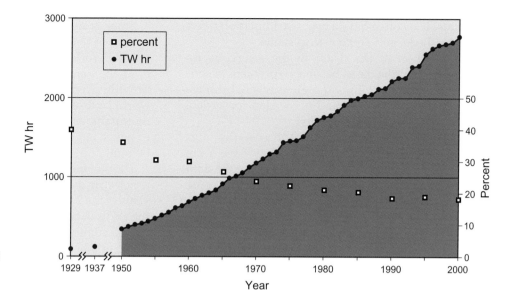

Figure 46. World hydroelectric power generation and percent of the world's total electricity generation.

HYDROELECTRIC POWER GENERATION

Hydroelectric power generation has grown steadily during the 20th century, as shown in Appendix Table 61 and Figure 46, but its percentage contribution to the world's total electricity generation has steadily decreased. During the 1920s, hydroelectric power represented about 40% of the total electricity generated worldwide; its contribution declined to about 30% by the mid-1950s and to about 20–21% by the mid-1980s. It has remained at about 18% during the 1990s (Appendix Table 61; Figure 46). Hydroelectric power is used in more than 150 countries, accounting for more than 90% of the total national supply of electricity in 24 countries and more than 50% in 63 other countries.

Appendix Table 61 and Figure 47 show the hydroelectric power generation of the United States, Canada, Brazil, the FSU, and China, the five principal contributors of the total hydroelectric power in the world. They have accounted for between 50 and 54% of the yearly total hydroelectric power generated in the world since 1950. These five countries plus Norway and Japan accounted for 50–63% of the world's hydroelectric power during the same period.

Whereas Brazil and China increased their generation very rapidly, particularly since 1970, the hydroelectric power generation by the United States, Norway, and Japan has remained fairly stable or has grown slowly since the mid-1970s. Canada's generation increased steadily until the mid-1990s but seems to have flattened since then. The slow growth of hydroelectric power production in the United States, Norway, and Japan can be attributed to the decreasing availability of suitable sites for large dams. The apparent slowing down in Canada may be caused by

the same reason. Hydroelectric power generation increased rapidly in the former Soviet Union until the breakup of the country in 1991. The generation by both Russia and by the combined countries into which the FSU split has begun to increase again during the late 1990s. Other countries with important generation of hydroelectric power are, in descending order, India, France, Sweden, Venezuela, and Italy.

The percentage of the total energy produced in the world represented by hydroelectricity has increased progressively from about 5% in the 1950s to about 7% in the 1980s and 1990s (Appendix Table 8; Figure 12A, B).

THE ROLE OF HYDROELECTRIC POWER IN THE 21ST CENTURY

Although hydroelectric power is the most efficient (80–90%) power delivery system and probably the cleanest major source of power yet developed, its future is clouded by its environmental, social, and economic drawbacks and by the limits in the availability of sites in which to build large hydroelectric plants.

The strong opposition to the building of dams that started in the 1970s has continued to grow, particularly in the developed countries. The negative environmental and social impacts of hydroelectric power have received an unprecedented level of attention in many countries. Except for China, Brazil, and India, the commissioning of new large hydroelectric plants has steadily diminished in the last 20–25 years. By the early 1990s, it became clear that the controversy was seriously affecting the future of hydropower as an important source of electricity.

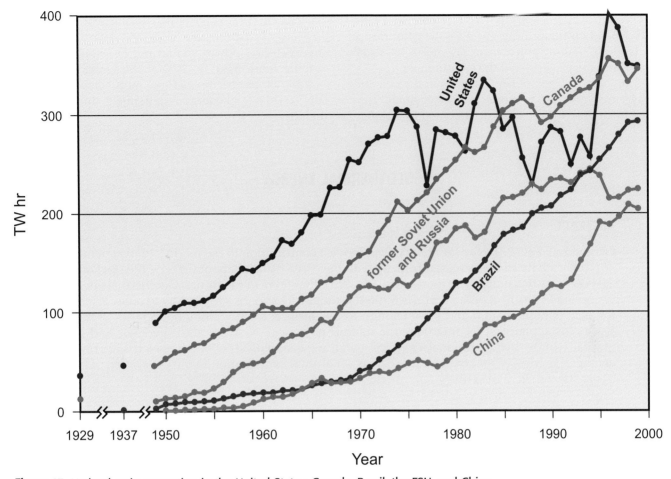

Figure 47. Hydroelectric generation in the United States, Canada, Brazil, the FSU, and China.

The decisions to build new dams are being increasingly contested to the point where the possibility of the building of large hydroelectric dams in the future in many countries is in question. It is obvious that the building of large hydroelectric dams will need to address in the future the social and environmental problems, inequities, and conflicts they would create. As pointed out by the World Commission on Dams (2000, p. 234), "Rivers, watersheds and aquatic ecosystems are the biological engines of the planet. They are the basis for life and the livelihoods of local communities. Dams transform landscapes and create risks of irreversible impacts."

However, even if the opposition to the building of dams was to vanish or at least diminish, site availability and economic viability may still remain as limiting factors in the building of hydroelectric plants and in the future contribution that hydroelectric power will make to the world's demand for energy, in general, and electricity, in particular, during the 21st century.

As mentioned above, few favorable sites for large hydroelectric dams remain in North America and Europe that do not involve conflicts of population displacement, land, water use, or aquatic biosystems. Latin America, Africa, and Asia, where the highest rates of population growth are being registered, have almost half of the world's undeveloped hydroelectric power capacity, but many of the potential sites are far from centers of electricity consumption, and the countries in which these sites are located often lack the necessary financial resources for the building of large hydroelectric plants. These plants require large initial investment and often involve long construction periods, which make them a poor choice for power generation when viewed only on the basis of short-term economic considerations.

Despite these serious drawbacks, hydroelectric power should be expected to continue to make a significant contribution to the world's supply of electricity, particularly if the countries in the developing world, which have the biggest share of potential sites for the building of medium and large hydroelectric plants, raise their standard of living, demand larger amounts of electricity, and acquire access to suitable funding.

The pace of hydroelectric power development should also depend on the priority given by the countries of the world to limiting greenhouse gas emissions, social issues, and the protection of the environment.

If the present trend of fast increases in the demand for electricity persists, hydroelectric power should continue to increase for yet a few decades and to maintain its present 18% contribution to the world's electricity supply. During the mid- and late 21st century, hydroelectric power should not be expected to contribute more than 15–20% of the world's total generation of electricity and no more than 2–2.5% of the world's total energy requirements during the 21st century.

GEOTHERMAL ENERGY

ABSTRACT

Geothermal energy has so far made only a modest contribution to the world's total generation of electricity: its share increased from 0.1% in the early 1970s to about 0.4% by the start of the 21st century. It is unlikely to increase much above that level in the future. Geothermal energy, generated by the heat of the Earth, is limited geologically and geographically to regions where the heat is high enough and close enough to the surface, mainly along boundaries of major Earth crustal plates (spreading ridges or subduction zones). Not only these geologic and geographic realities but operational, environmental, and economic drawbacks as well will prevent geothermal energy from making more than a marginal contribution to the generation of electricity during the 21st century.

INTRODUCTION

Geothermal energy, in the broadest sense, is the natural heat of the Earth. Temperatures in the Earth rise at different rates in different regions with increasing depth (geothermal gradient). Most of this heat, however, is far too diffuse ever to be recovered economically. Consequently, most of the heat in the Earth cannot be considered a potential economic energy source, unless it is concentrated enough and hot enough when close to the Earth's surface.

Economically significant concentrations of geothermal energy occur now where high temperatures (40° to more than 380°C; 104° to more than 716°F) are found in porous and permeable rocks at shallow depths (less than 3000 m; about 10,000 ft). The geothermal energy is stored in both the solid rock and the water or steam-filling pores and fractures. The steam or hot water are used mainly as the fuel for the operation of electricity-generating turbines or for space heating.

There are, therefore, three main requirements for the commercial development of geothermal resources: shallow high temperatures, rocks with good permeability, and sufficient volumes of water.

At present, there are two major types of commercial geothermal systems for the generation of electricity: water-dominated (hot-water) systems and vapor-dominated (dry-steam) systems. Hot-water systems are much more common than dry-steam systems, of which only four commercial fields are known at present: The Geysers, 115 km (70 mi) north of San Francisco, California; Larderello, in central Italy; Matsukawa in Japan; and possibly Kawah Kamojang in western Java, Indonesia.

A third type, the hot-dry-rock system (HDR), is designed to recover heat from impermeable hot dry rocks at depths of 1500–2000 m (5000–6500 ft) by drilling two closely located wells into them, artificially fracturing the rocks, and injecting water through one well and recovering it from the other after having been heated by the hot rocks. Experimental attempts to recover geothermal heat in this way have so far failed to demonstrate that hot-dry-rock systems are economically viable. In these systems, energy recharge is only by thermal conduction and, because of the slowness of this process, their geothermal energy should be considered exhaustible (Stefansson, 2000). Some experts, however, believe that the long-range future of geothermal energy may depend on HDR systems becoming a technological and economic reality (World Energy Council, *Survey of Energy Resources*, 2001).

The occurrence of favorable geothermal conditions for the commercial generation of electricity is limited geographically and geologically. Regions with geothermal potential are mainly located along belts of active magmatism, mountain building, and faulting

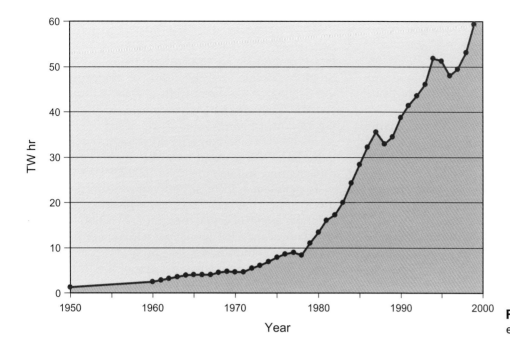

Figure 48. World geothermal electricity generation.

principally localized along the boundaries of major Earth crustal plates, belts where either new material from the mantle is being added to the crust (spreading ridges) or where crustal material is being dragged downward and consumed in the mantle (subduction zones). In both cases, molten rock is generated and moved upward into the crust and near the surface of the Earth. Geothermal energy for the commercial generation of electricity is absent in the stable continental shields, which are characterized by lower-than-average geothermal gradient.

USE OF GEOTHERMAL ENERGY

Geothermal resources throughout the world have been used for various purposes for centuries. Hot springs were used for bathing and medicinal purposes in ancient Greece, Rome, Babylonia, and Japan, among many other places. "Medicinal spas" are still popular resorts throughout the world.

Geothermal hot water has been used directly for space and district heating, for aquacultural and agricultural purposes, and for a variety of industrial applications. Much of Reykjavik, the capital of Iceland, located on the Mid-Atlantic spreading ridge, has been heated since the 1930s with geothermal hot water. Many other countries, such as Japan, New Zealand, Hungary, and the United States (Boise, Idaho, and Klamath Falls, Oregon, for example), use geothermal energy for space heating. Such direct use of geothermal energy, in fact, is possible virtually anywhere in the world.

Direct use of geothermal energy, although considerable now and possibly a sizable source of heat in the future, is not included in the following discussion. The discussion is restricted to the use of geothermal energy for the generation of electricity, which was accomplished for the first time at Larderello, Italy, in 1904. Although the first generator produced only enough electricity for five light bulbs, the Larderello geothermal field has been generating electricity on a commercial scale since 1913.

Elsewhere, generation of electricity from geothermal sources began in New Zealand in 1958 and a year later in Mexico. In the United States, it started in 1960 at The Geysers field in California. Japan followed in 1966 and the former Soviet Union in 1967. Many other countries began geothermal electricity generation shortly thereafter. The Geysers is, at present, the largest geothermal electric power installation in the world.

GEOTHERMAL-ENERGY ELECTRICITY GENERATION

Worldwide commercial generation of electricity from geothermal sources is shown in Appendix Table 62 and Figure 48. It increased slowly and steadily until the late 1970s and at a much faster rate in the 1980s and 1990s, mostly because of active expansion of geothermal programs in the United States, the Philippines, Mexico, Indonesia, and Japan.

The most important producers at present are, in the following order, the United States, the Philippines, Mexico, Italy, Indonesia, Japan, New Zealand, and several countries in Central America, namely, Nicaragua, El Salvador, Costa Rica, and Guatemala.

The percentage of the worldwide total electricity generated from geothermal sources has increased from 0.1% in the early 1970s and 0.3% in the late 1980s to about 0.4% in the late 1990s.

Significant geothermal resources have been identified in 142 countries. In 2000, 21 of them used geothermal power for the generation of electricity; an additional 45 used geothermal heat for other purposes, and most of the rest are now gathering data and assessing the possible use of geothermal heat for the generation of electricity or for direct uses.

Worldwide installed electricity-generation capacity from geothermal sources stood at 7974 MW in 2000, of which 2228 MW, or 28%, was in the United States. By 2002, it had increased to 8227 MW. The Philippines and Indonesia now lead in the aggressive addition of new capacity. The history of the development of geothermal installed electricity-generation capacity is similar to that of the geothermal electric generation: a slow increase until the late 1970s and a much faster increase in the 1980s and 1990s. A slowing down in the early 2000s is expected to end by 2004 or 2005, when slow increase may resume.

THE ROLE OF GEOTHERMAL ENERGY IN THE 21ST CENTURY

Geothermal energy was portrayed in the 1980s as a renewable source of commercial energy, socially, environmentally, and economically attractive, destined to provide an important share of the world's electricity in the future. These early forecasts of the potential of geothermal power now seem overly optimistic for economic, political, technical, and even environmental reasons.

Most damaging to the early optimistic predictions is, perhaps, the fact that geothermal power may not be renewable but is more likely a finite resource that will need to be exploited carefully. Twenty or thirty years of experience in the development of the known geothermal reservoirs indicates that the hot water or steam in these reservoirs can ultimately be depleted. The time of depletion in these reservoirs has been estimated to range between 40 and 100 years. Geothermal resources are only renewable if an equilibrium is maintained between the hot water or steam extracted through wellbores and the natural or artificial recharge of water into the system. The life of a geothermal system can be extended by alternating periods of production and shutdown.

At The Geysers, after less than 20 years of steam production, a pressure decline became evident in the late 1980s, which continued until the mid-1990s. This pressure decline was the result of overexploitation: the drilling of too many wells and the withdrawal of steam in quantities far exceeding the field's capacity. Of the 24 power plants, 6 had to be closed by 1995, and electricity-generation capacity decreased from 2000 MW in the late 1980s to about 1300 MW in 1997. Sustainable capacity may not be more than 1000 MW.

To maintain or increase reservoir pressure and steam production, it has been necessary to resort to a carefully planned water-injection program: injection into the reservoir of recycled municipal wastewater from several nearby communities. This program seems to have been successful in recovering part of the lost power-generation capacity and in providing the contributing communities with a solution to the disposal of their wastewater.

Similar water-injection programs may be necessary in other geothermal fields in the coming years, provided that the injection of cold water does not lower the temperature of the reservoirs.

Although geothermal power has some evident advantages—low or no emission of CO_2, the gas blamed for what has been called "global warming"—it also has some serious limitations:

- Geothermal steam and hot water, unlike coal, oil, and natural gas, cannot be transported over long distances from the producing well to a power plant. Geothermal power plants therefore need to be built in the vicinity of the geothermal source, and when the centers of consumption are a considerable distance away, long and expensive electricity transmission lines are necessary.

- Geothermal power may not be as socially and environmentally attractive as it was at first depicted; it can have some adverse environmental impacts: unless reinjected into the hot-water or steam reservoir, the spent water discharged from the electricity-generating plants can contaminate local streams and potable-water aquifers. Reinjection of spent water, of course, is desirable not only for environmental reasons, but for the maintenance of the pressure and the volume of fluid in the reservoir as well.

- Of lesser concern are the emission in some cases of H_2S, CO_2, and radon and the content of arsenic, boron, and other poisons in the discharged water. Also mentioned as shortcomings of the development of geothermal power are the noise of steam production, thermal pollution, and the possible subsidence of the land surface resulting from the removal of large quantities of underground fluids. Reinjection of spent water or of water from other sources can eliminate or reduce the land subsidence.

- Drilling for steam and hot water is, in many cases, more difficult and expensive than drilling for oil or natural gas because of high temperatures and very hard or fractured rocks.
- Finally, although geothermal dry-steam systems seem to be economically attractive, hot-water systems are, at present, economically marginal or unfavorable. Even if geothermal power plants can be built in increments of virtually any size depending on the need, front-end investments are quite heavy and not easily funded, which may make geothermal power economically prohibitive for many Third World countries, even for those heavily dependent on the importing of other fuels for the generation of electricity.

Geothermal power currently makes a modest contribution to the electricity supply of some countries. It will probably continue to do so in the future. However, considering its geologic and geographic limitations and its economic and environmental uncertainties, the function of geothermal power in the 21st century is expected to be only marginal; it is doubtful that the contribution of geothermal power to the total generation of electricity in the world will be higher in the future than the present 0.4%. The percentage of its contribution to the total worldwide demand for energy during the 21st century would be even smaller.

ENERGY FROM THE SUN, THE WIND, AND BIOMASS

ABSTRACT

Sunlight, the wind, and biomass (wood and agricultural, animal, and municipal wastes) supply, at present, a small fraction of the total commercial energy consumed in the world. Sunlight is used for passive household space and water heating and in the generation of electricity by the photovoltaic process (PV). For both purposes, the use of sunlight energy is limited by its intermittence. Photovoltaic electricity generation is not economically competitive at a large scale and is restricted to small-volume niches, remote areas without access to other sources of electricity.

Wind-propelled turbines offer a more favorable prospect as a future source of electricity. Wind power is now the fastest growing source of electricity in the world. The most undesirable aspect of both large-scale sunlight or wind electricity-generation installations, however, may be the great expanses of land that they require.

Biomass has been extensively used and is still used in the developing countries for home heating and cooking, for the small-scale generation of electricity, and for production of combustible gases and liquids (alcohols). As a large-scale source of energy, biomass has serious drawbacks. The use of wood aggravates the already serious problem of deforestation and massive soil erosion in many Third World countries; and the availability of any kind of wastes is limited. Biomass will continue to be used in the developing countries as before and may supply energy everywhere as a by-product of other activities (for instance, the forest products and the sugarcane industries), but in the long run, it will be replaced by more convenient sources of energy. Sunlight, the wind, and biomass are not expected to be major suppliers of commercial energy, in any form, during the 21st century.

INTRODUCTION

In its broadest sense, solar energy has been defined as all the sources of energy resulting from the reception of sunlight on Earth. In this sense, solar energy would include not only heating and cooling of buildings and the generation of electricity by the direct use of sunlight, the power of the winds, and the use of wood and other forms of biomass. It would also include oil, natural gas, and coal (what has been called "fossil sunshine"), because these fossil fuels are the product of the conversion of sunlight into organic matter through photosynthesis and the subsequent generation of oil, natural gas, and coal by the effects of heat, pressure, and time. This expanded definition would cover almost all sources of energy in the Earth since its birth, except for nuclear and geothermal power.

A more restricted use of the term "solar energy" is favored in this study. Solar energy is applied here to just the use of direct sunlight, the solar energy arriving on Earth at present. Wind power and the energy obtained from biomass will be discussed separately. Humans have used all these sources of energy in one way or another for many centuries: the heat of the sun, wood for fire, and the wind for moving sailing vessels and powering windmills.

The use of ocean tides and waves and the energy provided by ocean thermal energy conversion (OTEC) are believed to have only very small potential as future sources of energy and will not be discussed in this study.

The term "renewables" has been used, not always consistently, to include the energy provided by the sun, the wind, and biomass, as well as geothermal, hydroelectric, and "marine" (tides, waves, thermal, etc.) energy. Renewables will not be used in the following discussion to refer to any of these energy sources.

USE OF DIRECT SUNLIGHT

The amount of solar energy reaching the Earth is incomprehensibly vast, inexhaustible, and free, but it is intermittent; the sun shines only during the day and is often obscured by clouds, and at high latitudes, days are very short during the winter. Over an entire year, day and night, summer and winter, cloudy and clear, and at all latitudes, only a part of the sun's energy reaches the Earth's surface.

Furthermore, solar energy is diffuse and needs to be concentrated and stored. To collect sufficient amounts of energy, it is necessary in some cases to use large and expensive equipment and great expanses of land that can make the use of solar energy economically questionable and even environmentally objectionable.

Direct sunlight can be used for two main purposes: for heating and cooling and for the generation of electricity.

Heating and Cooling

Solar heating is the oldest and technically most mature use of direct sunlight. Passive solar heating has been used by humans for centuries for warming people and their dwellings, for drying food, and for growing plants in greenhouses. Proper architectural design of buildings to take advantage of sunlight's heat is receiving increasing attention.

Several devices (flat-plate collectors, parabolic trough collectors, and parabolic dish collectors) have been in use since the early 20th century for heating water and for space heating in countries and regions of the world where sunlight is prevalent. These devices, however, need some kind of backup for times when there is no sunlight. Space cooling is in a less advanced stage, both technologically and economically, because it requires considerably more complex and expensive systems.

Generation of Electricity

A more recent development in the use of direct sunlight is for the generation of electricity. Two different processes have been developed for this purpose: a thermal conversion process and a photovoltaic (PV) conversion process.

The solar thermal conversion process employs a large field of heliostats (two-axis movable mounted mirrors) that track the sun and reflect sunlight into a central receiver at the top of a "solar tower." Water in the receiver is converted to steam that operates a turbine generator on the ground, producing electric power. The first experimental electric power plant using this technology (Solar One, near Barstow, California) was shut down in the late 1980s. A second, larger, and more advanced plant, Solar Two, started to operate in 1996 but was shut down in 1999 because of lack of funding. No other plants using this technology are now in operation. The economic viability of this electricity-generation process remains to be proven.

The PV conversion process uses solar cells—solid-state semiconductor devices that convert sunlight directly into electricity, silently and without pollution. The PV process is the simplest, most dependable technology to harness the power of the sun. Solar cells or solar modules are easy to install and, as a result of their simplicity, require minimal maintenance.

Solar PV systems are successfully used today to generate electricity at a small scale in many countries and for many purposes. They are in particular demand where only modest amounts of power are required and in areas remote from electric power plants and without access to electricity grids: highway signs, rural water-pumping stations, remotely located communities and dwellings, space vehicles, and marine or air navigation beacons.

In the last 35 years, the worldwide small-scale generation of electricity by PV devices has increased substantially, particularly in the last few years, from less than 5 MW/yr in the late 1970s to about 130 MW/yr in the late 1990s; but this is still a minute percentage of the total world generation of electricity.

Several economic, geographic, and technical barriers still need to be overcome for large-scale PV electricity generation to achieve significant penetration in the world electricity market.

Central large-scale PV power plants cannot economically compete at present with other sources of electricity, even after important breakthroughs in PV technology, and continuing improvement in the efficiency of solar-cell manufacturing in the last 20 years have pushed down the cost of manufacturing PV devices. At present, the price of PV electricity is 5–10 times higher than that of electricity generated in conventional coal, natural gas, hydroelectric, or nuclear power plants. Further refinements in technology and possible increases in the scale of production could decrease the cost of solar cells and improve the potential for PV electricity generation, not only in large plants, but also in individual residences and other buildings. However, it is generally believed that large-scale generation of PV electricity will be more expensive than electricity from other sources for many decades to come, and it is this high cost of PV electricity that has been the major obstacle to its more widespread use (World Energy Council, *Survey of Energy Resources*, 1995, 1998).

Large-scale use of direct sunlight, either by thermal conversion or by PV devices, also has serious geographic limitations. It requires clear skies, sunny climates, and large expanses of land that would become unavailable for any other purpose. When the sun does not shine, solar PV power installations require backup electrical supplies from conventional sources. Use of sunlight for the generation of electricity is, therefore, better suited for low-latitude and sparsely populated regions. However, large electricity-generating plants in sparsely populated areas far from centers of electricity consumption require massive and expensive electricity storage or long-distance transport of the electricity generated, further hindering the plants' economic viability.

A scheme suggested to overcome these geographic limitations is the placement of large arrays of solar cells in synchronous-orbit satellites 30,000–35,000 km above the Earth. The PV arrays would convert the high solar radiation to electricity, which in turn would be fed to a microwave generation system for transmission to Earth. The microwave transmissions would be collected by large receiving antenna systems and converted to alternating or direct current electricity. The possibility of placing one of these solar power satellites in orbit has received no serious consideration.

An even more extravagant proposal would place large arrays of PV cells on the moon to transmit the solar power generated to Earth by microwave beams.

WIND POWER

Wind, generated by the sun's differential heating of the Earth, by its rotation and by differential solar absorption between land and sea, has been a source of energy for many centuries. Wind energy is plentiful, but like direct sunlight, it is diffuse and intermittent; it needs to be captured, concentrated, and, for some uses, transported or stored. Winds are variable in place, time, and intensity, but unlike sunlight, they can blow day and night, at all latitudes, and during any time of the year.

Although humans had used the power of the wind for centuries to propel their sailboats, it was not until the beginning of the 12th century that wind power became an important source of energy with the development of the windmill. Besides pumping water, windmills became invaluable for grinding wheat and corn, sawing wood and stone, and for producing oils, dyes, and later, paints, paper, and textiles.

The emergence of the steam engine in the 18th and 19th centuries, which brought the Industrial Revolution, ended the importance of the windmill as a source of mechanical power. The widespread availability of sources of electricity during the 20th century further reduced the need for windmills, except in remote regions where other sources of energy are not available.

Wind power has reentered the energy supply picture in the last 30 or 35 years and developed rapidly during the last 20 years, this time for the purpose of generating electricity by means of wind turbines. Wind power has become, over the last decade, the fastest growing source of electricity in the world. However, even at this fast growth rate, it will take a long time for wind to gain a significant share of the electricity market.

As a source of electricity, wind power has some definite advantages but also some important disadvantages.

Wind power is probably the cleanest and most environmentally harmless source of energy. It also has a much shorter lead time of construction compared to that of other electricity-generating plants. Wind turbines are easy to build and can be added progressively and quickly as needed.

Perhaps the greatest disadvantage of wind power is the great expanses of suitably located land needed for the installation of the scores of wind turbines that would be necessary for the development of large-scale "wind farms." Wind turbines cannot be sited too close to each other because of aerodynamic interference effects; no more than four to six 50-m (160-ft)-rotor-diameter turbines can be sited in 1 km² (0.4 mi²) of land, giving an energy yield of about 8–10 GW h/km² (World Energy Council, *Survey of Energy Resources*, 1998, p. 169). Wind farms, however, unlike PV power plants, do not preclude farming, ranching, or other activities in the land they occupy.

Other often mentioned, but less important, concerns about wind power are the possible icing of the turbine blades in cold weather, the unsightly appearance of the wind farms, the noise they make, the disruption of wildlife habitats and of the paths of migrating birds, and possible interference with telecommunication transmissions (World Energy Council, *Survey of Energy Resources*, 2001).

Wind farms typically include 10–100 wind turbines, but some already in operation have several hundred. Because the wind is intermittent, wind-diesel or wind-batteries systems need to be installed, or backup access to other sources of electricity need to be available to provide uninterrupted electricity supply.

Until recently, wind power could not compete economically with other sources of electricity. This has changed during the past 10 or 15 years. Important improvements in the reliability of the technology for the design and construction of wind turbines have significantly narrowed the gap between the cost of wind power and that of electricity generated by more conventional methods (coal, natural gas, hydroelectric, and nuclear power plants). However, many wind-power installations are not cost-competitive at present with other sources of electricity generation and, therefore, need to receive government subsidies or premium prices for the electricity they generate to be able to operate. In a variety of cases (remote communities, oceanic islands, and undeveloped regions), wind power is now the cheapest electricity-generation option.

Most favorable for the development of wind power are the coastal regions of the world, where the wind is more steady and has the preferable range of velocity—critical factors for obtaining the most profitable electricity-generation yields. For this reason, and as a possible solution to the problem of land availability, particularly in densely inhabited regions such as Western Europe, consideration has been given to the development of offshore wind-power installations, either in platforms, barges, or otherwise. The higher wind speed in offshore areas is expected to offset the higher costs of such installations. All offshore installations built so far are relatively close to shore. The advantage of stronger and steadier winds farther offshore needs to be balanced, however, with the higher cost of installations in deeper water. Structures must also be designed to withstand the impact of the waves and some complex wind and wave interactions.

Despite the substantial increases in the generation of wind power in the last 10 years, wind provides, at present, an insignificant percentage of the world's total electricity, although in some countries (such as Denmark), in certain remote areas, and for certain purposes, wind power makes a considerable contribution to the demand for electricity. Wind power has become firmly established as an energy option in many markets because of the rapid improvements in the technology of the construction of wind turbines and because of the important recent reductions in costs. Further improvements both in performance of the wind turbines and in the cost of wind power are expected. Governments, universities, and industry are increasingly paying greater attention to the future of wind power, particularly in the United States, Western Europe, India, and China, and financial commitments to research and development programs for better wind-power technology are becoming significant.

How much wind power will contribute to the supply of electricity during the 21st century will depend on the performance of the present wind-power installations and of those to be put in production during the next 10–20 years. Most wind farms have been in operation for only a few years, and there is still limited information concerning their longer term mechanical and economic performance (cost of maintenance and lifetime of the turbines, for instance). The large expanses of land needed for large-scale generation of electricity with wind turbines will probably limit wind power to relatively small-scale installations.

ENERGY FROM BIOMASS

Dictionaries define "biomass" as the amount of organic matter in a particular area. As a source of energy, biomass includes:

- Wood in different forms: forest residues; by-products of the forest products industry, namely, wood chips, sawdust, and wood pellets;
- Charcoal;
- Agricultural residues or wastes: corn cobs, stalks and husks, rice hulls, and sugarcane bagasse;
- Animal wastes: manure and other farm wastes;
- Municipal solid and liquid wastes: garbage and sewage.

Biomass is, by far, the oldest source of energy; it has been used by humans since they "discovered" fire and its many uses. Fire set man in the path toward control of energy. In addition, alcohol has been produced from various fruits and grains for centuries. Less than a century ago, wood and farm wastes were the principal fuels used, even in the presently called "developed countries." In these countries, both wood and farm waste have progressively been replaced by more convenient sources of energy, but in many Third World countries, where other sources of energy are either not available or not affordable, wood, charcoal, and agricultural wastes still provide a large portion of their energy requirements.

The most traditional use of biomass as a source of energy is by direct burning of wood (as fuel for space heating and cooking and as fuel in some industries) and its conversion to charcoal. Some of the more recent uses of wood and other kinds of biomass as a source of energy include the following:

- Agricultural waste, crop residues, and municipal sewage and garbage are used in the production of low-heat-content gas by bacterial action (biogas, mainly composed of methane), fuel alcohols (mainly ethanol), and for the generation of electricity by some thermal process. In Hawaii, for instance, sugarcane bagasse provides electric power for the sugar industry and contributes to part of the electricity supply in the islands. Sugarcane bagasse is also extensively used in Brazil for the generation of electricity and for the production of ethanol, used alone or mixed with gasoline as fuel for transportation vehicles. In the United States, ethanol produced from corn mixed with gasoline (gasohol) has also been used as vehicle fuel.
- Manure and other farm and cattle feedlot wastes have been used to generate combustible gas that also provides a source of heat or fuel for the generation of electricity.

These by-product processes are attractive, because they combine the production of energy with the disposal of unwanted wastes. The use of manure as a source of energy, of course, reduces its availability as farm fertilizer.

Biomass is the most poorly documented of all sources of energy, because accurate data on its production and use are very difficult to collect. The reason is that much of the biomass used to generate energy is on a small scale, in rural areas, primarily in developing countries, where most of the biomass is gathered and used locally in individual households. Small manufacturers, as well as some large commercial industries like the forest-products industry, use their residues as a fuel for the generation of internal heat and electricity, but also in these cases, the volumes of biomass used and the energy generated by it are very poorly documented. For this reason, the contribution of biomass to the world's energy consumption has not been considered in this study's discussion of the sources of energy in general or of the types of primary energy used for the generation of electricity.

Nevertheless, it may be of general interest to mention that the volume of wood used as a source of energy has been estimated (World Energy Council, *Survey of Energy Resources*, 1998, 2001) to have amounted to 1.7 billion t in 1996 and 1.4 billion t in 1999, which is equivalent to about 4 billion and 3.5 billion bbl of oil, respectively. This amount represents about 6–7% of the global primary energy production, the same order of magnitude as the contribution of nuclear power or hydroelectric power.

The total amount of wood used as a source of energy in the developing countries has increased steadily in the last 50 years as their population has increased. During that time, their share of the total world consumption of wood for energy uses has increased from 76 to 89%. At present, wood provides 10–15% of the consumed energy in the developing countries of Asia and Latin America, 33% of those in Africa as a whole, and reaches as much as 60% in sub-Saharan Africa. However, the share of the total energy consumption held by wood in most developing countries has progressively diminished, whereas the use of fossil fuels has increased.

In the developed countries, the volume of wood used as a source of energy has slowly declined. Currently, wood provides only 3% of the consumption of energy in Europe and North America, 2–3% in the FSU, and only 1.5% in the industrial countries of Asia and the Pacific region (World Energy Council, *Survey of Energy Resources*, 1998).

No estimates are available of the amount of other kinds of biomass used as sources of energy.

The use of biomass as a source of energy has the distinct advantage that all forms of biomass are renewable and that it involves well-known technology. It is also the only practical source of renewable energy that can be the source of liquid fuels, not limited to the generation of electricity. Large-scale use of biomass as an energy source can provide employment opportunities and help local economies, particularly in rural areas.

There are, however, serious limiting factors to the use of wood and other kinds of biomass as a source of energy. Often mentioned are their low energy efficiency and their health and environmental hazards because of their objectionable generation of smoke, haze, and air-polluting gases when burned. Wood, in addition, is burdensome to collect, transport, and store, and a possible sizable contribution of wood to the world energy demand would result in further extensive deforestation, which would aggravate a current, already serious problem in many developing countries of the world where wood is being burned at a faster rate than it is being replaced.

Reforestation and the creation of what have been called "energy plantations" have been proposed as a solution to the destruction of forests. Energy plantations of selected fast-growing plants and trees, not only on land, but also in the oceans and in fresh-water bodies, have been suggested as a source of biomass for the generation of energy. They have so far been developed on a very small scale with unsatisfactory economic results. Successful energy plantations, in any case, would have to be limited to the tropics where the fast-growing plants and trees would grow best.

Possible competition for fertile land with food production has also been a major concern about energy plantations, which, if they were to become an important source of energy, would require the use of vast expanses of land. Critics of the use of wood as a source of energy ask, "In a hungry world, who can afford to set aside land to grow gasoline?"

In the case of agricultural and municipal wastes, their possible large-scale use as a source of energy is limited by the supply of such wastes.

The economics of generating energy with biomass are not favorable at present. As a by-product of, or supplement to, other profitable ventures, like its use by the forest products and sugar industries as a fuel for the generation of internal electrical needs, waste recycling for energy is an attractive undertaking. Residue recovery and waste recycling, however, would probably not stand by themselves as economic sources of energy. Generation of energy from biomass as a large-scale, stand-alone industry is currently uneconomic. The production of ethanol for vehicular fuel in Brazil and the United States has been heavily subsidized by the respective governments. Ethanol cannot now and will probably never be able to compete in price with gasoline. In addition, the amount of energy used to produce ethanol from sugarcane bagasse or corn is greater than the energy obtained by burning the ethanol.

The use in Brazil of ethanol, either alone or mixed with gasoline as fuel in transportation, much talked about during the 1980s, began to lose acceptance in the early 1990s, because alcohol subsidies were gradually decreased, ethanol shortages recurred, and the problems in starting the vehicles in cold weather persisted. In 2002, ethanol represented only 4.2% of the transportation fuels, and the sale of cars operating with ethanol or ethanol-gasoline mixtures decreased to virtually zero. At present, the introduction of the so-called "flex-fuel" car, powered by either ethanol or gasoline, is mentioned as possibly bringing a renaissance of the use of ethanol in transportation in Brazil.

THE ROLE OF SOLAR, WIND, AND BIOMASS ENERGY IN THE 21ST CENTURY

As mentioned before, from the time when humans learned to use fire to warm themselves and their dwellings and to cook their food until the mid-1800s, the sun, the wind, and biomass were mankind's only sources of energy: first, wood and other combustible biomass, eventually the wind, and all along, the heat of the sun. Starting in the mid-1800s, coal progressively replaced wood as a fuel, and during the 20th century, oil and then natural gas gradually became the main sources of energy (see Figure 1).

At present, wood and other sorts of biomass provide about 6–7% of the world's primary energy, mainly in the developing countries. The use of direct sunlight for heating and cooling and that of sunlight and wind for the generation of electricity represent only an insignificant percentage of the world's total energy supply. This small contribution will probably not increase appreciably during the 21st century.

Direct sunlight has been in use during most of the 20th century in sunny regions of the world for water and space heating. Practical space cooling has not yet been achieved. Devices that would combine water and household heating, space cooling, and electricity generation (thermal-photovoltaic solar collectors) are a possible further advance in solar-power technology. Direct sunlight used for these purposes, however, still has geographic, economic, and technological limitations. It will likely continue to be used only as small-scale installations in sunny regions of the world and in more affluent countries.

Generation of electricity by direct sunlight (solar thermal or PV processes) is not expected to become a major source of energy in the 21st century. Large-scale PV electric plants have serious technical limits and commercial disadvantages that will keep them from becoming significant electricity sources for many years to come. They will require immense expanses of land in sunny regions of the world. If the regions with such critical geographic conditions are not near the centers of electricity consumption, long-distance transmission of the electricity generated would be necessary, which would considerably impair the economic feasibility of PV large-scale electricity generation. In addition, the cost of electricity generated by PV devices, which is now 5–10 times higher than that of electricity generated in thermal plants, is not expected to come down substantially in the foreseeable future.

The prospects of electricity generation in wind farms are more favorable. The economic picture is better, and there are fewer geographic restrictions. Wind power is now the fastest growing source of electricity in the world, but it still represents a very small percentage of the electricity market.

This does not mean that solar PV and wind electricity generation will not be a factor in the supply of electricity during the 21st century. Their development should be encouraged. In remote locations hard to reach by conventionally sourced electricity, in undeveloped regions, or other special cases, solar and wind electricity will be invaluable, the best or only option. But while they can fill these special niches, solar and wind electricity, particularly solar PV, are not expected to become a major contributor to the world's future fast-increasing demand for electricity.

Wood, charcoal, and other combustible biomass (agricultural, animal, and municipal wastes) will continue to be an important source of energy in the developing countries during the 21st century. Their total volume, however, will remain stable or grow slowly, but its share of the total world energy supply will decline from the present 6–7% as the population of the developing countries becomes more urban, its living standards improve, and its consumption of energy from other more efficient and higher quality fuels increases at a faster rate. A possible larger contribution would face serious economic and environmental constraints: a limit in the supply of wastes, undesirable extensive deforestation, geographic restrictions, low efficiency, and unfavorable economics. The use of industrial, agricultural, and municipal wastes as a source of energy is attractive as a by-product of or supplement to other activities, but it is, at present, uneconomic as a stand-alone source of income. Energy plantations and the production of ethanol from corn, sugarcane, or other biomass require economic subsidies, and large-scale development of both would require too much of the land needed for other purposes.

More optimistic estimates anticipate a more efficient use of biomass as a source of energy, the result of more research, new ideas, and new technology. However, more pessimistic outlooks predict a decline in the use of biomass, particularly of wood, for energy generation, as continuing deforestation in the developing countries drastically reduces the sources of supply of wood and wood wastes.

Technological and manufacturing advances resulting from future research on the uses of direct sunlight, wind, and biomass will contribute to the continuing demand for these sources of energy. However, although the sun and the wind are free and inexhaustible, the high cost of the equipment needed to convert them into useful energy and their serious geographic limitations will restrain them from becoming significant contributors to the supply of energy during the 21st century. The sun does not always shine, and the wind does not always blow, so that to provide reliable power, both solar and wind electric plants require backup by conventional sources of primary energy, namely, coal, natural gas, and hydroelectric or nuclear power.

Energy from the sun, the wind, and biomass will fill gaps left open by other sources of energy, but optimistic predictions of a major future role for them are probably unrealistic. They should be viewed as supplementary to fossil fuels and nuclear and hydroelectric energy in the generation of electricity and not as an important replacement, and they will have an insignificant share of the transport fuel market.

HYDROGEN AND FUEL CELLS

ABSTRACT

For all the reasons given below, hydrogen is not expected to supply a significant share of the world's energy demand before the mid-21st century at the earliest. Hydrogen is an environmentally clean, inexhaustible, efficient, and potentially major source of energy, available worldwide—the "universal fuel of the future." Unlike other clean sources of energy (nuclear and hydroelectric power, sunlight, and wind) that are capable only of generating electricity, hydrogen can be a source of both electricity and fuel for ground and air vehicles. At present, the best prospect for hydrogen making substantial inroads as a major source of energy is in fuel-cell-driven vehicles. However, a hydrogen-based transportation system faces severe technological, operational, and economic barriers. The problem of how to place hydrogen on board fuel-cell vehicles has not been solved; hydrogen can only be stored now in either pressurized or liquefied form in large, thick-walled, heavy, expensive containers. Current prototype fuel-cell vehicles, therefore, run on gasoline or methanol that is converted to hydrogen in the vehicle. Still, even if this problem was to be solved, large-scale use of hydrogen-fueled vehicles would require the replacement of millions of the vehicles currently running on gasoline, diesel, or jet fuel, and the establishment of a new and widespread infrastructure that would make possible their refueling. This will require a long time and enormous investments.

INTRODUCTION

Growing concern about the rising concentration of carbon dioxide (CO_2) in the atmosphere has encouraged more and more countries, especially the technologically strong, developed ones, to study the possibility of using hydrogen as a replacement for the fossil fuels. Their concern is caused by:

- The effects of the combustion of oil, natural gas, and coal on the environment, especially the potential contribution to global warming;
- The recognition of the uneven geographic distribution and possible eventual depletion of the fossil fuels;
- The realization that during the 21st century, cleaner sources of energy (nuclear, hydroelectric, geothermal, wind, and solar power) may not be adequate to replace the fossil fuels in supplying the anticipated increase in the demand for energy in general and for transportation fuels in particular.

Hydrogen-energy programs are being undertaken in Japan, several European countries, Canada, and Iceland. In the United States, President Bush called in early 2003 for a $1.7 billion research and development program to promote the development of hydrogen-powered fuel-cell cars. Some companies have started to commercialize hydrogen technology, hydrogen know-how, and hydrogen energy systems, and the scientific community has started to seriously study the prospects of a "hydrogen economy."

On the plus side, hydrogen is seen as an environmentally nonpolluting, truly inexhaustible, efficient, and potentially economical source of energy for electricity generation and as transport fuel, available everywhere in the world from several sources. It can be transported over long distances and stored and used in gaseous or liquid form in industry, households, power stations, and land and air vehicles—the "universal fuel of the future." It is now generally accepted, however, that it will be the widespread use

of hydrogen in fuel cells, mainly in transportation vehicles, that will enable hydrogen to become a major source of energy in the 21st century. However (and this is a critical disadvantage), today's prototype fuel-cell cars actually use either gasoline or methanol (wood alcohol) as the source of hydrogen. Refueling directly with elemental hydrogen is not yet possible.

In addition, on the minus side, the use of hydrogen is not free from potential problems:

- Hydrogen is not a primary source of energy. In nature, it does not occur in its elemental state, and a source of energy is required to produce it.
- When burned in air, it causes the emission of nitrogen oxides (NO_x) derived from the air itself, the amount depending on the flame temperature and the duration of the combustion.
- The low density of hydrogen, either as gas or liquid, makes its energy density on a volume basis so low that considerable space is required to store it either in stationary power-generating plants or onboard in transportation vehicles.
- Hydrogen ignites readily, raising safety fears, magnified by memories of the fiery explosion of the German zeppelin Hindenburg in 1937. It is feared that, in the case of a leak, to which hydrogen is clearly vulnerable, it may more easily ignite than gasoline, natural gas, or other fuels. This fear, however, may be exaggerated—safety is not the problem it is perceived to be. Experience so far seems to indicate that the potential danger of hydrogen in gaseous or liquid form is not greater than that of conventional hydrocarbon fuels (Van Vorst, 1995, p. 227–228; Dunn, 2001, p. 19).
- Finally, and perhaps most important, hydrogen is now expensive to produce, store, transport, and distribute.

To become a major source of energy, hydrogen will have to compete with—and, in fact, be dependent for its production on—gasoline and other liquid fuels in transportation and with coal, natural gas, and hydroelectric power in the generation of electricity. It will face formidable technologic, economic, political, and institutional barriers.

Hydrogen was first used as a fuel in a combustion engine in 1820, but during the rest of the 19th century and most of the 20th century, interest in hydrogen as a fuel was purely academic. No commercial applications were investigated (Hoffmann, 2001, p. 27–37). In the early 20th century, oil became the choice of transportation fuel, whereas coal and hydroelectric power were the dominant fuels in the generation of electricity. Some interest in hydrogen grew in Europe in the 1920s and 1930s, but no commercial development occurred. In the 1950s, hydrogen was used in the development of fuel cells for space applications. Interest in hydrogen as a fuel has increased since then (Dunn, 2001, p. 18–21; Hoffmann, 2001, p. 37–51). Some of the early supporters predicted that a ''hydrogen economy'' would begin to develop before the end of the 20th century, and that hydrogen would become a major supplier of energy by the middle of the 21st century. The early years of this century have not yet witnessed any measurable contribution of hydrogen to the world's energy consumption.

HYDROGEN PRODUCTION

Hydrogen is mainly found in combination with oxygen in water and with carbon in hydrocarbons and most organic compounds. It has to be produced from these sources through the use of three basic methods (Dunn, 2001, p. 28–35):

1) Catalytic steam reforming of natural gas (methane), other hydrocarbons, methanol, and other alcohols or gasoline. At present, this is the most common, least expensive, and most energy-efficient way to produce hydrogen. About 95% of the hydrogen produced commercially now is produced by this method and will probably remain as the main source of hydrogen well into the 21st century.
2) Electrolysis of water, the use of electricity to split water into hydrogen and oxygen. In the long-term future, this is ideally the most promising source of hydrogen (because water can be regarded as essentially inexhaustible) particularly if the source of electricity can be some form of renewable, clean energy (solar, wind, and hydroelectric). However, as discussed earlier, these sources of electricity cannot be expected to be available in major volumes, now or in the future. Manufacture of hydrogen by the electrolysis of water using nuclear power, favored by some supporters of a hydrogen economy, will have to wait for a renaissance of the nuclear industry, now in a period of definite decline.

Electrolysis of water is also too expensive at this time to produce hydrogen in large volumes; only about 5% of the hydrogen produced at present is obtained this way, and the problem of what to do with the oxygen generated by this method has not yet been resolved. Electrolysis of water is now a poor choice for the production of hydrogen. We will have to rely on the reforming of hydrocarbons, alcohols, or gasoline for quite some time. It is possible, therefore, that until hydrogen can be obtained economically by the electrolysis of water, the use of hydrogen as a source of energy will be much restrained.
3) Thermal reforming of coal through gasification. This method is not competitive now, because it requires the sequestration of the carbon released by the gasification. In the future, with the advent of new technology, it could become an important source of hydrogen by reason of the large reserves of coal in the world.

A variety of other methods of producing hydrogen have been or are currently being investigated: gasification of biomass and various methods using renewable solar energy—direct thermal decomposition (thermolysis), photolysis, biolysis, radiolysis, and thermochemical cycles. None of them is now considered practical.

HYDROGEN TRANSPORTATION

Hydrogen will most likely be transported as a gas under high pressure in pipelines that may require only small adaptations to those currently employed for natural-gas transmission. The possible embrittlement of the pipelines' steel by the hydrogen may be a problem that will need to be studied. Liquid hydrogen can be transported by sea using expensive cryogenic tankers like those now used for LNG.

HYDROGEN STORAGE

Hydrogen's low density makes its storage a severe problem. At present, there are three means of storing hydrogen: as a gas at high pressure, as a liquid at low temperature, or as a solid metal hydride. Each option has serious drawbacks.

In the gaseous state, even when highly pressurized, a large amount of space per unit of caloric value is required; to store hydrogen as a gas under pressure requires thick-walled, heavy vessels and can be a safety hazard. Compressed-gas vessels are now made from very expensive materials such as carbon fiber and are relatively large, making it difficult to fit them in vehicles where storage space is an issue. Onboard containers that are safe, light, and inexpensive are still in development.

The problem of storing hydrogen in liquid state is that of maintaining extremely low temperatures and, as a consequence, the great amount of energy needed to chill the gas and the high costs involved. Cryogenic hydrogen storage tanks are being operated experimentally in Germany and Japan, but they are heavy, very expensive, potentially dangerous, and difficult to maintain.

Solid hydrides are costly, and they are unlikely to become economically competitive anytime in the foreseeable future.

Safe, effective, and economical storage of hydrogen may prove to be the main barrier to the rapid introduction of hydrogen as a fuel, and it is certainly the critical element in the introduction of hydrogen-fueled vehicles. It is likely that it will take some time to develop onboard hydrogen-supply systems that are smaller, lighter, and cheaper than those available at present.

HYDROGEN USE

At present, hydrogen is principally used in the refining of petroleum and the manufacture of ammonia fertilizers, resins, plastics, solvents, and other industrial commodities. Very little is currently used as a source of energy. Approximately 550 billion m^3 (20,000 ft^3) of hydrogen per year is produced worldwide now, primarily reformed from natural gas, about one-fifth of it in the United States.

In the future, the use of hydrogen represents a large potential for reducing carbon dioxide emissions. The most promising are the uses of hydrogen in stationary or portable electricity-generating plants, as a means of highly efficient transmission of energy over long distances and, most significantly, as a promising nonpolluting alternative fuel for both ground and air transportation.

The potential of hydrogen as the preferred fuel for ground and air vehicles has long been recognized, and its use will depend on how critical the need for reducing the air pollution caused by the combustion of fossil fuels will become in the future. Hydrogen's failure so far to attain its potential is caused by the high cost of its production and use and by the abundance,

availability, and low cost of gasoline, diesel, and jet fuel, which are deeply entrenched and excellent transportation fuels in every respect except for their contribution to air pollution.

Until now, the use of hydrogen as a fuel for vehicles has been limited to studies, especially in aircraft design, and to the production and demonstration of experimental road vehicles. Some proponents of hydrogen as a fuel believe that in the future, hydrogen is likely to become a more dominant factor in aviation than as a fuel for road vehicles.

The use of hydrogen as fuel in adapted internal combustion engines has been investigated for quite some time, but although probably technically practicable, it does not seem to be the most efficient use of hydrogen as fuel for vehicles. Research on hydrogen-fueled internal combustion engines has not settled whether it is preferable to develop an engine especially designed to use hydrogen or if it will only be necessary to modify the currently used engines.

The replacement of internal combustion engines by fuel cells seems, at present, to offer the best prospect for hydrogen to make substantial inroads in the transportation market, because hydrogen performs more efficiently in fuel cells than in internal combustion engines.

There are, however, some obstacles to the large-scale use of hydrogen as the primary transportation fuel, either using internal combustion engines or fuel cells. Most critical are the establishment of an adequate infrastructure that would allow the production, transportation, storage, and ready availability for the delivery of hydrogen and the development of an onboard source of hydrogen in the hydrogen-fueled vehicles.

The development of an entirely new and full-fledged hydrogen infrastructure will require a long time and enormous investments, not only to replace the more than 750 million gasoline, diesel, and jet-fuel vehicles now in use in the world with hydrogen-fueled vehicles, but also to have in place an adequate number of refueling stations to supply the demand for hydrogen. The availability of widespread hydrogen refueling stations will require a significant number of hydrogen-fueled vehicles in operation, but the existence of a large fleet of such vehicles will depend on the access to the refueling stations. The development of a hydrogen infrastructure will take an even longer time in rural areas and in the developing countries.

Because of the undesirable weight, volume, and cost of currently available systems of onboard hydrogen storage in transportation vehicles and the lack of an adequate extensive refueling infrastructure, it is likely that the use of hydrogen as a fuel will be limited in space-constrained passenger cars. It will, therefore, be restricted in the near term to urban buses, trucks, and other commercially operated and centrally refueled fleets. It is not likely that the railroad industry will switch from diesel to hydrogen unless mandated by new and strict emission standards.

FUEL CELLS

Introduction

Fuel cells are devices that use electrochemical reactions to convert the chemical energy of a fuel directly into electricity—quietly, efficiently, and without moving parts. Fuel cells run on hydrogen or some other fuel, like natural gas, gasoline, or methanol, that can be converted to hydrogen with the aid of catalysts. In the fuel cell, hydrogen is oxidized at the anode with a concurrent reduction of oxygen at the cathode.

The fuel cell was invented by Sir William Grove in 1839, but it remained little more than an academic curiosity until the mid-20th century, when its potential as a practical source of energy was realized (Hoffmann, 2001, p. 145–154). Its high cost effectively precluded its use for some time except for restricted special applications, like the space program, where cost is not an issue. In the last few decades, encouraged by environmental concerns, new technology has developed smaller, more powerful, cheaper, and more useful fuel cells. Further advances are certainly on the horizon.

The direct use of hydrogen as the fuel for fuel cells, if developed, will offer the possibility of completely clean energy, because the by-product from such fuel cells would only be hot water. Fuel cells are recognized as being able to potentially provide the basic technology that would allow hydrogen to become an environmentally harmless (except, perhaps, in large cities, where the huge output of hot water will certainly raise the humidity) and inexhaustible major source of energy for both transportation and electricity generation in the 21st century.

The technology of fuel cells is now available and is fairly straightforward, although it still needs improvements to make fuel cells economically attractive. Progress in catalysts and electrode technology, along with cost reductions, has to be made before fuel cells can be considered economically competitive for general use. For instance, some fuel cells now use platinum as a catalyst to operate at reasonable temperatures. The substitution of platinum as a catalyst will be essential to reduce the cost of fuel cells and to make a major replacement of the current transport fleet by fuel-cell vehicles possible.

Six main types of fuel cells exist, classified according to the electrolyte used in the system (Yildiz and Pekmez, 1995; Dunn, 2001, p. 40–43; Hoffmann, 2001, p. 156–158): alkaline fuel cells (AFC); molten carbonate fuel cells (MCFC); solid oxide fuel cells (SOFC); proton exchange membrane fuel cells (PEMFC) (also called solid polymer fuel cells [SPFC], or polymer electrolyte fuel cells [PEFC]); phosphoric acid fuel cells (PAFC); and direct methanol fuel cells (DMFC).

Fuel-cell systems are also classified according to their working temperatures: low-temperature fuel cells (AFC, PEMFC, PAFC) that operate at 80–200°C and are of particular interest for space and transportation-vehicle applications; and high-temperature fuel cells (MCFC, SOFC) that operate at 600–1000°C and are mainly suitable for large-scale production of electricity and heat.

The PEMFC have received particular attention because they seem to be most suitable as a power source for transportation (as well as for stationary applications). They are quiet, compact, and low weight, have no moving parts, and have a simple electrochemical design; they operate at relatively low temperatures (about 90°C), are very efficient (about 50%), and do not produce any polluting emissions. A vehicle powered primarily by a PEMFC using gaseous hydrogen as fuel has been designed, assembled, and satisfactorily tested. Currently, more than 100 organizations plan to commercialize fuel cells for a broad number of purposes.

Use of Fuel Cells in Transportation

The technology of fuel-cell vehicles, although offering the best promise for the future replacement of internal combustion engines in ground and air vehicles, needs further development as far as reliability, performance, practicability of manufacture, and cost are concerned. At present, a good number of automobile manufacturers (Ford, General Motors, BMW, Mercedes Benz, Daimler Chrysler, Mazda, Renault, Volvo, and Toyota) are engaged in the development of fuel-cell cars. Numerous experimental cars and buses have been built since 1993 (Hoffmann, 2001, p. 48–51, 113–131), but the commercial production of such vehicles is not yet seriously considered. Other companies (ExxonMobil, Shell) are working with the automobile manufacturers in determining the choice of the preferred fuel for the fuel cells that will propel the new breed of vehicles. Onboard conversion of gasoline or methanol to hydrogen is generally favored over the direct use of hydrogen because of the serious problems of onboard storage of hydrogen discussed before: large, heavy, thick-walled, and expensive containers to store either pressurized or liquefied hydrogen. Both governments and industry are therefore devoting more time and resources to the development of these fuels systems than to the direct use of hydrogen in fuel-cell vehicles—for instance, development of the best form of gasoline to fuel the fuel cells. They reason that, even under pressure, the energy density of hydrogen is too low to provide sufficient driving range, and that to develop processors that would generate hydrogen from gasoline (or methanol) in the car itself would allow the transition to fuel-cell vehicles to occur much sooner than if a widespread hydrogen-refueling infrastructure had to be put in place. As a result, all prototype fuel-cell cars so far developed run on gasoline or methanol that are reformed into hydrogen in the car. Supporters of a hydrogen economy are disheartened by this choice, because they proclaim that it would thwart the desired progress toward the decreasing use of fossil fuels.

The choice of fuel for fuel-cell vehicles, a critical issue if such vehicles are to become acceptable, therefore remains controversial (Dunn, 2001, p. 47–62). Each fuel has its advantages and drawbacks that are stressed by their supporters and detractors. Direct use of hydrogen, in compressed-gaseous or liquid form, would be the simplest and least environmentally damaging long-term choice for fuel-cell vehicles, but it still faces major technological (onboard storage), economic, and marketing barriers. Methanol is the easiest fuel to reform onboard, but gasoline, although more difficult to convert to hydrogen, has the advantage of having an existing infrastructure and being familiar to the public. The use of gasoline as fuel for fuel-cell vehicles, however, would slow down the move toward a hydrogen-based transportation system. It is evident that if the onboard conversion of either methanol or gasoline were to attain market dominance, it would block for a long time the establishment of a hydrogen system, missing the important environmental benefits that hydrogen-fueled fuel-cell vehicles could provide.

Use of Fuel Cells for the Generation of Electricity

Electricity-generating plants using fuel cells are also under consideration by several companies for both residential and commercial consumption. General Motors has recently unveiled a small, stationary fuel-cell unit that would generate enough electricity to run a house, mentioning that it can build more powerful units capable of supplying electricity to whole residential subdivisions and factories. Many other companies in North America, Europe, and Japan are also developing small stand-alone fuel-cell electricity-generating plants. If these small fuel-cell power plants are to run

on hydrogen, a widespread infrastructure that would make hydrogen easily available will be necessary. Major power plants using fuel cells seem to be a more distant possibility. Nevertheless, it is likely that stationary fuel-cell electricity-generating units, small or large, will be commercially available for businesses and homes before fuel-cell vehicles are widely in use.

THE ROLE OF HYDROGEN AS A SOURCE OF ENERGY IN THE 21ST CENTURY

If the concerns about the atmospheric pollution and potential global warming caused by the burning of fossil fuels reach an alarming and urgent level throughout most of the world during the next few decades, consideration will have to be given to replacing the fossil fuels with cleaner, environmentally benign sources of energy. Hydrogen used as the fuel for fuel cells can become the desired alternative, if not the only alternative, because, unlike other clean sources of energy (nuclear, hydroelectric, geothermal, solar, and wind) that are only capable of generating electricity, hydrogen-fueled fuel cells can supply both electricity and the fuel for transportation vehicles.

In addition, hydrogen offers a possible solution to the storage of electricity, and it could provide highly efficient transmission of energy, invaluable in situations where the transmission of electricity is inefficient, impractical, or impossible.

In addition, hydrogen is an inexhaustible source of energy.

For these reasons, the potential of hydrogen as an important source of energy in the 21st century is receiving increasing attention. However, the possible large-scale production, delivery, storage, and use of hydrogen as a source of energy will face enormous technologic and economic obstacles.

For one thing, the cost of producing, storing, transporting, and distributing hydrogen needs to be lowered considerably.

Fuel-cell technology is now available, but despite important advances and rapid progress toward commercialization in the last few years, it still requires considerable improvements, particularly in lowering the cost of fuel cells, now a serious barrier to their extensive and economically acceptable use. A long step still needs to be taken from theoretical possibility to operational reality.

Predictions made a few years ago that by 2004 fuel-cell cars, not just prototypes, would be in limited production and running on the streets and highways have not turned out to be correct. Similar predictions that it is possible that hundreds of thousands of hydrogen fuel-cell vehicles would be on the road by 2010 now seem unrealistic.

Even if the cost of large-scale production, storage, and transport of hydrogen and the cost of vehicles powered by fuel cells can be reduced considerably, the new vehicles need to prove their operational reliability, gain the acceptance of the general consuming public, and prove to be successful in the free market, but most importantly, (1) an appreciable number of the hundreds of millions of vehicles in the world fueled by gasoline, diesel, or jet fuel need to be replaced by fuel-cell-powered vehicles; and (2) an entirely new, huge, and widespread (both urban and rural) infrastructure for the production, storage, and distribution of hydrogen needs to be put in place for the refueling of hydrogen fuel-cell vehicles. This will require a very long time and an enormous investment. A substantial replacement of the current vehicle fleet by hydrogen fuel-cell vehicles and the availability of adequate numbers of hydrogen-refueling stations are unlikely in the foreseeable future, even in the developed countries. It will not occur until much further in the future in the developing countries for obvious economic reasons.

The probably lengthy conversion to a hydrogen-transportation system determined by the need of its acceptance by the consuming public, the replacement of the current transportation vehicle fleet, and the development of a hydrogen infrastructure in a free-market economy could, however, be shortened by specific government actions. Progress in a free-market environment is often slow and sometimes capricious, because it needs to overcome natural inertia and the skepticism of the public toward something entirely new. Government encouragement, incentives, legislation, directives (selective taxation), and regulations could facilitate and hasten the transition to a hydrogen-transportation system, particularly if guided and encouraged by logical reasons based on sound concerns for the environment and the possibility of appreciable global warming.

However, with or without government action, even the most optimistic supporters of a new global hydrogen economy (or as sometimes called a "hydrogen civilization") do not believe that hydrogen will be economically and operationally competitive before 2020 or 2030 and agree that hydrogen will not supply a significant percentage of the world's energy demand before the mid-21st century at the earliest. Nakicenovic et al. (1998, p. 87), for instance, state that "in the absence of very stringent environmental policies, hydrogen is unlikely to play more than a marginal role before 2050." Less optimistic prognosticators see the development of a large-scale hydrogen-energy infrastructure as an insurmountable technical and economic barrier to the establishment of a hydrogen-transportation system. To replace a technologically advanced energy system with an entirely new one would be a vast, expensive, and time-consuming undertaking. The long lifetimes of power plants, refineries, and other components of the existing energy system also need to be considered.

Hydrogen may eventually become a factor as a pollution-free source of energy, but it will be slow to make major inroads, probably not until the second half of the 21st century. Meanwhile, hydrogen should perhaps be considered as a means of transferring and storing energy and not as a primary source of energy.

How important a factor hydrogen will be will depend on the extent of the public acceptance, or lack thereof, major technological improvements particularly in hydrogen storage, the cost reductions that would make hydrogen competitive with other sources of energy, the success in developing an adequate hydrogen infrastructure, government actions, and how critical the protection of the environment becomes during the 21st century.

5

Electricity: Generation and Consumption

ABSTRACT

Worldwide consumption of electricity has increased steadily during the second half of the 20th century, in the developed and developing countries alike. At present, 40% of the primary energy in the world is used to generate electricity, and 14% of the energy consumed is in the form of electricity. It is expected that the consumption of electricity will continue to increase during the 21st century. During the middle part of the 20th century, coal and hydroenergy contributed 80% of the primary energy for the generation of electricity. By the start of the 21st century, coal had, in part, been replaced by oil, nuclear energy, and particularly by natural gas, which had become the preferred fuel for new power plants. However, coal remained the main source of energy for the generation of electricity. During the last two decades of the 20th century, the shares of oil and hydroenergy decreased, whereas that of nuclear energy, after rapidly increasing in the 1970s and 1980s, stabilized during the 1990s. Other sources of energy made only small contributions to the generation of electricity. During the first half of the 21st century, it is likely that coal, natural gas, and hydroenergy will be the main sources of primary energy for electricity generation. The share of coal will probably decrease, whereas that of natural gas will substantially increase. The share of oil and nuclear energy will decrease. Other sources of energy, with the possible exception of wind, will remain minor contributors. During the second half of the century, nuclear power may make a comeback, and coal and natural gas will be the main fuel for electricity generation. Oil and wind probably will make small contributions. Hydrogen may enter the electricity-generation field if some of the obstacles to its use can be eliminated.

INTRODUCTION

At present, about 40% of the world's primary energy is used to generate electricity, and 14% of the energy consumed in the world is in the form of electricity—more in some industrialized societies. Electricity use consistently increased worldwide during the 20th century as electric power became essential to sustain a high standard of living and an advanced level of industrialization in the developed countries and some of the developing countries of the world.

All evidence indicates that the consumption of electricity will continue to increase during the 21st century and to be a very important factor in the supply of energy, especially if the technological advances to be expected increase the efficiency of its use and lower its generation cost, making it possible for the present developing countries to increase the consumption of electricity and thereby raise their standard of living. Long-term projections indicate that although global total energy consumption may more than double in the next 50 years, electricity consumption will more than triple because it is such a convenient form of energy.

A discussion of the consumption and possible sources of energy during the 21st century, therefore, needs to consider the role that electricity has played and will most likely continue to play, during this new century.

Usable electricity is generated through several different processes using various sources of primary energy: coal, oil, natural gas, uranium, flowing water, geothermal, wind and solar energy, and biomass.

Until recent years, electricity has been generated either by steam-driven turbines, the thermal processes, or by water-driven turbines, the hydroelectric process. Thermal plants may use coal, oil, natural gas, or heat produced by a nuclear reaction to generate the steam that activates the turbines. Hydroelectric plants use the flow of water in rivers or from a lake created by a dam to power the turbines. Recently, electricity is also being generated by geothermal heat, wind-driven turbines, or by the photovoltaic (PV) conversion process. These last three processes, however, supply a very small percentage of the total electricity generated in the world today.

Electricity consumption involves two main problems: It is difficult to store in large quantities, and it has, therefore, to be consumed as it is generated; and it suffers losses as it is transported over long distances.

As a solution to these problems, it has been proposed that electricity could be used to generate hydrogen by electrolysis of water, and the hydrogen could then be stored and transported in gaseous or liquid form. However, the commercial development of these processes is still far in the future.

A BRIEF HISTORY OF ELECTRICITY AND ITS SOURCES

Public supply of electricity first became available in some of the more developed countries during the 1880s. The first hydroelectric plant was built at God-alming, England, in 1881. The first thermal power plant was placed in operation in London in January 1882, followed shortly after, in September 1882, by another public power plant in New York City. Supply, small in plant capacity and in range of distribution, was first limited to localized urban areas in the form of direct current (DC) and at low voltages. As demand increased, with the resulting increases in the distances from generating plants to consumers, the difficulty of transmitting low-voltage direct current for long distances became evident. High-voltage alternating current (AC) systems, first used about the end of the 19th century (1898–1899), rapidly became the common means of transporting electricity during the 20th century. Particularly after World War II, when the demand for electricity grew apace, increasingly higher voltages were used in transporting the growing amounts of electricity.

The rapid increase in the demand for electricity during the early and mid-20th century forced an increase in the construction of power plants. During the first half of the century, the abundance and ready availability of coal and of undeveloped waterpower in the developed countries favored the preponderance of these primary energy sources in the generation of electricity. The discovery of large oil accumulations during the mid-20th century provided another source of primary energy for electricity generation; and in the 1960s nuclear power was added as a promising new source of energy for electric thermal plants. Finally, natural gas, a source of primary energy cleaner than coal and oil, has progressively become the preferred energy source for the generation of electricity in the last few decades.

Appendix Table 63 and Figure 49 show the percentages of the various primary energy sources used

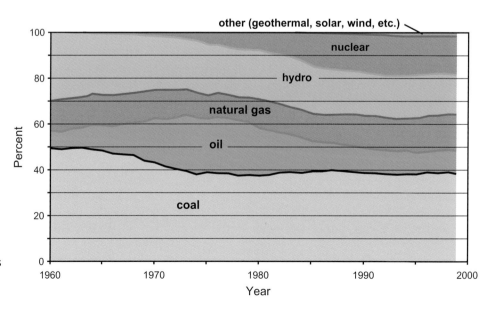

Figure 49. Primary energy sources used in electricity generation (percentage).

worldwide in the generation of electricity since 1960. Coal has remained the largest single source during this period. Its use has declined in the developed countries during the second half of the 20th century but has increased in the developing countries, particularly in those with appreciable coal reserves. The use of coal to generate electricity has declined from 49% in 1960 to about 38% in 1999. The contribution of oil reached a maximum of about 20–24% in the 1970s but has steadily declined since then to about 10% in the late 1990s. The share of natural gas, which had remained at about 10–13% from 1960 to 1990, has increased during the 1990s to 16%. Hydropower decreased from 30% in 1960 to about 20–21% by the mid-1980s and to 18% during the 1990s. The contribution of nuclear power to the generation of electricity increased since its introduction in the late 1950s to about 17% in 1990 and has remained at that level since then.

Other primary energy sources (geothermal, PV, wind, and hydrogen) have been actively investigated as potential sources of energy in the generation of electricity but have so far made very small contributions (totaling less than 2%) to the electricity generated worldwide, and it does not seem at this time that any of them, with the possible exception of wind, will contribute appreciably to the worldwide demand for electricity, at least not during the first half of the 21st century.

The proportions of the various sources of primary energy used to generate electricity vary from country to country, depending on their respective availability of primary energy, local economics, and government policies. For instance, coal provides 80% of the primary energy in the generation of electricity in Australia, 70–75% in China, and about half in North America. India and the FSU also use coal as a major source of energy in the generation of electricity. In France, Lithuania, Slovakia, and Belgium, more than 50% of the electricity is generated now in nuclear plants. Canada, the United States, Brazil, China, and the FSU have accounted for between 50 and 54% of the yearly total hydroelectric power generated in the world since 1950. About one-third of the countries of the world rely now on hydroelectric power for more than half of their consumption of electricity.

GENERATION AND CONSUMPTION OF ELECTRICITY: HISTORICAL DATA

The world's generation of electricity steadily increased during the 20th century (Appendix Table 64; Figure 50). The amount of electricity generated in 1999 was 15 times the amount generated in 1950. Growth was exponential in the 1950s and 1960s and essentially linear since 1970. This increase has been caused by the continuous growth of the world's population, the rapid rural electrification, the increasing and widespread urban illumination, intensified industrialization, and improving of the standard of living of the developed and many developing countries during the second half of the 20th century. The last few decades of the 20th century have witnessed the widespread ownership

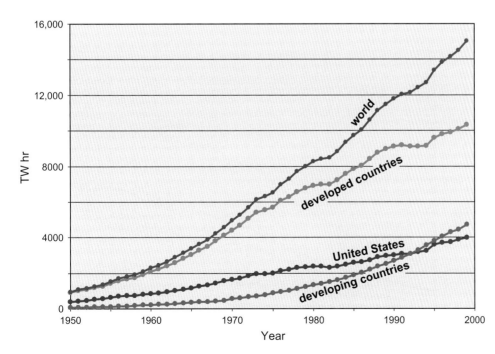

Figure 50. Electricity generation.

and use of refrigerators, radios, television sets, air conditioners, electric heaters, computers, and many other electric appliances.

Because electricity is difficult to store in large quantities and is therefore consumed as it is generated, the amounts of electricity consumed are essentially equivalent to the amounts generated, only lowered by the losses suffered during transportation, particularly over long distances.

Appendix Table 64 and Figure 50 also show the growth in electricity generation in the developed and the developing countries and in the United States. Appendix Table 64 includes the percentage of the total electricity generated worldwide that is produced by the developed and the developing countries and the United States.

Electricity generation has steadily increased in all countries, developed and developing. The developed countries have accounted for the major share of the electricity generated in the world: 90–94% during the 1950s and 1960s, decreasing to 80% in the mid-1980s and to 69–70% in the late 1990s. The share of the developing countries increased from 6% in 1960 to about 30% in the late 1990s. The share of the United States has decreased from about 40% in the 1950s to about 25–27% in the 1990s.

The patterns of electricity generation and consumption in individual countries and regions of the world, of course, depend not only on their stage of industrial development but also on the distribution of the population, the general level of prosperity, and the primary energy sources available for electricity generation.

The percentage of the total primary energy used worldwide for the generation of energy has increased from about 14–16% in the early 1950s to 30% in the late 1970s and close to 40% in the late 1990s. As the generation of electricity rises during the 21st century, probably at a faster rate than other sources of energy, the percentage of primary energy devoted to the generation of electricity will also most likely increase.

Electricity provided about 5% of the energy consumed in the world in the early 1950s. This share grew to 10% by the late 1970s and is now about 14–15%. It will undoubtedly continue to grow during the 21st century.

CONSUMPTION OF ELECTRICITY PER CAPITA

Appendix Table 65 and Figure 51 show the consumption of electricity per capita for the world as a whole, for the developed and the developing countries, the United States, and the developed countries excluding the United States.

Worldwide, electricity consumption per capita has increased slowly from about 379 kW hr/yr in 1950 to about 2500 kW hr/yr in 1999. A great disparity, however, exists in the consumption of electricity per capita between the developed and the developing countries. In 1950, the consumption of electricity per capita in the developed countries was 5460 kW hr/yr, 150 times the per-capita electricity consumption of 35 kW hr/yr in the developing countries. By the late

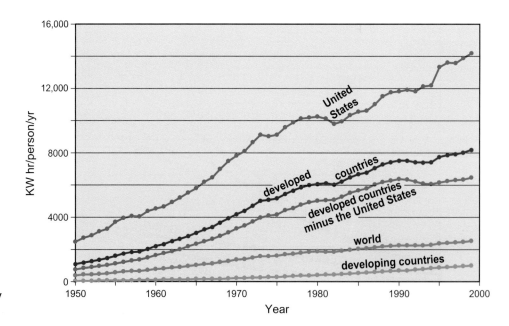

Figure 51. Per-capita electricity consumption.

1990s, the developed countries still consumed eight to nine times more electricity per capita than the average of the developing countries: 8000 kW hr/yr vs. 900–1000 kW hr/yr.

The consumption of electricity per capita in the United States is considerably higher than that in the rest of the developed countries: 2500 kW hr/yr vs. 700 kW hr/yr in 1950 and 13,000–14,000 kW hr/yr vs. 6000–6500 kW hr/yr in the late 1990s, about three and two times higher, respectively. Consumption of electricity per capita in the United States was 70 times higher than that in the developing countries in 1950 and 13–15 times higher in the late 1990s, and although the per-capita consumption of electricity in the developed countries other than the United States has leveled off during the late 1980s and 1990s, it has continued to increase in the United States (Appendix Table 65; Figure 51).

GENERATION AND CONSUMPTION OF ELECTRICITY IN THE 21ST CENTURY

It is safe to say that the generation and consumption of electricity, which has grown steadily during the 20th century, will continue to increase during the 21st century. Consumption of electricity and its share of the total world energy consumption should be expected to increase linearly for yet a few decades and then begin to flatten as the growth rate of the world's population also decreases and improvements in efficiency in the use of electricity are developed. Conservation may also become a factor. Consumption of electricity is not anticipated to be restrained by insufficient sources of primary energy. However, the share of each of these sources will undoubtedly vary considerably, depending on economic, environmental, political, and emotional circumstances, as well as on the local or regional availability of the source or sources of primary energy.

Coal was the main fuel used in the generation of electricity during the 20th century. Despite its serious drawbacks as a source of environmental pollution and its progressive replacement by oil, natural gas, and nuclear energy during the middle decades of the century, coal's share has remained essentially unchanged since 1970; it still provides about 35% of the primary energy for the generation of electricity at the beginning of the 21st century (Appendix Table 63; Figure 49). The use of coal in the generation of electricity may decline in the future in most developed countries but may increase in the developing countries, particularly in those with ample reserves of coal.

At least in the first half of the 21st century, coal will probably maintain its position as the largest source of primary energy for the generation of electricity.

If coal, because of its environmental drawbacks, is to be eliminated as the main fuel for the generation of electricity, or at least its role is to be considerably reduced, a substitute or substitutes must be selected and developed during the next few decades.

Oil is also environmentally objectionable, and its use as a fuel in the generation of electricity has steadily decreased during the last three decades of the 20th century. The sharp increase in price during the 1970s (the oil price shocks of 1973 and 1979) effectively eliminated oil's advantage as the cheapest source of primary energy for the generation of electricity. In addition, concerns about its long-range possible exhaustion have contributed to reduce its desirability. Its contribution, therefore, will undoubtedly continue to decrease in the future.

Natural gas, however, is less of a source of environmental pollution and is progressively becoming more the chosen fuel for new power plants (Appendix Table 63; Figure 49). This shift toward the increasing use of natural gas, which started in the 1990s, will probably continue during the 21st century. During the first half of this century, natural gas will become the world's second largest source of fuel for the generation of electricity, surpassing both hydropower and nuclear power. The developed countries will account for more than half of this increase. The growth of the natural-gas share will, however, depend on its price remaining relatively low. If the price of natural gas increases excessively, coal may have a clear advantage—utilities will always tend to choose the most cost-effective fuels.

Two potential large-scale suppliers of primary energy for the generation of electricity that do not contribute polluting combustion products to the environment, nuclear and hydroelectric energy, face strong opposition as possible replacements for the fossil fuels.

Nuclear power increased rapidly in the 1970s and 1980s, but its share of the total electricity generation remained essentially unchanged during the 1990s (Appendix Table 63; Figure 49). The more extensive use of nuclear power has many problems, some real, some perceived, many of them political and emotional. Its short-range future does not look good. If the consumption of electricity continues to increase, as public resistance slows the building of new nuclear power plants, and as more and more older plants are progressively decommissioned, the share of nuclear power will inevitably decrease during the first half of the 21st century. To retain an appreciable share of the electricity generation, nuclear power needs to overcome the present public resistance that was considerably strengthened by the disastrous 1986 Chernobyl accident.

A "nuclear renaissance" during the mid- or late 21st century is still uncertain. It will depend on government policies and on an evolution of the nuclear industry that will make possible the solving of the problems that now make nuclear power unpopular and even feared: safety, economics, and disposal of radioactive waste (see section on nuclear power in Chapter 4).

Hydroelectric power, once representing as much as 40% of the total electricity generated in the world, has provided only about 20–21% since the mid-1970s. Hydroelectric power, despite being probably one of the cleanest sources of electricity and the only clean one used on a large scale, faces a future, as the nuclear power does, clouded by environmental, social, and economic drawbacks and by the limits in the availability of sites in which to build large hydroelectric plants (see section on hydroelectric power in Chapter 4). In the developed countries, hydropower will increase little, if at all, during the 21st century. It is in these countries that the opposition to the building of dams, which started in the 1970s and has continued to grow, is stronger. In addition, in most of these countries, particularly in Europe, the best sites for large hydroelectric dams have already been developed, and the commissioning of new large hydroelectric plants has steadily diminished in the last 20–25 years. Most of the development of new hydroelectric power will occur in the developing countries, namely, China (where the huge Three Gorges Project is scheduled to be commissioned between 2003 and 2009), India, and Latin America, but only if such a source of electricity is found to be economically competitive with other sources. As in the case of nuclear power, if the consumption of electricity continues to increase and no new large hydroelectric plants are built, the share of hydroelectric power in the generation of electricity will inevitably decrease during the 21st century.

Even if the opposition to hydroelectric power vanished, or at least diminished, site availability and lack of economic attractiveness may still limit the future contribution that hydroelectric power will make to the world's demand for electricity during the 21st century to no more than 15–20%.

Other potential sources of primary energy for the generation of electricity (geothermal energy, PV processes, wind, biomass, and hydrogen) have so far made only minimal contributions to the world's generation of electricity. Although their contribution will undoubtedly increase, mainly in the developed countries, it is believed that they will not become major sources of electricity during the 21st century, at least not during the first half of the century. Their most important role may be in providing electricity to remote locations. Wind power appears now to have the best possibilities.

So, how will electricity be generated in the 21st century?

During the first half of the century, it is most likely that coal, natural gas, and hydropower will remain as the prime sources of primary energy for electricity generation. The share of coal will probably decrease, whereas that of natural gas will substantially increase, particularly in countries and regions where gas is available, and if its price remains low. Hydroelectric power will supply no more than 15–20% of the electricity consumed in the world. The three (coal, natural gas, and hydropower) will, in all probability, contribute between 75 and 80% of the primary energy used in electricity generation during the first half of the 21st century. The share of oil and nuclear power will decrease to an average of about 10% each. The "non-hydro" renewables (geothermal, solar, wind, and biomass) will contribute less than 5% to the generation of electricity.

As discussed earlier, resource distribution, economic variables, political resolutions, and many other factors will determine, in different regions and in different countries, the proportions of the basic primary energy sources used in the generation of electricity. Coal, for instance, will remain the principal fuel for electricity generation in China, India, and Australia.

Also influential in determining the share of the various sources of primary energy will be the rate of development of new technologies for electricity generation and for the development of more environmentally harmless use of the fossil fuels. Specifically, if new technologies could be developed for the non-polluting, or less polluting, use of coal, coal would remain as the main fuel of power plants for many decades to come, given its abundance in many regions of the world.

During the second half of the 21st century, nuclear power may make a comeback; the contribution of coal will decrease, unless new technologies allow its use in more environmentally acceptable ways; oil will no longer be used in the generation of electricity; hydropower will maintain its contribution at about the same level; and natural gas will continue to be the preferred choice for the fuel of power plants. The contribution of other sources of primary energy, principally wind power and hydrogen, will depend on public acceptance and on major technological, political, environmental, and economic developments that will be discussed in the next section of this study. It does not seem likely, however, that these and other "non-hydro" renewables will make a major contribution to the generation of electricity during the second half of the 21st century.

But even if the use of wind power, hydrogen, or an entirely new, environmentally acceptable source of primary energy for the generation of electricity materialized during the 21st century, a considerable period of time and an enormous amount of capital would be necessary to convert from the currently used processes of power generation and current energy-consumption infrastructure to new ones.

For most of the 21st century, it is likely that coal, natural gas, and hydroenergy will remain the main sources for the generation of electricity.

6

Energy Consumption and Probable Energy Sources During the 21st Century

ABSTRACT

Five main factors have guided the preparation of five scenarios of energy consumption and sources of supply in the world during the 21st century:

1) The world's energy resources are estimated to be more than sufficient to supply the demand for energy during the whole century. The issue will be how and where they will be available and how they will be used; efficiently, one hopes, and with minimal damage to the environment.
2) The world human population will grow more slowly but still reach 10–12 billion by 2100. The population of the developing countries will account for 80–90% of the total.
3) A great discrepancy exists in economic development between the developing and developed countries, as testified by the profound divergence of their respective per-capita energy consumption. Because the vast majority of the world population will be in the developing countries, how their economies develop will be critical in estimating the total world energy consumption.
4) No radically new technology for energy generation and use is anticipated; however, important advances in known technologies are expected to make major contributions: clean use of coal, gas-to-liquids conversion, more efficient transport engines and electricity-generation plants, and development of fuel cells powered by hydrogen for transport and electricity generation.
5) Most important, finally, will be how widely governments, industries, and the people recognize the potential damage to the environment by the use of the fossil fuels and how decisively they act to protect it.

During the 21st century, energy will continue to be used principally as transportation fuels and for electricity generation. The fossil fuels (oil, natural gas, and coal) will be the major source of energy supply during the entire century. Other sources are forecast to contribute in different but lesser proportions in the five scenarios.

INTRODUCTION

The past is not always a sure guide to the future, but history can provide an important foundation on which to base predictions of what the future may bring. The previous sections of this study have attempted to provide a review of the past worldwide consumption of energy and the sources that supplied it—the basic factors that have determined consumption and supply, and, as complete as possible, a factual history of these factors. It is hoped that this thorough review of the past will make forecasts of energy consumption in the 21st century and its possible sources more realistic, well grounded, and defensible. If past developments and

DOI:10.1306/1032710SP226

trends leading up to the present do not point to the future, what other guide do we have?

Several historical factors and potential future developments influence the forecasting of consumption and possible energy sources during the 21st century. Different combinations of these historical factors and different choices of possible future developments yield different forecasts of energy consumption in the 21st century and its sources of supply. What happens, of course, will be subject to possible great changes because of unpredictable events such as wars, famines, political manipulations of the availability and cost of energy sources (such as the oil shocks of the 1970s), disruption of the routes of transportation, and many other factors.

One thing is certain: The forecasts will be wrong. The questions are "how wrong?" and "wrong in what direction?" We will have to wait and see. The future cannot be predicted with any degree of certainty. As Francis Bacon put it, "Dreams and predictions ought to serve but for winter talk by the fireside." But we can try. Predictions need not be perfect to be valuable.

Five forecasts (scenarios) will be included as the final output of this study. For each forecast, the choices of basic information and assumptions of possible future developments will be clearly stated and discussed.

Readers, however, can decide not to agree with any of these forecasts. They will then be able to choose their own historical data and possible future developments and make their own forecasts based on their own assumptions about population trends, future economic and technological developments, etc. The historical information, previously scattered through countless publications (not all of them easy to access), will be readily available for their use. That, in fact, was one of the objectives of this study: to compile in one place all the factual historical information that will provide the basis to forecast the consumption and probable sources of energy during the 21st century (world population-growth trends; patterns of energy consumption; statistics of past production and reserves of possible sources of energy by countries or groups of countries; history of the development of estimates of the ultimate recovery of these potential energy sources; and the sources from which all this information was obtained).

PRINCIPAL FACTORS IN FORECASTING THE CONSUMPTION AND SOURCES OF ENERGY SUPPLY IN THE 21ST CENTURY

Energy consumption and the sources that supply the demand for energy are determined by five closely related principal factors.

Availability of Sources of Energy

The previous review of the magnitude and distribution of possible sources of energy (the energy resource base) conclusively indicates that they are more than sufficient to supply the expected demand during the 21st century. Predictions of possible scarcity of energy resources, in general, are not realistic. On the contrary, the estimated volumes of most worldwide energy resources have increased with each successive assessment. The sources of energy are there, although irregularly distributed. The issue is how they will be made available and how and in what proportions they are going to be used, hopefully economically, efficiently, and with minimal damage to the environment.

Energy in the 21st century will continue to be consumed in two principal forms: as liquid and gaseous fuels for transportation and as electricity.

No scarcity of ground and air transportation fuels is anticipated. Oil will be the main source of liquid fuels for many years yet. The production of conventional oil may peak sometime during the 21st century, but large resources of unconventional oil are known in tar sands, oil shales, and as natural-gas liquids in many natural-gas accumulations. The availability of liquid fuels for transportation will be further assured if improvements in the technology to convert coal and natural gas to liquids make these fuel sources economically competitive; gas resources are great and growing, and coal is known in vast amounts in many countries of the world. Liquid fuels from these sources will be there to fill the gap when conventional oil production begins to decline.

The expected rapidly increasing demand for electricity also will be readily supplied mainly from plants fueled by coal, natural gas, and hydropower. Nuclear power may become a factor in the future generation of electricity, but only if the problem of the disposal of spent fuel is satisfactorily resolved, if the public's concerns about proliferation and safety are overcome, and if measures to protect the environment and counteract possible global warming become an important factor in the selection of energy sources. Geothermal, solar, and particularly, wind power may also contribute to the future generation of electricity, but it does not seem now that their share will be important.

Whereas the availability of energy sources will not limit the supply of energy at the global level during the 21st century, several geographic, environmental, political, and financial factors may limit the availability and the consumption and supply of energy at certain times and in certain regions and countries of the world.

One such factor is the uneven geographic distribution of some sources of energy: 65% of the oil reserves are in the Persian Gulf region; 71% of the conventional natural-gas reserves are in the former Soviet Union (FSU) and the Persian Gulf region; 60% of the coal reserves are in the United States, the FSU, and China; and the location of favorable sites for hydroelectric plants has definite geographic restrictions.

Difficulty of access to and high cost of production of certain sources of energy will increase the price of the commodity. Distance from the source of energy to the places of energy consumption, which will increase as the energy sources closer to the centers of demand are progressively exhausted, may require costly transport with resulting increases in price. These high costs of access, production, and transportation, instead of the availability and magnitude of the global sources of supply of energy, may be the determining factors in the levels of energy consumption in some regions of the world.

The future prices of the different forms of energy could be considered a sixth factor in determining energy demand and consumption in the 21st century. Oil and gas prices could increase when the lowest-cost accumulations are depleted, and it becomes necessary to develop higher-cost reserves: accumulations in very deep water, in hostile environments, or in regions remote from centers of consumption. Prices could also increase whenever dominant producers seek higher prices. On the other hand, prices could decrease because of lower production costs brought by technological advances. The Law of Supply and Demand, to be sure, will have much to say. To predict future energy prices, however, is extremely difficult and has not been attempted in this study. Prices can be affected by too many unpredictable events.

The previously mentioned uneven geographic distribution of some energy sources and the potential high cost of certain forms of energy in certain regions of the world indicate that, to supply the energy demand throughout the world during the 21st century, it will be necessary to deal with conflicting technological, economic, political, and environmental circumstances and values and to rely on a balance of diverse energy sources, mainly oil, natural gas, coal, hydroelectric power, and perhaps, wind power.

Increased globalization of the energy market will also be necessary to ensure an adequate supply of energy to all countries. If the supply of energy became fully integrated geographically, all regions and countries in the world would be able to secure their requirements for energy from all potential sources; and, all energy-producing countries would be free to supply the demand for energy anywhere in the world—the ideal and the most efficient system of energy supply and consumption.

Human Population Growth

The growth rate of the total world human population, which was exponential during the first and middle thirds of the 20th century, slowed considerably during the last third of the century, during which the growth became linear. Rates of population growth seem to have peaked everywhere in the world except in the Middle East and in Africa, particularly in sub-Saharan Africa, where the population is still increasing at a very fast rate. More and more developing countries have taken steps to slow their increases in population, and as a result, fertility rates have generally declined in most countries much faster than anticipated.

The slowing of the rates of population growth, however, has been far from uniform throughout the world; whereas in the developed countries, the population has been growing very slowly, even decreasing in some countries, in many developing countries, the population is still growing rapidly. Similar trends should be expected in the future. As a result, the percentage of the world population in the developing countries is predicted to increase from about 79% at the beginning of this century to about 87–89% during the second half of the century and reach 90% by 2100.

In the developed countries, the population will probably remain stable or grow slowly, and in the second half of the 21st century, it will represent only about 10–15% of the world's population. However, their high consumption of energy per capita, particularly in the United States, will make their share of the total world consumption of energy considerable, certainly during the first half of the century, even if very desirable efforts are made to increase the efficiency in the use of energy.

For the developing countries, the future rate of population growth, as well as the relationship between fertility rates and population growth on the one hand and economic development and poverty levels on the other, have been the subject of lasting and contentious debate for many years. Will the rate of growth continue to slow down during the 21st century? Does rapid population growth have a negative, positive, or little impact on economic development and the reduction of poverty? What effects will population growth have on the environment?

Most recent studies, using more extensive and much improved data sets and more advanced methodologies, seem to conclude that in general, rapid population growth has an adverse effect on the pace of economic development in the developing countries and is an important cause of poverty and of the present economic disparity between developed and developing countries (Birdsall et al., 2001). Excessive population growth will have a very negative impact on the environment. Declining fertility, if coupled with the proper government policies, has a strong positive effect on economic growth and can contribute to the decline of poverty. High fertility increases the inequalities in income distribution and makes the decrease of poverty more difficult and less likely.

Population numbers alone, of course, are not the only determining factor in predicting the demand for energy during the 21st century in both the developed and the developing countries. Rates of migration between regions of the world and the increasing urbanization of the population will also be important elements in determining the demand for energy.

Low incomes and high rates of unemployment are common in Third World countries. As a result, attempts to migrate to developed countries, where manual-labor jobs that few citizens of the developed countries want are available, are common in many regions of the world. Over the long run, however, there will be a limit to the number of these jobs available and a limit, therefore, to the portion of the rapidly increasing population of the developing countries that can find employment in the industrialized world.

Increasing urbanization will also become an important factor. It has been estimated that by 2050, 75% of the world's population will live in urban areas where per-capita consumption of energy, particularly in the developed countries, is much higher than the national averages.

Because the great majority of the population of the world during the 21st century will be in today's developing countries, long-term forecasts of global energy consumption will depend critically on estimates of the future rates of their population growth and their future trends in energy consumption per capita.

Other factors, apart from population numbers, that contribute to the prediction of energy demand are the degree of economic development and the resulting improvement or deterioration of the standard of living, the efficiency in the consumption of energy, and the concern, or lack thereof, for the protection of the environment that governments may consider in establishing their energy policies.

Economic Development

Given that there are more than enough sources of energy in the world to supply adequately its overall energy demand for a long time to come, to forecast the future economic development of both the present developed and developing countries is essential in determining the demand for energy in the 21st century, and to do it keeping in mind the historical trends of population growth and energy consumption.

Estimates of the economic development and consumption of energy in the developed countries can be made with reasonable assurance. Their populations should be expected to vary within fairly narrow limits during the 21st century, as shown in Appendix Tables 27–29 and Figure 16. To forecast future economic development in the developing countries, given that they will contain the great majority of the world population during the 21st century, will be much more difficult but particularly important.

At present there is a great disparity between the developed and developing countries in economic development, per-capita income, per-capita energy consumption, and general standard of living.

The disparity in energy consumption between the developed and developing countries is illustrated by the following historical trends discussed earlier in this study.

Per-capita consumption of energy in the developed countries increased from 13.7 BOE/p/yr in 1950 to 32.5 BOE/p/yr in 1979, and it has remained at 31–34 BOE/p/yr since then (see Appendix Table 10; Figure 13). It has been estimated that the consumption of energy per capita in the developed countries will remain within this range or decrease somewhat during this century as a result of low rates of population growth and increasing efficiency in the use of energy.

The consumption of energy per capita in the developing countries has been much lower than in the developed countries; it increased from 0.7 BOE/p/yr in 1950 (about 1/20 of that of the developed countries) to about 5 BOE/p/yr (about 1/7 of that of the developed countries) during the 1990s (Appendix Table 11; Figure 13). However, because the population of the developing countries during those years grew from 1.7 billion to 4.5 billion, the total consumption of energy increased 20-fold from 1.2 billion to about 24.5 billion BOE/yr, whereas in the developed countries, it increased only about 4-fold from 11 billion to 42 billion BOE/yr.

Can this large disparity in per-capita energy consumption and, as a consequence, in economic development and standard of living between the present developing and developed countries substantially decrease or vanish during the 21st century?

Several constraints will make it difficult, at least during the next few decades: the continuing high rate of population growth in some developing countries (despite recent indications of an apparent slowdown); insufficient access to the investment capital necessary to establish a widespread energy infrastructure; possible future increases in the price of transportation fuels and electricity caused by increasing costs of producing and transporting energy sources to centers of consumption;

and possible governmental policies and taxes intended to reduce combustion gases to protect the environment.

Enormous financial investments will be necessary to create the infrastructures that would make available increasing amounts of transportation fuels and electricity to the population of the present developing countries. However, these financial investments may not be forthcoming if the potential consumers are not able to pay for the products of the investments. How much capital is available to be invested in the creation of energy infrastructure in the developing countries and when it will be available are critical issues. Major changes in economic development do not occur overnight; they come gradually and generally take a long time. They require an educated population, political stability, a favorable investment climate, and free trade. Globalization of the energy industry may become a necessary factor.

Because of all these variables, the estimates of the future consumption of energy by the developing countries can vary widely, depending on the assumptions made concerning their future economic development and their per-capita consumption of energy. Will they maintain their rate of increase of energy consumption per capita of the last 50 years? Will they increase it, or will they decrease it? How much? Will most, or at least some, of the developing countries be able to afford increasing their per-capita consumption of perhaps increasingly expensive fuels for transportation vehicles and electricity that would provide them a higher standard of living? Will they have access to domestic or foreign capital to build the necessary infrastructure to make available affordable sources of energy? Will they have access to new energy technology, and perhaps more important, how much will the rates of population growth in the still-developing countries increase or decrease during the 21st century? All are very difficult questions to answer, given the fragility of the present economic, political, and social infrastructures of many developing countries.

As shown in Appendix Table 34 and Figure 18, depending on the options chosen, the total consumption of energy in the developing countries may surpass that of the developed countries as early as 2020, if they rapidly improve their economic development, or not until the mid-21st century, if they do not. The total consumption of energy of the developing countries at the end of the century, therefore, may range from as little as 50 billion to as much as 150 billion BOE/yr.

At least in some of the developing countries, a substantial increase in per-capita energy consumption is less than certain because of excessive increases in their population, their insufficient access to the capital necessary to establish an adequate energy infrastructure, and lack of means to acquire energy supplies from abroad.

More than 2 billion people in the developing countries now live below the poverty level, have no access to commercial energy, and have to depend on locally available sources of energy such as traditional wood, agricultural wastes, and coal. And, the total number of poor continues to grow in many regions of the world. Half of the world population (3 billion) still lives on less than $2 per person per day, and one quarter (1.5 billion) subsists in deep deprivation on less than $1 a day.

Nearly 50% of the current 4.8 billion Third World population (2.4 billion) has no access to basic sanitation, adequate education, or health care; almost one third (1.6 billion) has no access to clean fresh water; one quarter (1.2 billion) lacks adequate housing; 1 billion live in slums, and more than 100 million are homeless. Electricity is unavailable for 2 billion people around the world, and this problem probably will worsen as the population of the developing countries continues to grow. In Africa, the share of the total population with access to minimal amounts of electricity has dropped from 12 to 8% in the last 10 years.

Some 14 million babies and young children under age 4 starve to death each year in the developing countries, and millions of children are being left orphaned in the worst affected countries by the increasing spread of AIDS.

Twenty percent of the world's population controls 80% of the resources, and 80% of humanity has to make do with the remaining resources. This economic gap between the prosperous millions and the poor billions seems to be widening. It would be unrealistic and confusing to believe that this disparity in economic development and in per-capita consumption of energy between the present developed countries and most of the developing countries will be greatly reduced or essentially disappear sometime during the 21st century, or that the descendants of the billions of people now living in poverty or near poverty will attain a comfortable standard of living in the next 50 or 100 years.

Technological Progress

The potential development of better, more efficient, or entirely new technologies for the generation, distribution, and use of energy will have critical importance in estimating the consumption and possible sources of energy during the 21st century. A forecast based on past trends can be made meaningless by an unforeseen, radically new technological development.

Whereas the possible appearance of such entirely new technologies that would radically change the energy supply and consumption picture cannot be ignored—it is not impossible—it does not seem likely in view of past experience.

If the past is any guide, a review of the evolution of the systems of energy generation, supply, and use during the 20th century does not give much hope for the development, introduction, and acceptance of radically new and different systems of transportation and electricity generation, the two principal forms of energy use. The internal combustion engine running on liquid fuels produced principally from oil, although enormously improved over the years, has been the main source of propulsion of land transportation vehicles since the early 20th century. Over the same period, electricity has been generated and continues to be generated by thermal or hydroelectric processes: by putting steam, water, or air through turbines.

Predictions made in the 1960s and 1970s about future development of new technologies in the production and use of energy now seem too optimistic and unrealistic. Technological advances in the energy field have been nowhere as fast as forecasted; radically new technologies have not been developed and put in widespread use. Although impressive and revolutionary advances have been made in other fields (medicine, telecommunications, space exploration, to name a few), comparable innovations have not been seen in the energy field.

Some authors have commented that new energy technology has not produced more spectacular results, because research on energy sources has been neglected during the last few decades. Fusion is commonly mentioned in this respect. Perhaps if more attention is given and more funds are devoted to energy technologies in the 21st century, better results, revolutionary and unexpected findings, may be obtained. What these results may be in the next 100 years is very difficult to predict.

But even if such revolutionary findings were to occur, it is necessary to keep in mind that the lifetimes of present energy supply systems and their infrastructures are very long; that the introduction and commercialization of a new technology of energy supply and consumption, particularly if it is radically new, would require enormous initial investments; and that the necessary replacement of previously established infrastructures by entirely new ones would take many years. For China to reduce its dependence on coal, for example, would take decades.

The introduction of a totally new system of energy supply would be especially difficult in the developing countries where the desirable local technical expertise and, particularly, the necessary vast investment capital to put in place an entirely new infrastructure may not be readily available. This limitation of financial resources will also reduce the ability of the developing countries to take full advantage of new energy technologies that could improve efficiency and provide access to new sources of energy.

How many transportation vehicles (cars, buses, trucks, trains, tractors, and other agricultural machinery) will be running on something other than internal combustion engines fueled by liquid fuels (gasoline and diesel fuel) produced from oil (or natural gas or coal) by the middle of the 21st century? How many by the end of the century? If most vehicles were to be powered by electric engines, for example, how much additional electricity would have to be generated in the world and by what means? Hybrid vehicles, a potentially important development in transportation technology, still run in part on internal combustion engines fueled by gasoline.

All things considered, it now looks like radically new technologies will not be as important a factor in the future consumption and supply of energy as some authors and organizations have predicted. Some conceptions of future energy systems are too farfetched to be seriously considered, and it is unrealistic and dangerous to rely blindly on radically new technology that will make available totally new sources of energy.

Improvements in presently known technologies, however, will hopefully become an important factor. Such technological improvements should contribute not only to an ample supply of energy but also to the economic growth and to a better quality of life for a larger part of the population of the developing countries.

Technological advances can be expected in the design of more efficient internal combustion, jet, and hybrid engines and in the development of major improvements in the presently known systems for the generation of electricity. However, the internal combustion engine and electricity-generating turbines will remain the principal means of transportation and of generating electricity, respectively, during the 21st century.

A particularly important objective of future pursuits for improvements in the production and use of energy will be the development of technologies that will provide cleaner sources of energy, sources that will help protect the environment and prevent a potential global warning (for example, less polluting means of using the vast coal resources of the world in the generation of electricity—a clean coal-burning technology). The development and adoption of clean-coal technologies will be crucial for the future use of coal.

Equally important will be the development of economically competitive means of converting coal and natural gas to liquid fuels. The technology to do this is available now, but it needs to be improved to be more efficient and cheaper. Converting natural gas to liquids at commercially competitive prices would greatly reduce the long-distance transportation problem. Some large natural-gas accumulations, located a great distance from consumption markets, cannot be economically developed if the gas needs to be transported in gaseous form in long, expensive pipelines, or in liquid form (LNG) by means of equally expensive cryogenic systems. Transporting the natural gas as a liquid at atmospheric temperature permits the development of large, now-uneconomic, or "stranded" natural-gas resources. Such a development will have great consequences in any energy-supply scenario during the 21st century.

Oil is now the main source of liquid fuels for transportation (gasoline, diesel, and jet fuel). When and if oil becomes insufficient to provide these fuels, coal and natural gas will still be available in large volumes as the main potential sources of liquid fuels; as mentioned before, other sources of energy (hydroenergy, nuclear, geothermal, solar [PV], and wind) can only be used to generate electricity.

Other desirable technological advances would be:

- The eventual development of hydrogen-fueled fuel cells as the source of propulsion of transportation vehicles and in the generation of electricity.
- The development of effective carbon-sequestration technology (principally the capture and storage of CO_2 from exhaust combustion gases), because the use of fossil fuels cannot be avoided for many decades to come.
- The generation of improvements in the technologies of exploration for oil and natural gas. Major improvements have been made in the last few decades, and more can be made.
- Improvements in oil- and gas-production technologies, particularly those that could bring profitability to now-uneconomic accumulations (production enhancement of oil fields, lowering the costs of production).
- The development of technology to recover the very large volumes of gas in hydrates.

To be accepted and used, new energy technology (either radically new or improvements to presently known technology) needs to provide efficient and reliable energy, preferably to be protective of the environment and, most importantly, to supply energy at a reasonable cost. Environmentally favorable energy technologies could be promoted by governments through taxation policies and research funding, although they may not represent the preferred choice of consumers.

Protection of the Environment and Means to Attain It

Probably nothing could change the patterns of demand, consumption, and supply of energy in the 21st century as radically as the extent to which the world's nations recognize the potentially dangerous environmental effects of the emission of the combustion products of fossil fuels (particularly CO_2) and the apparent resulting global warming, along with international resolve to do something to counteract them.

Environmental concerns have been high on the agenda of organizations and governments for a long time, but not much has been done. What is done, or not done, to attempt to protect the environment will strongly influence the choice of energy sources and the initiatives and policies of the societies and governments of the world's nations. Such initiatives and policies could also have an important influence on the rate of growth of the population and on the emphasis given to the development of new technology.

Serious concerns about the environment are likely to result in attempts to shift, as much as possible, from fossil fuels (oil, natural gas, and particularly coal) to other cleaner sources of energy and to try to develop means of using fossil fuels in ways as environmentally benign as possible. New technology would be critical in this last respect, as previously discussed.

Encouraging energy conservation in the countries that have high rates of energy consumption per capita, particularly the United States (see previous sections on energy consumption and electricity), and more efficient use of energy would also reduce the consumption of fossil fuels and lessen the resulting pollution of the atmosphere. New technology would also help these objectives. Conservation, in fact, can be considered as an energy resource, because it extends the life of known sources of energy.

Finally, governmental energy policies and regulations, such as limiting CO_2 emissions, requiring increases in the efficiency of the use of transportation fuels, increasing fuel and electricity taxes, and creating economic incentives to conserve and use all types of energy more efficiently, could be necessary for more effective protection of the environment. Such policies would encourage investment in improved or new technology and the initiation of a trend toward a flexible, convenient, and clean supply of energy at local, regional, and global levels.

There is, however, a limit as to how much the environment can be protected and how much the amount of global warming can be minimized. People need to consume energy to maintain an adequate standard of living or just to survive, and as the population of the world continues to grow, and particularly if the standard of living of the developing countries is hopefully to improve, an increase in the global consumption of energy is inevitable. Transportation fuels and electricity will be the principal forms of energy consumption during the 21st century, and there is no way to avoid the use of fossil fuels to provide them, particularly if two of the cleanest and potentially more readily available sources of electricity, nuclear and hydroelectric power, continue to be strongly and vociferously opposed by environmental organizations and other supporters of a clean environment. For the provision of transportation fuels, there are few and not promising alternatives to the products of oil, natural gas, and coal.

FIVE ENERGY SCENARIOS: CONSUMPTION AND SUPPLY DURING THE 21ST CENTURY

Introduction

Some general guidelines and basic assumptions have been used in designing the scenarios of consumption and supply of energy during the 21st century.

First of all, the importance of the factual historical information discussed in the first part of this study has been kept clearly in mind—historical trends are essential in predicting the future.

The selection of the basic parameters for designing each scenario has been based on the five closely interrelated factors discussed earlier in this section. However, in choosing these parameters, not all factors have been given the same importance.

The differences between the scenarios have been based principally on the prediction of future rates of population growth and economic progress of the developing countries, because they will represent about 80–90% of the world's population during the 21st century.

It is clear that the population of the world will continue to grow, although at a slower rate, and that this growth will mainly be in today's developing countries. The current great disparity between the developing and developed countries in economic development, income, and energy consumption per capita, and their general quality of life, will probably decrease during the 21st century; but the persistent population increase in the developing countries will make inevitable the disparity that will still remain throughout the century. Nevertheless, the bulk of the increase in energy demand will come from the developing countries. How large this increase will be is an essential element in estimating the demand for energy in the 21st century, and different estimates produce the basic differences between the different scenarios.

Next in importance in designing the scenarios is the degree of concern that all countries of the world will have about protection of the environment. If measures to reduce the degree of environmental damage are established and seriously enforced and if the current wasteful use of energy is no longer tolerated, the consumption of energy and the sources of supply would have to change radically.

New technologies or advances in the technologies known today will be given less importance in the selection of the design parameters of the different scenarios. Radically new technology, of course, could completely change or strongly modify the patterns of supply and demand of energy. However, the introduction and use of new technology, in particular radically new technology, takes a very long time, and it is not expected that any such radically new technology will be an important factor in the consumption and supply of energy in the 21st century. The potential importance of revolutionary new technology has repeatedly been exaggerated in the past. New technological innovations have frequently been oversold. Novelty has too commonly been mistaken for progress. Improvements in currently known technology, however, could be most important: more efficient transport vehicles, cleaner internal combustion engines, more efficient and cleaner power plants, cleaner use of coal, gas-to-liquids conversion, and many others.

In all scenarios, it will be assumed that, worldwide, energy resources will be more than sufficient to supply the expected demand during the 21st century, although their uneven geographic distribution, which may require long-distance transportation, and possible price increases may limit the availability of energy sources in certain regions or countries of the world. Other economic and political factors may also determine the ready local availability of energy sources.

World energy consumption will continue to increase during the 21st century, certainly during its first half and probably during its entirety. Electricity will supply a growing share of the total energy consumption. Fossil fuels (oil, natural gas, and coal) will remain the main sources of the energy consumed in the world well into the 21st century, both as liquid and gaseous motor fuels and as fuels for the generation of electricity, the two biggest general uses of energy. Only in the last few decades of this century, with increasing concern about the environment, other sources of energy (nuclear, hydroelectric, and wind power) may claim a somewhat larger share in the generation of electricity. Transportation fuels will still be produced from fossil fuels, because it is not anticipated now that other sources of energy will be developed to replace them. A possible exception could be the use of hydrogen in fuel-cell-driven vehicles. However, hydrogen supplying a substantial share of the transport fuel is unlikely before the end of the 21st century.

It is not expected that entirely new sources of energy will become available, or that the function of those known and used today will change significantly. There will be a shift toward electricity and toward high-quality, more convenient, flexible, efficient, and environmentally cleaner fuels such as natural gas. How electricity is to be generated will become a particularly important issue.

Any major change in the distribution of the sources of energy will have to overcome the inertia of choice and acceptance by the consumers and the very long time necessary to establish entirely new infrastructures. The more radical the change, the longer the time needed for these adaptations.

Given the long lives of the current suppliers of energy (power plants and refineries), it is unlikely that major changes will occur during the first half of the 21st century.

Oil will be the dominant source of motor fuels during most of the 21st century, an important energy source during the entire century. It is easier to produce, transport, store, and use than natural gas or coal. Its use will be only limited by the amount of the recoverable resources of both conventional and unconventional (tar sands and oil shales) oil. Improvements in the present technology for discovering new reserves and for increasing recovery and maintaining low prices will assure the availability of oil during the 21st century.

Oil products are now the main source of transportation fuels and enjoy a worldwide infrastructure for their distribution. A limited probability exists that oil (or natural gas) may be replaced by other fuels in the transportation field. The use of oil as the fuel in power plants has steadily decreased in the last three decades and will probably essentially vanish during the first half of this century, except perhaps in the main oil-producing countries.

Natural gas is now considered the fuel of the future, the transition fuel of choice. It is cleaner than oil and coal, abundant in many regions of the world, and capable of highly efficient electric power generation. An increasing percentage of recently built power plants are fueled by natural gas at the expense of coal, oil, and other fuels. Future new power plants will provide the bulk of the incremental use of natural gas.

Natural gas resources, both conventional and unconventional [coalbed methane, tight sands, hydrates(?)], are very large, and improved technologies are expected to be influential in increasing the efficiency of gas production and transportation.

Natural gas will become a major factor in the supply of energy during the 21st century if its price does not increase excessively, and particularly, if the technology to convert gas to liquids progresses to make the liquids economically competitive as a transporta-tion fuel. Compressed natural gas may also become a widely used motor fuel.

Coal will remain an important source of energy in the 21st century, although its share of the energy market will most likely decline progressively during the first half of the century. Reasons for this decline are the environmental drawbacks of the mining, transporting, and burning of coal. Increases in the uses of coal would require the opening of new mines and the development of new transportation infrastructures, all involving long periods of time and very large investments. However, coal resources are vast and widely distributed geographically, a secure supply of energy that provides a powerful incentive to develop the technology to convert coal to gas and, particularly, to liquids, making coal a potential commercially competitive and environmentally acceptable major source of energy. The development of this technology for the clean use of coal (clean coal technology) will make coal available during the second half of the 21st century to replace natural gas in the generation of electricity and oil as the main source of transportation fuels, as oil and gas production begins to decline. In any case, the direct burning of coal is expected to diminish during the 21st century.

The future of nuclear power as a source of energy is much in question. It depends on the resolution, or lack of resolution, to the current controversies concerning the safety of nuclear plants, the disposal of spent fuel, and the fears of a potential nuclear proliferation. A large expansion of nuclear power may require the development of new waste disposal approaches, including credible and safe recycling processes that would reduce the volume of waste.

The contribution of nuclear power to the generation of electricity, despite its great potential, will inevitably decrease progressively during most of the first half of the 21st century because of the current great political and social opposition. Many of the currently active nuclear plants will be decommissioned in the next few decades, and the decisions about their replacement will determine the future of nuclear power. A renaissance during the second half of the 21st century will depend on the general recognition of nuclear power as a clean source of energy, the resulting general realization that the reduction of the contribution of nuclear power to the generation of electricity would inevitably mean a greater dependence on fossil fuels, and governmental policies seeking to expand the use of nuclear power to reduce combustion gases in their countries.

The development of a new generation of nuclear plants (cheaper, safer, more efficient, and perhaps smaller) and the development of new, safer technology for the disposal of spent fuel could also help in improving the social acceptability of nuclear power and its more important, long-term role in the generation of electricity.

Hydroelectric power will remain a moderate source of electricity during the 21st century despite the strong opposition to the building of new, large hydroelectric dams, the large initial capital investments that such dams require, and the limits to the availability of favorable sites for large hydroelectric plants.

The "non-hydro" renewables, geothermal, solar (photovoltaic [PV]), and wind, are mainly sources of electricity and will have a limited role as future energy suppliers. Wind is, by far, the most promising, but in relative terms, it is not expected that even wind power will become a significant source of electricity during the 21st century. The future of these "non-hydro" renewables will depend, to a great extent, on the development of technological advances that would make them price competitive with other sources of electricity.

Use of wood will continue in the developing countries, in some of them as the main fuel for heating and cooking. Other uses of biomass (generation of electricity and production of methane and alcohols) are minimal now and are expected to remain at that level in the future. Use of biomass as a source of energy has some very serious limiting factors: low efficiency and health and environmental hazards. In the case of wood, for example, a sizable contribution to the world energy demand would result in competition with food production, extensive deforestation, and prohibitive pollution. Would extensive burning of wood and other kinds of biomass be allowed if the world gets serious about protecting the environment?

Land-use conflicts may become major obstacles not only for the use of all forms of biomass, but also for the possible increase in the use of solar (PV) devices and wind turbines for the generation of electricity.

Finally, hydrogen as the fuel for fuel cells is unlikely to become more than a marginal factor in the supply of energy during most of the 21st century. Even if stringent governmental policies to protect the environment are instituted in the foreseeable future, hydrogen is not expected to make more than a meager contribution until very late in the century.

In summary, the contribution of the various sources of energy to the world's consumption of energy during most of the first half of the 21st century is not expected to change appreciably from that of today: the fossil fuels (oil, natural gas, and coal) will still be the main source of energy in the production of motor fuels and the generation of electricity. It is very unlikely that they will be replaced, to an important extent, by any other source of energy. The availability of liquid fuels could be considerably increased if improvements in the technology to convert natural gas and coal to liquids make these fuels cost competitive. Hydroelectric and nuclear power will supply most of the rest of the energy to be consumed during the first half of the century in the form of electricity.

During the second half of the century, the situation is more uncertain. It will depend on technological developments, industrial strategies, consumer choices, population and economic growth rates in the developing countries, and governmental polices.

If most of the world's nations recognize the serious need to protect the environment against the damaging effects of combustion gases, the second half of the 21st century could witness a comeback of nuclear power, significant development of hydroelectric power, ready supply of clean liquid fuels derived from natural gas and coal, increase in the development of wind power, the initiation of a hydrogen economy, and perhaps the introduction of some revolutionary technologies for the supply and use of energy that we cannot even imagine today.

Governments can contribute greatly to the protection of the environment by encouraging and directly funding research on, and development of, clean and efficient means of producing and using energy, by stimulating conservation of energy, and by enacting pricing and taxing policies that would favor restraint in the use of energy (high gasoline taxes, limitations on CO_2 emissions, improvement of vehicle fuel efficiency, etc.) and the choice of cleaner and more environmentally benign sources of energy.

However, no matter how many technological advances and governmental environmental policies become a reality, an adequate supply of energy that would assure a comfortable standard of living and a reduction of the levels of poverty in the developing countries will not be possible unless their population stabilizes in the second half of the 21st century at a reasonable level or even begins to decrease from the numbers it will inevitably reach in the first part of the century.

The assumptions and predictions made in planning and structuring the scenarios that follow are based on demographic, economic, technological, and environmental considerations and are tied to a common set of historical data; all scenarios incorporate the lessons learned from history. The assumptions and predictions range from fairly evident and well founded to uncertain and debatable. They all become increasingly questionable as the scenarios extend further and further into the future.

As discussed above, there is little doubt that there will be ample energy resources to supply the demand of energy during the 21st century, although their uneven geographic distribution and possible future price fluctuations may limit the availability of energy in certain regions of the world. For these reasons, it is easier to anticipate the forms in which energy will be used by future consumers (transportation fuels and electricity) than to estimate the absolute amounts of energy demand that the energy sources will need to supply or the geographic distribution of the consumption of energy during the 21st century.

Trends of population growth are also predictable within fairly close ranges, given the absence of catastrophic events such as famines, wars, or plagues and pandemics. Changes in fertility occur slowly over long periods of time.

However, there is considerable uncertainty in predicting future energy-price fluctuations, the possible development of radically new technology, the emphasis to be given by governments worldwide to the protection of the environment and the reduction of global warming, their enactment of environmental policies, and the effect that these policies may have on the choice of energy sources—a renaissance of nuclear power, for example.

The energy scenarios reflect a viewpoint that is basically optimistic. However, a review of the events that occurred during the 20th century and a look at a newspaper or television news program today tells us that

political and economic upheavals, wars, and pestilence will not be entirely avoided during the 21st century. On the other hand, favorable geopolitics and the globalization of the energy markets, as well as the general acceptance of the need to protect the environment, can be expected. Improvements to present technologies will make possible the switch to clean fuels and safe, less-polluting transportation vehicles. Plenty of energy sources exist. As a result, the world will not suffer any major scarcity of energy to satisfy the demand. The rate of population growth in the developing countries will diminish and perhaps stabilize during the second half of the 21st century, and their standard of living may improve but will not reach that of the present developed countries.

Only the development of radically new technology for the clean, cheap production and use of energy is placed in doubt.

The Five Scenarios

For *Scenario 1*, I chose a rather negative (probably unrealistically negative) outlook for the world during the 21st century. I forecasted that the human population of the world will increase to only 10 billion by 2100, which would actually be a positive factor, of which 9 billion will live in the developing countries and 1 billion in the developed countries. On the negative side, I projected

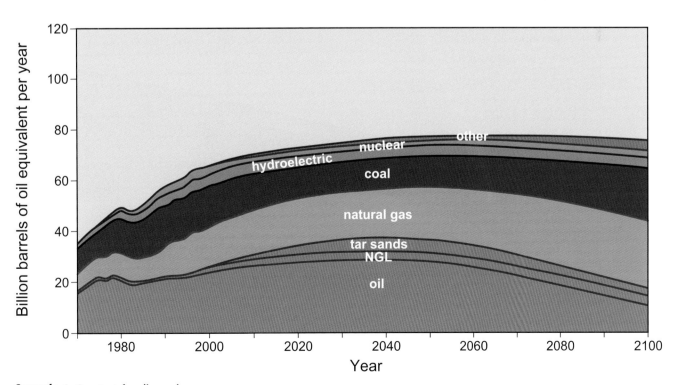

Scenario 1. See text for discussion.

that the per-capita energy consumption of the developing countries will not increase during the 21st century, remaining at the level of the 1990s: 5 BOE/p/yr. The per-capita energy consumption of the developed countries is projected to decrease moderately from 33 BOE/p/yr in the 1990s to 30 BOE/p/yr in 2100 because of expected increasing conservation, voluntary or by legislation, and more efficient use of energy. The forecast of the per-capita energy consumption in the developing countries may appear to be unrealistically low. It is based, however, on numerous indications that a large percentage of the population of the developing countries is making little progress in its economic development, and that this situation may not improve as their populations continue to increase, albeit, for the most part, at a lower rate. The effect of the estimated energy consumption in the developing countries is, of course, much stronger than that of the developed countries, because the former will represent 80–90% of the world population during the 21st century. Under these assumptions, total world energy consumption in 2100 reaches only 75 billion BOE.

The fossil fuels—oil (including bitumen from tar sands, extra-heavy oil, and natural-gas liquids), natural gas, and coal—are expected to provide 85–86% of the primary energy throughout the 21st century. The contribution of oil is based on an estimated ultimate recovery of 3.5 trillion bbl, and the gas production schedule is based on an estimated ultimate recovery of 31,000 tcf.

Oil production is forecasted to peak around 2040 and to decline during the second half of the 21st century, while natural-gas production increases consistently throughout the century. Bitumen and extra-heavy-oil production is projected to peak at about 5.5 billion bbl/yr in 2050. Coal's share of the world total energy consumption decreases during the first part of the 21st century as the share of oil and natural gas expands but increases during the second half to replace the diminishing contribution of oil, bitumen, and extra-heavy oil. This predicted increase in coal's production is based on the belief that advances in the technology to convert coal to liquid and gaseous fuels will diminish the objections to its use.

Hydroenergy has been projected to remain at the same level during the 21st century, no major comeback of nuclear power has been forecasted, and the remaining sources of energy (oil shales, geothermal, sunlight, wind, and biomass energy), are not expected to make important contributions to the energy supply. Coal and natural gas, and to a lesser extent oil and hydroenergy, are expected to provide the primary energy for the generation of electricity during the 21st century.

Scenario 2 is probably more realistic. The population of the world is projected to reach 11 billion in 2100, of which 9.9 billion will be in the developing countries

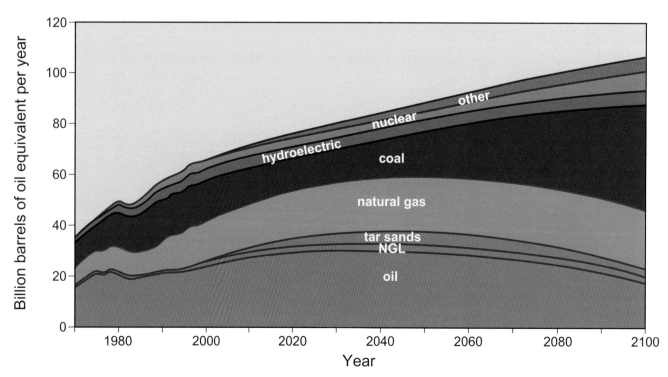

Scenario 2. See text for discussion.

and 1.1 billion in the developed world. The per-capita consumption of energy in the developing countries is estimated to increase gradually throughout the 21st century and reach 7.5 BOE/p/yr by 2100. The per-capita consumption of energy in the developed countries, as in Scenario 1, is projected to decrease moderately from 33 BOE/p/yr in the 1990s to 30 BOE/p/yr in 2100. Total world energy consumption under these assumptions will reach 107 billion BOE in 2100.

Essentially similar levels of yearly energy consumption by 2100 (102 billion BOE) are reached if it is predicted that the per-capita consumption of the developed countries decreases to 25 BOE/p/yr. As mentioned before, the forecasts of population and energy consumption for the developing countries are what determines the total world energy consumption. Changing the per-capita energy consumption (or the levels of population) of the developed countries makes small changes in the total amounts of energy consumption.

As in Scenario 1, the fossil fuels are expected to provide about 80–85% of the primary energy consumed in the world throughout the 21st century. The oil contribution is based on an estimated ultimate recovery of 4 trillion bbl, and that of gas is based on an estimated ultimate recovery of 24,400 tcf. Annual oil production peaks at about 30 billion bbl in 2030–2040. Coal production declines moderately during the first three decades of the 21st century but then increases continuously during the

rest of the century. By 2100, coal provides about 38–39% of the energy consumed in the world. As in Scenario 1, this increase of coal production becomes necessary to replace the diminishing contribution of the other fossil fuels, if it is predicted, as is done in this scenario, that other sources of energy (hydroenergy, nuclear, geothermal, sunlight, and wind) will not make important contributions to the world's energy consumption during the 21st century. The primary energy for the generation of electricity, as in Scenario 1, will be provided primarily by coal and natural gas, with lesser contributions by oil and hydroenergy. A mild revival of nuclear power is predicted for the second half of the 21st century.

In *Scenario 3*, the population of the world is projected to grow faster during the 21st century and to reach 12 billion in 2100, of which 10.8 billion will be in the developing countries and 1.2 billion in the developed countries. This represents the highest world population growth rate contemplated in a previous section of this study. The trends of per-capita energy consumption for developing and developed countries predicted in this scenario are the same as those of Scenario 2: in the developing countries, it will increase to 7.5 BOE/p/yr, and in the developed countries, it will decline to 30 BOE/p/yr by 2100. The total world energy consumption under these forecasts will reach 117 billion BOE in 2100.

As in Scenarios 1 and 2, the fossil fuels are assumed to provide 80–85% of the primary energy consumed in

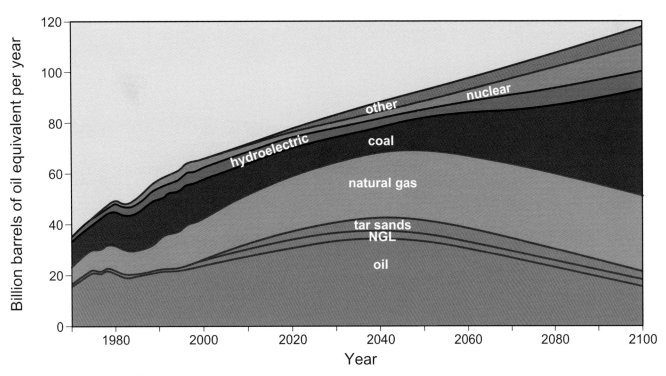

Scenario 3. See text for discussion.

the world during the 21st century. Oil production, based on an estimated ultimate recovery of 4 trillion bbl, increases faster than in Scenario 2 during the first four decades of the century, peaking at about 34 billion bbl in 2040. Predicted natural-gas production is based on the same estimate of ultimate recovery as in Scenario 1: 31,000 tcf. Coal production, as in the first two scenarios, is projected to decrease during the first three decades of the 21st century and then increase gradually during the rest of the century. Hydroenergy and nuclear power increase modestly during the second half of the 21st century, contributing 6 and 9%, respectively, of the total world consumption of energy by 2100. Other sources of energy, as in the previous scenarios, are not expected to make important contributions to the world energy consumption.

Forecasts of higher rates of population growth or higher increases in the per-capita energy consumption in the developing countries (to 10 BOE/p/yr, for example) would result in a higher estimate of the world's consumption of energy in 2100 and in a necessary higher contribution of coal to it.

In Scenarios 1–3 it has been assumed that the efforts of governments, the energy industry, and the population of the world to protect the environment from air pollution and to avoid a potential harmful global warming will be limited to increasing the use of natural gas, particularly for the generation of electricity; developing the technology that would allow the clean use of the vast coal resources of the world; encouraging the more efficient and frugal use of energy in the developed countries; and accepting that hydroenergy and nuclear power, despite some negative features, are potential environmentally benign sources of electricity.

Scenario 4 assumes that governments and society will take stricter measures to protect the environment from the combustion products of the fossil fuels and to slow the growth of the world population. How strict these measures can be, as mentioned before, has a limit. People need to consume energy to maintain an adequate standard of living, and the population of the world will certainly increase, to a greater or lesser extent, during the 21st century.

It has been forecasted that the main aim of such measures will be to reduce, as much as possible, the use of coal as a source of energy, especially as the fuel for the generation of electricity, and to develop the technology that will make possible the launching of hydrogen as the fuel for both transportation and electricity generation. A modest renaissance of nuclear power and the rapid increase in the use of sunlight, wind, oil shales, and geothermal energy as sources of energy are also given strong support in this scenario.

The human population of the world is assumed to increase to only 9 billion by 2100, 8.1 billion in the developing countries and 0.9 billion in the developed

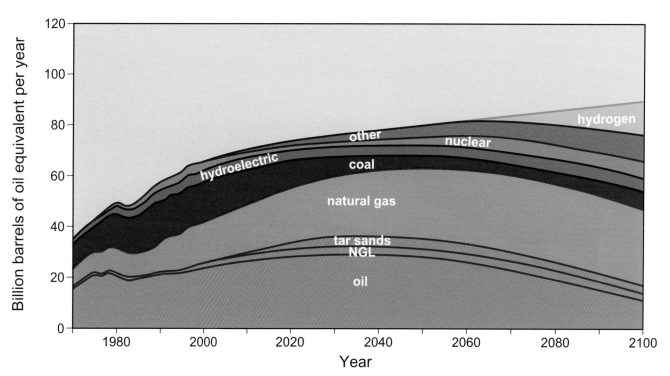

Scenario 4. See text for discussion.

world, the result of a comprehensive campaign of education and family planning in the Third World. This lower level of population is hoped to bring about a higher standard of living in the developing countries. Per-capita energy consumption, as in Scenarios 2 and 3, is assumed to increase by 2100 to 7.5 BOE/p/yr in the developing countries and to decrease to 30 BOE/p/yr in the developed countries. The total world energy consumption will reach only 88 billion BOE by 2100.

In this scenario, oil is expected to remain as an important source of transportation fuels during the 21st century, and natural gas is expected to remain as the main energy source, in gaseous or liquid form. A fully developed and economically favorable technology for gas-to-liquids conversion will make it possible for gas to be not only the principal fuel for electricity generation, but to replace oil to a considerable extent as a transportation fuel as well. These predictions of oil and natural-gas production during the 21st century are based on estimated ultimate recoveries of 3.5 trillion bbl of oil and 31,000 tcf for natural gas.

The use of coal will be progressively reduced during the first half of the 21st century and projected to remain at that level during the second half, limited to countries like China and India that have large coal resources but lack sufficient oil and natural-gas deposits. Hydroenergy and nuclear power, viewed with disfavor by the defenders of the environment, remain as modest sources of electricity during the entire century. Hydrogen enters the energy picture shortly after the middle of the century, and by 2100, it provides about 15% of the total energy consumed in the world, both in hydrogen-fueled fuel-cell vehicles and in electricity generation. Other sources of energy are optimistically forecast to provide 10–11% of the energy consumed in the world by the end of the 21st century.

Scenario 4 is probably overly optimistic and unrealistic, with the reduction of the growth rate of the population being too high and the limited use of coal during the whole century and the early entrance of hydrogen as a sizeable source of energy being questionable.

Scenario 5 is considered more realistic and is my preferred choice, the most likely scenario within the limits of reality. The population and per-capita energy consumption parameters are the same as those of Scenario 2: the world population is projected to reach 11 billion by 2100 (the middle of the three projections previously proposed) of which 9.9 billion will live in the developing countries and 1.1 billion in the developed world. The per-capita consumption of energy in the developing countries is estimated to increase gradually throughout the 21st century and reach 7.5 BOE/p/yr in 2100; that of the developed countries is estimated to decrease moderately from 33 BOE/p/yr at the beginning of the century to 30 BOE/p/yr in 2100. The total world

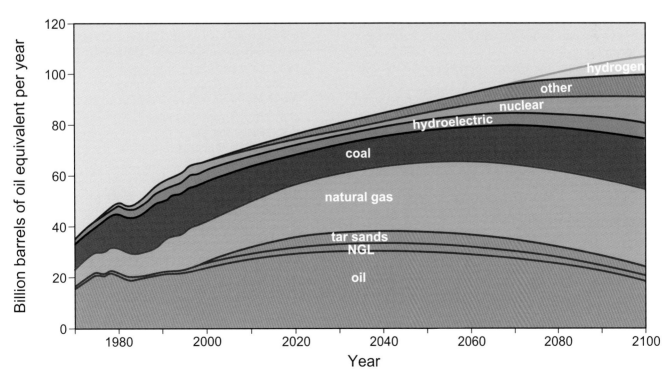

Scenario 5. See text for discussion.

energy consumption under these assumptions will reach 107 billion BOE in 2100.

The fossil fuels (oil, natural gas, and coal) that, during the first half of the 21st century provide 87–88% of the total world energy consumption, decrease their contribution gradually during the second half and, by 2100, provide only 70%. The estimated ultimate recovery of oil is 4 trillion bbl, and that of natural gas is 31,000 tcf. Coal, despite the possible development of a clean-coal technology, does not increase its share of the total world energy consumption during the second half of the 21st century as much as in Scenarios 1–3, and by the end of the century, coal provides only 19% of the primary energy consumed in the world. The contribution of hydroenergy remains at about 5–6%; hydrogen initiates its participation in the energy market late in the 21st century; and nuclear power and the other sources of energy, particularly wind power, increase gradually their share in the electricity generation after midcentury. Total world energy consumption in 2100 is 107 billion BOE. Natural gas is expected to be the principal source for the generation of electricity, with lesser contributions by coal, hydroenergy, and nuclear power.

Although the assumptions concerning population growth, per-capita energy consumption, and total world energy consumption at the end of the 21st century are the same in Scenario 5 and in Scenario 2, the contribution of the various energy sources, except for oil, is different in 2100: higher for natural gas, hydroenergy, nuclear power, and the other possible sources of energy. Hydrogen, which is not contemplated as a source of energy anytime during the 21st century in Scenario 2, is predicted to enter the energy market late in the century in Scenario 5.

The assumptions concerning population growth and distribution and per-capita energy consumption, as well as the estimates of the share of the different sources of energy at the end of the 21st century used in the preparation of the five scenarios, are summarized in Table 17.

Where Do We Go from Here?

Assembling the basic historical information for this study, and particularly devising the five scenarios of energy consumption and probable sources of supply, lead one inevitably to give thought to what the living conditions of the human population of planet Earth may be during the 21st century and beyond. How will the future unfold? What should governments, industries, institutions of higher learning and research, and the people do concerning consumption and supply of

Table 17. Assumptions and Estimates in the Year 2100.

	Scenarios				
	1	**2**	**3**	**4**	**5**
Population (Billions)					
Total world	10	11	12	9	11
Developing countries	9	9.9	10.8	8.1	9.9
Developed countries	1	1.1	1.2	0.9	1.1
Per-capita Energy Consumption (BOE/p/yr)					
Developing countries	5	7.5	7.5	7.5	7.5
Developed countries	30	30	30	30	30
Total world energy consumption (billion BOE)	75	107	117	88	107
Percent of World Energy Consumption					
Oil + NGL + tar sands	24	22.5	18	19	22.5
Natural gas	33	21	25.5	34	28
Coal	28	38.5	36	8	19
Hydropower	5.5	5.5	6	5.5	6
Nuclear power	4	7	9	8	9
Hydrogen	–	–	–	14.5	7
Other	5.5	5.5	5.5	11	8.5

NGL = natural-gas liquids.

energy to make these conditions better? Some obvious suggestions come to mind:

- Encourage the reduction of the rate of growth of the human population, particularly in the countries of the Third World, where a high percentage of the inhabitants now live in poverty. Not an easy task. Uncontrolled population growth in many countries is perhaps the number-one problem that the world faces today. Overpopulation breeds poverty, hunger, unemployment, wars, and pestilence; it increases the faster depletion of the nonrenewable sources of energy and harms the environment.

- Promote better use of the abundant energy resources now available by encouraging conservation of transport fuels and electricity and more efficient, less wasteful use of energy in general through: more efficient vehicles; better, more available public transportation; and cities planned to make public transportation more convenient and acceptable.

- Encourage and finance research on technologies that make possible the efficient and environmentally less harmful consumption of the presently used energy sources, as well as of sources little used or not used at present: conversion of coal to gaseous or liquid fuels, conversion of gas to liquids, use of hydrogen for transport fuels and electricity

generation, even the development of fusion nuclear power, and of the recovery of gas from hydrates.
- Encourage the comeback of safe nuclear power and the full development of hydroelectric power.
- Promote wider recognition of the situation that we face and acceptance of the steps to be taken.

Some of these suggestions could be facilitated and hastened by the proper governmental policies—perhaps taxes on energy consumption? Effective education in the developing countries would also help enormously.

Over the long range, when oil and natural gas become scarcer and their price higher, entire life patterns will have to change. Coal and hydrogen may become the main sources of transportation fuel, since other energy sources can be used only to generate electricity. Over the even longer range, we have to believe that new technology will resolve the current obstacles for the production and use of coal and hydrogen, and that these two, along with nuclear and hydroelectric power, will become the principal energy sources in the world.

Appendix Tables

Appendix Table 1. Population of the world, developed countries, and developing countries (millions at midyear).

Year	World	Annual Increase	Developed	%	Developing	%
1950	2519.49		834.13	33.11	1685.36	66.89
1951	2565.71	46.22	843.84	32.89	1721.87	67.11
1952	2612.01	46.30	854.37	32.71	1757.64	67.29
1953	2658.74	46.73	865.12	32.54	1793.62	67.46
1954	2706.21	47.47	876.19	32.38	1830.02	67.62
1955	2754.72	48.51	887.67	32.22	1867.05	67.78
1956	2804.50	49.78	899.07	32.06	1905.43	67.94
1957	2855.78	51.28	910.62	31.89	1945.16	68.11
1958	2908.73	52.95	922.23	32.83	1986.50	67.17
1959	2963.49	54.76	934.18	31.52	2029.31	68.48
1960	3020.18	56.69	945.99	31.32	2074.19	68.68
1961	3078.86	58.68	957.72	31.11	2121.14	68.89
1962	3139.57	60.71	969.75	30.89	2169.82	69.11
1963	3202.32	62.75	983.62	30.71	2218.70	69.29
1964	3267.06	64.74	995.09	30.46	2271.97	69.54
1965	3333.72	66.66	1005.87	30.17	2327.85	69.83
1966	3402.25	68.53	1016.28	29.87	2385.97	70.13
1967	3472.52	70.27	1025.63	29.53	2446.89	70.47
1968	3544.27	71.75	1036.09	29.23	2508.18	70.77
1969	3617.18	72.91	1044.68	28.88	2572.50	71.12
1970	3690.92	73.74	1052.74	28.52	2638.18	71.48
1971	3765.51	74.59	1062.43	28.21	2703.08	71.79
1972	3840.80	75.29	1072.51	27.92	2768.29	72.08
1973	3916.29	75.49	1081.56	27.62	2834.73	72.38
1974	3991.33	75.04	1090.81	27.33	2900.52	72.67
1975	4065.51	74.18	1099.70	27.05	2965.81	72.95
1976	4138.58	73.07	1108.03	26.77	3030.55	73.23
1977	4210.78	72.20	1115.88	26.50	3094.90	73.50
1978	4282.78	72.00	1123.51	26.23	3159.27	73.77
1979	4355.53	72.75	1131.10	25.97	3224.43	74.03
1980	4429.75	74.22	1138.72	25.71	3291.03	74.29
1981	4505.50	75.75	1146.27	25.44	3359.23	74.56
1982	4582.66	77.16	1153.52	25.17	3429.14	74.83
1983	4661.43	78.77	1160.99	24.91	3500.44	75.09
1984	4742.03	80.60	1168.37	24.64	3573.66	75.36
1985	4824.51	82.48	1175.84	24.37	3648.67	75.63
1986	4909.05	84.54	1184.10	24.12	3724.95	75.88
1987	4995.39	86.34	1190.61	23.83	3804.78	76.17
1988	5082.60	87.21	1198.32	23.58	3884.28	76.42
1989	5169.38	86.78	1206.17	23.33	3963.21	76.67
1990	5254.82	85.44	1214.22	23.11	4040.60	76.89

Copyright ©2005 by The American Association of Petroleum Geologists.
DOI:10.1306/1032711SP226

Appendix Table 1. Population of the world, developed countries, and developing countries (millions at midyear) (cont.).

Year	World	Annual Increase	Developed	%	Developing	%
1991	5338.53	83.71	1220.56	22.86	4117.97	77.14
1992	5420.68	82.15	1226.72	22.63	4193.96	77.37
1993	5501.61	80.93	1232.87	22.41	4268.74	77.59
1994	5581.86	80.25	1238.38	22.19	4343.48	77.81
1995	5661.86	80.00	1243.72	21.97	4418.14	78.03
1996	5741.67	79.81	1248.20	21.74	4493.47	78.26
1997	5821.13	79.46	1252.53	21.52	4568.60	78.48
1998	5900.18	79.05	1256.73	21.30	4643.45	78.70
1999	5978.73	78.55	1260.43	21.08	4718.30	78.92
2000	6056.71	77.98	1264.33	20.87	4792.38	79.13

Source: United Nations Population Division.

Appendix Table 2. Population of developed countries (millions at midyear).

Year	Total	United States	Total Minus the United States	Canada	Australia	New Zealand	Japan	Western Europe	Eastern Europe	Former Soviet Union
1950	834.13	157.81	676.32	13.74	8.22	1.91	83.62	283.70	104.88	180.98
1951	843.84	160.02	683.82	14.13	8.40	1.95	85.12	285.07	105.95	183.20
1952	854.37	162.51	691.86	14.52	8.59	1.99	86.47	286.83	107.06	186.40
1953	865.12	165.23	699.89	14.91	8.79	2.04	87.70	288.73	108.22	189.50
1954	876.19	168.11	708.08	15.31	8.99	2.09	88.81	290.73	109.45	192.70
1955	887.67	171.07	716.60	15.74	9.20	2.14	89.81	293.53	110.80	196.15
1956	899.07	174.09	724.98	16.17	9.41	2.18	90.74	294.94	111.89	199.65
1957	910.62	177.13	733.49	16.62	9.62	2.23	91.60	297.41	112.86	203.15
1958	922.23	180.16	742.07	17.07	9.84	2.27	92.42	299.72	113.89	206.86
1959	934.18	183.18	751.00	17.50	10.06	2.32	93.25	302.35	115.02	210.50
1960	945.99	186.16	759.83	17.91	10.28	2.37	94.10	305.54	116.04	214.33
1961	957.72	189.09	768.63	18.28	10.50	2.42	94.98	307.69	116.61	218.15
1962	969.75	191.95	777.80	18.63	10.72	2.48	95.92	311.06	117.49	221.50
1963	983.62	194.71	788.91	18.96	10.95	2.53	96.89	316.36	118.12	225.10
1964	995.09	197.34	797.75	19.31	11.17	2.58	97.88	319.32	119.34	228.15
1965	1005.87	199.80	806.07	19.68	11.39	2.63	98.88	323.04	120.29	230.94
1966	1016.28	202.01	814.27	20.08	11.61	2.67	99.88	325.46	121.04	233.53
1967	1025.63	204.19	821.44	20.50	11.83	2.70	100.89	327.69	121.84	235.99
1968	1036.09	206.19	829.90	20.93	12.05	2.74	101.94	329.81	124.13	238.30
1969	1044.68	208.15	836.53	21.34	12.28	2.77	103.08	331.46	125.09	240.51
1970	1052.74	210.11	842.63	21.72	12.53	2.82	104.33	333.34	125.51	242.78
1971	1062.43	212.10	850.33	22.05	12.81	2.87	105.70	335.41	126.39	245.10
1972	1072.51	214.11	858.40	22.34	13.11	2.93	107.18	338.11	127.23	247.50
1973	1081.56	216.14	865.42	22.61	13.40	2.99	108.69	339.88	128.05	249.80
1974	1090.81	218.16	872.65	22.88	13.67	3.05	110.16	341.68	129.01	252.20
1975	1099.70	220.16	879.54	23.14	13.90	3.08	111.52	343.05	129.96	254.50
1976	1108.03	222.15	885.88	23.42	14.07	3.10	112.76	344.81	130.92	256.80
1977	1115.88	224.13	891.75	23.69	14.21	3.11	113.88	345.93	131.93	259.00
1978	1123.51	226.14	897.37	23.98	14.31	3.11	114.91	347.04	132.82	261.20
1979	1131.10	228.22	902.88	24.25	14.43	3.11	115.87	348.17	133.65	263.40
1980	1138.72	230.41	908.31	24.52	14.57	3.11	116.81	349.32	134.47	265.49
1981	1146.27	232.71	913.56	24.77	14.74	3.13	117.71	350.40	135.21	267.60
1982	1153.52	235.11	918.41	25.02	14.94	3.16	118.57	351.21	135.61	269.90
1983	1160.99	237.58	923.41	25.27	15.17	3.19	119.38	351.82	136.18	272.40

Appendix Table 2. Population of developed countries (millions at midyear) (cont.).

Year	Total	United States	Total Minus the United States	Canada	Australia	New Zealand	Japan	Western Europe	Eastern Europe	Former Soviet Union
1984	1168.37	240.07	928.30	25.54	15.40	3.22	120.14	352.40	136.70	274.90
1985	1175.84	242.53	933.31	25.84	15.64	3.25	120.84	353.06	137.18	277.34
1986	1183.10	244.95	938.15	26.18	15.88	3.27	121.47	353.98	137.59	279.78
1987	1190.61	247.35	943.26	26.55	16.13	3.29	122.05	354.97	138.05	282.22
1988	1198.32	249.75	948.57	26.94	16.39	3.31	122.58	356.24	138.45	284.66
1989	1206.17	252.22	953.95	27.33	16.64	3.33	123.08	357.90	138.57	287.10
1990	1214.22	254.78	959.44	27.70	16.89	3.36	123.54	360.15	138.42	289.54
1991	1220.56	257.43	963.13	28.06	17.13	3.40	123.97	378.00*	122.14*	290.43
1992	1226.72	260.18	966.54	28.40	17.37	3.45	124.37	379.69	121.94	291.32
1993	1232.87	262.99	969.88	28.73	17.61	3.50	124.75	381.38	121.69	292.22
1994	1238.38	265.85	972.53	29.04	17.84	3.56	125.11	382.99	121.51	292.48
1995	1243.72	268.74	974.98	29.35	18.07	3.60	125.47	384.44	121.31	292.74
1996	1248.20	271.67	976.53	29.65	18.29	3.65	125.83	385.53	121.20	292.38
1997	1252.53	274.61	977.92	29.94	18.51	3.68	126.17	386.47	121.13	292.02
1998	1256.73	277.55	979.18	30.22	18.72	3.74	126.51	387.24	121.09	291.66
1999	1260.43	280.43	980.00	30.49	18.93	3.75	126.82	387.85	121.03	291.13
2000	1264.33	283.23	981.10	30.76	19.14	3.78	127.10	388.42	120.96	290.94

Source: United Nations Population Division.
*East Germany reunited with West Germany.

Appendix Table 3. Population of the Western European countries (millions at midyear).

Country	1950	1955	1960	1965	1970	1975	1980	1985	1990	1995	2000
Austria	6.93	6.97	7.05	7.25	7.39	7.52	7.55	7.58	7.73	8.05	8.08
Belgium	8.64	8.87	9.15	9.46	9.66	9.80	9.86	9.86	9.97	10.14	10.25
Denmark	4.27	4.44	4.58	4.76	4.93	5.06	5.12	5.11	5.14	5.23	5.32
Finland	4.01	4.24	4.43	4.61	4.61	4.71	4.78	4.90	4.99	5.11	5.17
France	41.83	43.43	45.68	48.76	50.77	52.70	53.88	55.28	56.73	58.14	59.24
Federal Republic of Germany	47.85	50.17	53.22	59.04	60.65	61.83	61.56	60.97	63.23	81.66	82.02
Greece	7.57	7.97	8.33	8.55	8.79	9.05	9.64	9.93	10.16	10.45	10.61
Iceland	0.14	0.16	0.18	0.19	0.20	0.22	0.23	0.24	0.25	0.27	0.28
Ireland	2.97	2.92	2.83	2.88	2.95	3.18	3.40	3.54	3.51	3.64	3.80
Italy	47.10	48.63	50.20	52.11	53.82	55.44	56.43	56.59	56.72	57.30	57.53
Luxembourg	0.29	0.30	0.31	0.33	0.34	0.36	0.36	0.37	0.38	0.41	0.44
Netherlands	10.11	10.75	11.48	12.29	13.03	13.65	14.15	14.49	14.95	15.46	15.86
Norway	3.26	3.43	3.58	3.72	3.88	4.01	4.09	4.15	4.24	4.36	4.47
Portugal	8.40	8.61	8.86	9.00	8.68	9.09	9.77	10.01	9.90	9.92	10.02
Spain	28.01	29.20	30.45	32.06	33.78	35.60	37.54	38.47	39.30	39.74	39.91
Sweden	7.01	7.26	7.48	7.73	8.04	8.19	8.31	8.35	8.56	8.83	8.84
Switzerland	4.69	4.98	5.36	5.95	6.19	6.41	6.32	6.54	6.83	7.12	7.17
United Kingdom	50.62	51.20	52.37	54.35	55.63	56.23	56.33	56.68	57.56	58.61	59.41
Total	283.70	293.53	305.54	323.04	333.34	343.05	349.32	353.06	360.15	384.44	388.42

Source: United Nations Population Division.

Appendix Table 4. Population of developing countries (millions at midyear).

Year	Total	China	India	Africa	Latin America	Total of Four	Percent of Total
1950	1685.36	554.76	357.56	220.89	166.99	1300.20	77.15
1951	1721.87	568.73	364.13	225.63	171.40	1329.89	77.23
1952	1757.64	580.53	371.17	230.53	176.00	1358.23	77.27
1953	1793.62	590.82	378.69	235.58	180.76	1385.85	77.26
1954	1830.02	600.15	386.67	240.82	185.66	1413.30	77.23
1955	1867.05	609.00	395.10	246.26	190.69	1441.05	77.18
1956	1905.43	617.79	403.93	251.90	195.86	1469.48	77.12
1957	1945.16	626.82	413.12	257.76	201.17	1498.87	77.06
1958	1986.50	636.33	422.63	263.84	206.65	1529.45	76.99
1959	2029.31	646.51	432.38	270.15	212.32	1561.36	76.94
1960	2074.19	657.49	442.34	276.69	218.21	1594.73	76.88
1961	2121.14	669.39	452.48	283.46	224.31	1629.64	76.83
1962	2169.82	682.36	462.78	290.49	230.62	1666.25	76.79
1963	2218.70	696.55	473.29	297.76	237.10	1704.70	76.83
1964	2271.97	712.14	484.07	305.31	243.69	1745.21	76.81
1965	2327.85	729.19	495.16	313.12	250.35	1787.82	76.80
1966	2385.97	747.75	506.55	321.24	257.07	1832.61	76.81
1967	2446.89	767.67	518.22	329.64	263.85	1879.38	76.81
1968	2508.18	788.51	530.18	338.32	270.71	1927.72	76.86
1969	2572.50	809.67	542.41	347.22	277.67	1976.97	76.85
1970	2638.18	830.67	554.91	356.34	284.75	2026.67	76.82
1971	2703.08	851.42	567.69	365.65	291.94	2076.70	76.83
1972	2768.29	871.85	580.75	375.20	299.24	2127.04	76.84
1973	2834.73	891.63	593.99	385.06	306.65	2177.33	76.81
1974	2900.52	910.37	607.33	395.31	314.18	2227.19	76.79
1975	2965.81	927.81	620.70	406.03	321.82	2276.36	76.75
1976	3030.55	943.79	634.06	417.24	329.59	2324.68	76.71
1977	3094.90	958.44	647.45	428.93	337.46	2372.28	76.65
1978	3159.27	972.14	660.95	441.10	345.41	2419.60	76.59
1979	3224.43	985.47	674.72	453.75	353.37	2467.31	76.52
1980	3291.03	998.88	688.86	466.87	361.33	2515.94	76.45
1981	3359.23	1012.40	703.37	480.45	369.25	2565.47	76.37
1982	3429.14	1026.01	718.24	494.48	377.16	2615.89	76.28
1983	3500.44	1039.98	733.42	508.94	385.04	2667.38	76.20
1984	3573.66	1054.64	748.85	523.80	392.93	2720.22	76.12
1985	3648.67	1070.17	764.46	539.02	400.82	2774.47	76.04
1986	3724.95	1086.77	780.26	554.59	408.73	2830.35	75.98
1987	3804.78	1104.27	796.22	570.51	416.64	2887.64	75.89
1988	3884.28	1122.06	812.34	586.68	424.54	2945.62	75.83
1989	3963.21	1139.28	828.57	603.03	432.45	3003.33	75.78
1990	4040.60	1155.30	844.89	619.48	440.35	3060.02	75.73
1991	4117.97	1169.91	861.27	636.00	448.25	3115.43	75.65
1992	4193.96	1183.27	877.69	652.60	456.15	3169.71	75.58
1993	4268.74	1195.66	894.15	669.34	464.04	3223.19	75.51
1994	4343.48	1207.57	910.63	686.29	471.92	3276.41	75.43
1995	4418.14	1219.35	927.10	703.49	479.78	3329.72	75.36
1996	4493.47	1231.07	943.56	720.95	487.63	3383.21	75.29
1997	4568.60	1242.61	959.99	738.67	495.46	3436.73	75.22
1998	4643.45	1253.90	976.36	756.68	503.27	3490.21	75.16
1999	4718.30	1264.77	992.69	774.99	511.05	3543.50	75.10
2000	4792.38	1275.13	1008.94	793.63	518.81	3596.51	75.05

Source: United Nations Population Division.

Appendix Table 5. Energy consumption of the world, developed countries, and developing countries (million barrels of oil equivalent per year).

Year	World Total	Developed	%	Developing	%
1950	12,563.12	11,420.44	90.90	1142.68	9.10
1951	13,703.02	12,270.60	89.50	1432.42	10.50
1952	14,033.19	12,452.22	88.70	1580.97	11.30
1953	14,484.32	12,860.40	88.80	1623.92	11.20
1954	14,876.75	13,128.66	88.20	1748.09	11.80
1955	16,345.74	14,331.30	87.70	2014.44	12.30
1956	17,293.96	15,067.36	87.10	2226.60	12.90
1957	17,911.07	15,495.07	86.50	2416.00	13.50
1958	18,871.32	15,659.52	83.00	3211.80	17.00
1959	19,930.42	16,208.93	81.30	3721.49	18.70
1960	21,383.77	17,134.83	80.10	4248.94	19.90
1961	21,189.01	17,663.13	83.40	3525.88	16.60
1962	22,325.54	18,660.34	83.60	3665.20	16.40
1963	23,736.27	19,832.16	83.50	3904.11	16.50
1964	25,053.50	20,880.89	83.30	4172.61	16.70
1965	26,270.52	21,872.14	83.30	4398.38	16.70
1966	28,372.18	23,104.93	81.40	5267.25	18.60
1967	29,298.54	23,920.26	81.60	5378.28	18.40
1968	31,402.41	25,343.01	80.70	6059.94	19.30
1969	33,445.96	27,013.47	80.80	6432.49	19.20
1970	35,659.17	28,507.61	79.90	7151.56	20.10
1971	37,200.32	29,458.72	79.40	7741.16	20.80
1972	38,884.23	30,590.35	78.70	8293.88	21.30
1973	41,063.82	31,904.48	77.70	9159.34	22.30
1974	41,873.91	32,382.33	77.30	9491.58	22.70
1975	41,995.33	32,243.11	76.80	9752.22	23.20
1976	44,480.98	34,313.24	77.10	10,167.74	22.90
1977	46,026.68	34,939.89	75.90	11,086.79	24.10
1978	47,942.26	35,829.09	74.70	12,113.17	25.30
1979	49,394.40	36,814.59	74.50	12,579.81	25.50
1980	49,105.09	36,409.24	74.10	12,695.85	25.80
1981	48,631.84	35,333.39	72.60	13,298.45	27.30
1982	47,999.82	35,054.22	73.00	12,945.60	27.00
1983	47,462.85	35,247.51	74.30	12,215.34	25.70
1984	49,619.92	36,619.91	73.80	13,000.01	26.20
1985	51,163.99	36,953.52	72.20	14,210.47	27.80
1986	52,225.22	37,361.57	71.50	14,863.65	28.50
1987	54,287.23	38,436.93	70.80	15,850.30	29.20
1988	56,100.52	41,378.40	73.70	14,722.12	26.30
1989	57,290.55	41,697.32	72.80	15,593.23	27.20
1990	57,802.20	41,181.12	71.20	16,621.08	28.80
1991	59,718.98	41,042.91	68.70	18,676.07	31.30
1992	59,897.09	40,179.51	67.10	19,717.58	32.90
1993	60,267.99	40,140.11	66.60	20,127.88	33.40
1994	60,995.13	40,006.99	65.60	20,988.14	34.40
1995	62,389.29	40,303.28	64.60	22,086.01	35.40
1996	64,405.78	41,060.25	63.70	23,345.53	36.30
1997	64,803.06	40,739.56	62.90	24,063.50	37.10
1998	64,892.49	40,950.31	63.10	23,942.18	36.90
1999	65,118.25	41,518.89	63.80	23,599.36	36.20
2000	66,670.75	42,212.87	63.30	24,457.88	36.70
2001	66,884.78	42,067.88	62.90	24,816.69	37.10

Sources: United Nations Department of Economic and Social Affairs (*World Energy Supplies*, 1950–1978; *Yearbook of World Energy Statistics*, 1979–1981; *Energy Statistics Yearbook*, 1982–2001).

Appendix Table 6. Energy consumption of developed countries (million barrels of oil equivalent per year).

Year	Total	United States	Canada	Australia	New Zealand	Japan	Western Europe	Eastern Europe	Former Soviet Union	Developed Minus United States
1950	11,420.44	5737.26	350.67	121.85	18.56	220.01	2710.91	815.49	1445.69	5683.18
1951	12,270.60	6022.18	370.34	130.66	17.95	259.64	3021.86	884.35	1563.62	6248.42
1952	12,452.22	5984.15	376.51	134.20	19.86	276.87	3033.92	949.80	1676.91	6468.07
1953	12,860.40	6167.69	380.25	137.02	19.05	307.16	3029.38	1034.02	1785.83	6692.71
1954	13,128.66	5985.55	394.56	149.18	20.29	302.28	3188.28	1103.83	1984.69	7143.11
1955	14,331.30	6597.56	407.48	159.85	21.15	313.63	3433.95	1174.26	2223.42	7733.74
1956	15,067.36	6875.37	408.39	165.16	21.92	352.87	3639.16	1241.14	2363.35	8191.99
1957	15,495.07	6875.16	444.00	163.82	23.45	397.05	3671.07	1319.50	2601.02	8619.91
1958	15,659.52	6872.28	441.22	174.11	23.55	386.56	3608.76	1364.63	2788.41	8787.24
1959	16,208.93	7160.23	466.06	181.82	23.74	428.82	3603.38	1415.59	2929.29	9048.70
1960	17,134.83	7441.60	492.81	190.39	25.84	522.49	3894.22	1501.51	3065.97	9693.23
1961	17,663.13	7542.53	487.45	196.56	26.61	588.72	4027.46	1590.78	3203.02	10,120.60
1962	18,660.34	7870.77	526.27	204.89	26.80	663.10	4271.21	1688.48	3408.82	10,789.57
1963	19,832.16	8233.08	543.45	216.33	28.91	702.39	4616.49	1799.84	3691.67	11,599.08
1964	20,880.89	8575.89	621.22	237.29	30.01	776.15	4760.26	1910.55	3969.52	12,305.00
1965	21,872.14	8985.86	666.26	253.94	30.68	851.67	4938.14	1932.63	4212.96	12,886.28
1966	23,104.93	9593.21	700.62	260.69	33.88	928.15	5081.43	1975.40	4531.55	13,511.72
1967	23,920.26	9784.95	738.72	278.83	33.79	1101.11	5193.36	2004.76	4784.74	14,135.31
1968	25,343.01	10,390.77	789.59	294.05	34.17	1244.22	5524.88	2109.62	4955.71	14,952.24
1969	27,013.47	10,976.32	844.35	310.61	35.56	1441.11	5949.00	2257.03	5199.49	16,037.15
1970	28,507.61	11,379.70	918.58	325.69	38.95	1591.06	6461.54	2378.88	5413.21	17,127.91
1971	29,458.72	11,631.28	982.66	339.52	40.61	1741.91	6544.81	2472.10	5705.83	17,827.44
1972	30,590.35	12,186.18	1026.20	353.88	44.49	1809.15	6854.50	2520.27	5795.68	18,404.17
1973	31,904.48	12,620.34	1072.08	381.06	48.67	2042.90	7250.62	2597.02	5891.79	19,284.14
1974	32,382.33	12,867.10	1119.14	382.88	50.72	2014.91	7122.34	2682.19	6143.05	19,515.23
1975	32,243.11	12,547.50	1129.26	392.80	49.18	1990.83	6878.54	2813.03	6441.97	19,695.61
1976	34,313.24	13,216.70	1140.03	402.86	53.36	2052.99	7373.54	2953.18	7120.58	21,096.54
1977	34,939.89	13,505.50	1178.74	418.84	56.15	2083.63	7262.34	3061.52	7373.17	21,434.39
1978	35,829.09	13,876.40	1232.39	417.44	53.80	2150.99	7380.80	3026.63	7690.64	21,952.69
1979	36,814.59	13,990.00	1275.93	435.69	49.55	2214.69	7857.98	3061.30	7929.45	22,824.59
1980	36,409.24	13,457.90	1290.01	459.74	51.24	2192.04	7614.62	3201.23	8142.46	22,951.34
1981	35,333.39	12,776.20	1236.86	467.51	51.60	2121.67	7332.71	3118.40	8228.44	22,557.19
1982	35,054.22	12,162.70	1217.81	488.40	58.13	2102.61	7386.44	3194.78	8443.35	22,891.52
1983	35,247.51	12,085.00	1210.99	489.94	59.45	2162.79	7478.65	3186.06	8574.63	23,162.51
1984	36,619.91	12,732.90	1269.04	508.19	64.72	2347.95	7570.94	3262.95	8863.22	23,887.01
1985	36,953.52	12,746.90	1291.47	555.39	69.56	2344.21	7771.56	3271.30	9173.13	24,206.62
1986	37,361.57	12,779.10	1281.28	542.49	72.64	2282.93	7724.13	3350.18	9328.82	24,582.47
1987	38,436.93	13,277.60	1318.59	568.81	77.04	2284.61	7787.90	3415.41	9706.97	25,159.33
1988	41,378.40	13,941.70	1525.01	581.78	89.43	2699.49	8714.71	3351.42	10,474.86	27,436.70
1989	41,697.32	14,130.80	1557.48	615.87	94.19	2749.12	9030.05	3277.90	10,241.91	27,566.52
1990	41,181.12	14,163.00	1497.66	652.37	94.19	2894.84	9123.72	2909.79	9845.55	27,018.12
1991	41,042.91	14,290.60	1504.85	638.22	102.55	3000.17	9840.60	2084.80	9581.12	26,752.31
1992	40,179.51	14,488.50	1539.45	655.30	105.70	3023.92	9820.07	1905.87	8680.70	25,691.01
1993	40,140.11	14,813.20	1606.08	675.53	103.13	3041.51	9745.75	1947.51	8207.40	25,326.91
1994	40,006.99	15,099.10	1647.78	696.42	105.18	3190.60	9677.52	1901.91	7688.48	24,907.89
1995	40,303.28	15,406.20	1665.45	716.36	104.38	3277.90	9892.35	1962.90	7277.74	24,897.08
1996	41,060.25	15,925.90	1707.74	751.18	110.98	3363.59	10,285.02	2048.88	6866.96	25,134.35
1997	40,739.56	16,035.10	1738.24	740.70	114.13	3376.27	10,157.77	1993.54	6583.81	24,704.46
1998	40,950.31	16,096.70	1743.44	773.61	113.25	3386.53	10,397.24	1910.56	6528.98	24,853.61
1999	41,518.89	16,418.50	1794.38	772.58	120.21	3428.80	10,491.06	1916.06	6577.30	25,100.39
2000	42,212.87	16,766.60	1801.71	792.37	123.14	3481.50	10,655.25	1914.60	6677.70	25,446.27
2001	42,067.88	16,399.40	1792.92	805.57	121.68	3471.30	10,776.93	1940.98	6759.10	25,668.48

Sources: United Nations Department of Economic and Social Affairs (*World Energy Supplies*, 1950–1978; *Yearbook of World Energy Statistics*, 1979–1981; *Energy Statistics Yearbook*, 1982–2001).

Appendix Table 7. Energy consumption of developing countries (million barrels of oil equivalent per year).

Year	Total	China	India	Africa	Latin America	Total of Four	Percent of Total
1950	1142.68	232.85	178.84	207.39	322.26	941.34	82.40
1951	1432.42	268.19	188.14	219.40	367.07	1042.80	72.80
1952	1580.97	335.89	197.45	239.89	397.11	1170.34	74.00
1953	1623.92	352.14	201.62	242.69	395.82	1192.27	73.40
1954	1748.09	427.67	207.49	254.94	424.68	1314.78	75.20
1955	2014.44	508.65	221.87	283.13	467.90	1481.55	73.50
1956	2226.60	576.84	228.03	296.45	503.38	1604.70	72.10
1957	2416.00	682.47	256.37	312.22	551.80	1802.90	74.60
1958	3211.80	950.00	267.04	324.32	584.80	2126.16	66.20
1959	3721.49	1218.00	283.74	328.28	611.33	2441.35	65.60
1960	4248.94	1487.33	310.11	347.85	647.75	2793.04	65.70
1961	3525.88	1335.87	336.41	356.09	682.49	2710.86	76.90
1962	3665.20	1331.74	370.29	362.90	724.78	2789.71	76.10
1963	3904.11	1437.84	396.97	382.67	746.45	2963.93	75.90
1964	4172.61	1542.15	385.99	417.62	800.42	3146.18	75.40
1965	4398.38	1596.72	419.86	445.11	841.95	3303.64	75.10
1966	5267.25	1758.73	427.00	462.36	903.93	3552.02	67.40
1967	5378.28	1737.39	434.00	467.55	965.70	3604.64	67.00
1968	6059.94	1726.20	441.00	495.46	1042.23	3704.89	61.10
1969	6432.49	1874.30	448.00	514.74	1126.97	3964.01	61.60
1970	7151.56	2099.26	455.00	549.28	1228.26	4331.80	60.60
1971	7741.16	2054.00	461.86	602.16	1328.78	4446.80	57.40
1972	8293.88	2009.00	485.54	614.11	1397.98	4506.63	54.30
1973	9159.34	1964.00	493.97	654.93	1520.92	4633.82	50.60
1974	9491.58	1988.33	528.71	683.01	1574.70	4774.75	50.30
1975	9752.22	2189.10	559.35	737.40	1612.60	5098.45	52.30
1976	10,167.74	2334.38	585.23	788.56	1683.04	5390.83	53.00
1977	11,086.79	2611.68	616.75	825.80	1788.74	5842.97	52.70
1978	12,113.17	2914.92	632.21	870.88	1936.07	6354.08	52.50
1979	12,579.81	2991.23	673.92	927.32	2046.61	6639.08	52.80
1980	12,695.85	2837.08	714.60	1007.51	2161.25	6720.44	52.90
1981	13,298.45	2802.55	750.66	1069.32	2162.72	6782.25	51.00
1982	12,945.60	2990.71	808.43	1122.44	2208.53	7130.11	55.10
1983	12,215.34	3146.62	875.13	1184.75	2218.64	7425.14	60.80
1984	13,000.01	3447.96	906.57	1187.39	2253.53	7795.45	60.00
1985	14,210.47	3734.41	985.15	1209.38	2289.89	8218.83	57.80
1986	14,863.65	3934.38	1059.48	1249.76	2361.43	8605.05	57.90
1987	15,850.30	4103.70	1119.00	1285.31	2464.13	8972.14	56.60
1988	14,722.12	4462.80	1219.12	1319.47	2559.05	9560.44	64.90
1989	15,593.23	4551.42	1324.60	1415.63	2614.32	9905.97	63.50
1990	16,621.08	4584.25	1381.34	1446.43	2721.04	10,133.06	61.00
1991	18,676.07	4787.30	1485.86	1475.23	2840.74	10,589.13	56.70
1992	19,717.58	4991.66	1560.63	1458.45	2963.67	10,974.41	55.70
1993	20,127.88	5189.93	1632.32	1462.85	3067.24	11,352.34	56.40
1994	20,988.14	5605.98	1771.59	1524.57	3257.67	12,159.81	57.90
1995	22,086.01	6008.33	1847.16	1622.79	3330.46	12,808.74	58.00
1996	23,345.53	6289.73	2057.16	1667.87	3414.68	13,429.44	57.50
1997	24,063.50	6202.57	2085.82	1708.26	3619.26	13,615.91	56.60
1998	23,942.18	5845.38	2168.80	1742.85	3593.31	13,350.34	55.80
1999	23,599.36	5765.80	2220.70	1789.40	3651.00	13,426.90	56.90
2000	24,457.88	5898.40	2338.50	1829.70	3782.20	13,848.80	56.60
2001	24,816.69	6155.00	2348.70	1856.80	3747.70	14,108.20	56.80

Sources: United Nations Department of Economic and Social Affairs (*World Energy Supplies*, 1950–1978; *Yearbook of World Energy Statistics*, 1979–1981; *Energy Statistics Yearbook*, 1982–2001).

Appendix Table 8. World energy consumption by source (million barrels of oil equivalent per year).

Year	Coal	%	Oil	%	Natural Gas	%	Hydroenergy	%	Nuclear	%
1950	6971.80	55.50	3809	30.30	1242.80	9.90	617.00	4.90		
1951	6981.90	50.90	4276	31.20	1473.40	10.70	673.20	4.90		
1952	6948.60	49.50	4504	32.10	1594.10	11.40	718.90	5.10		
1953	7310.00	50.50	4793	33.10	1713.60	11.80	743.40	5.10		
1954	7311.50	49.10	4943	33.20	1795.90	12.10	787.50	5.30		
1955	7892.60	48.30	5526	33.80	1860.70	11.40	851.20	5.20		
1956	8331.10	48.20	6065	35.10	2003.60	11.60	927.00	5.40		
1957	8482.30	47.40	6399	35.70	2185.70	12.20	990.00	5.50		
1958	8815.00	46.70	6733	35.70	2376.20	12.60	1098.00	5.80		
1959	9193.00	46.10	7068	35.50	2656.00	13.30	1142.80	5.70		
1960	9780.10	45.70	7650	35.80	2916.70	13.60	1240.20	5.80		
1961	8986.30	42.40	8149	38.40	3170.30	15.00	1307.20	6.20		
1962	9193.00	41.20	8815	39.50	3464.10	15.50	1381.10	6.20	12.60	0.06
1963	9676.80	40.80	9477	39.90	3624.00	15.30	1444.90	6.10	19.30	0.10
1964	9949.00	39.70	10,217	40.80	3953.60	15.80	1509.80	6.00	27.00	0.10
1965	10,054.80	38.30	10,946	41.70	4230.50	16.10	1648.80	6.30	54.00	0.20
1966	10,276.60	36.20	11,938	42.10	4559.60	16.10	1779.80	6.30	62.30	0.20
1967	9671.80	33.00	12,822	43.80	5099.00	17.40	1820.50	6.20	75.40	0.30
1968	10,226.20	32.70	14,035	44.70	5639.00	18.00	1899.00	6.00	94.00	0.30
1969	10,274.20	30.70	15,062	45.00	6179.00	18.50	2028.10	6.10	111.90	0.30
1970	10,332.50	29.00	16,392	46.00	6719.00	18.80	2117.90	5.90	141.70	0.40
1971	10,432.00	28.00	17,479	47.00	7263.60	19.50	2214.40	5.90	195.20	0.50
1972	10,532.00	27.10	18,140	46.60	7791.50	20.00	2326.00	6.00	266.60	0.70
1973	10,624.40	25.90	20,153	49.10	7955.00	19.40	2374.20	5.80	356.80	0.90
1974	10,730.10	25.60	20,703	49.40	8159.30	19.50	2579.40	6.20	459.70	1.10
1975	10,817.30	25.80	19,381	46.10	8092.80	19.30	2625.30	6.20	633.20	1.50
1976	12,086.40	27.20	20,882	46.90	8379.60	18.80	2632.00	5.90	738.20	1.70
1977	12,340.40	26.80	21,728	47.20	8833.30	19.20	2725.20	5.90	914.40	2.00
1978	12,499.70	26.10	21,897	45.70	9017.90	18.80	2930.40	6.10	1071.00	2.20
1979	13,008.70	26.30	22,870	46.30	9390.70	19.00	3098.20	6.30	1118.20	2.30
1980	13,264.80	27.00	21,781	44.30	9608.40	19.60	3159.50	6.40	1226.20	2.50
1981	13,352.50	27.50	20,458	42.10	9668.30	19.90	3196.40	6.60	1441.10	3.00
1982	13,553.60	28.20	19,442	40.50	9636.90	20.10	3300.00	6.90	1566.50	3.30
1983	13,920.00	29.30	19,329	40.70	9494.10	20.00	3435.50	7.20	1812.60	3.80
1984	14,484.40	29.20	19,841	40.00	10,333.10	20.80	3550.50	7.10	2198.70	4.40
1985	15,314.50	29.90	19,567	38.20	10,814.00	21.10	3592.10	7.00	2607.10	5.10
1986	15,487.40	29.60	20,391	39.00	10,979.80	21.00	3641.60	7.00	2790.00	5.30
1987	16,219.70	29.90	20,465	37.70	11,753.10	21.60	3678.80	6.80	3069.00	5.60
1988	16,896.60	30.10	21,173	37.70	12,237.70	21.80	3800.00	6.80	3307.50	5.90
1989	16,940.40	29.60	21,762	38.00	12,797.40	22.30	3820.50	6.70	3414.60	6.00
1990	16,322.00	28.20	22,094	38.20	13,035.90	22.50	3975.70	6.90	3572.10	6.20
1991	16,046.80	26.90	21,879	36.60	13,183.30	22.10	4049.50	6.80	3724.20	6.20
1992	16,269.60	27.20	21,914	36.60	13,140.70	21.90	4061.50	6.80	3809.70	6.40
1993	16,158.20	26.80	21,837	36.20	13,255.50	22.00	4346.10	7.20	3879.00	6.40
1994	16,636.00	27.30	22,080	36.20	13,281.40	21.80	4306.90	7.10	3961.80	6.50
1995	17,125.90	27.40	22,455	36.00	13,529.10	21.70	4558.50	7.30	4082.00	6.50
1996	17,655.60	27.20	23,172	36.00	14,243.40	22.10	4586.30	7.10	4104.00	6.40
1997	17,386.50	26.80	23,896	36.90	14,010.70	21.60	4653.70	7.20	4082.40	6.30
1998	17,049.30	26.30	24,144	37.20	14,248.80	21.90	4705.70	7.20	4128.70	6.40
1999	16,176.40	24.80	23,615	36.30	14,307.30	22.00	4738.00	7.30	4288.50	6.60
2000	15,968.20	23.90	24,528	36.80	14,551.10	21.80	4861.90	7.30	4388.05	6.60
2001	16,880.00	25.20	24,342	36.40	15,110.80	22.60	4697.70	7.00	4506.80	6.70

Sources: United Nations Department of Economic and Social Affairs (*World Energy Supplies*, 1950–1978; *Yearbook of World Energy Statistics*, 1979–1981; *Energy Statistics Yearbook*, 1982–2001); U.S. Department of Energy, Energy Information Administration, *International Energy Annual*; *Oil & Gas Journal*.

Appendix Table 9. Per-Capita Energy Consumption (P-C EC): world.

Year	Energy Consumption (million BOE/yr)	Population (millions)	P-C EC (BOE/yr)
1950	12,563	2519	4.99
1951	13,703	2566	5.34
1952	14,033	2612	5.37
1953	14,484	2659	5.45
1954	14,877	2706	5.50
1955	16,346	2755	5.93
1956	17,294	2804	6.17
1957	17,911	2856	6.27
1958	18,871	2909	6.49
1959	19,930	2963	6.73
1960	21,384	3020	7.08
1961	21,189	3079	6.88
1962	22,326	3140	7.11
1963	23,736	3202	7.41
1964	25,053	3267	7.67
1965	26,271	3334	7.88
1966	28,372	3402	8.34
1967	29,299	3473	8.44
1968	31,402	3544	8.86
1969	33,446	3617	9.25
1970	35,659	3691	9.66
1971	37,200	3766	9.88
1972	38,884	3841	10.12
1973	41,064	3916	10.49
1974	41,874	3991	10.49
1975	41,995	4066	10.33
1976	44,481	4139	10.75
1977	46,027	4211	10.93
1978	47,942	4283	11.19
1979	49,394	4356	11.34
1980	49,105	4430	11.08
1981	48,632	4505	10.79
1982	48,000	4583	10.47
1983	47,463	4661	10.18
1984	49,620	4742	10.46
1985	51,164	4825	10.60
1986	52,225	4909	10.64
1987	54,287	4995	10.87
1988	56,101	5083	11.04
1989	57,291	5169	11.08
1990	57,802	5255	11.00
1991	59,719	5339	11.18
1992	59,897	5421	11.05
1993	60,268	5502	10.95
1994	60,995	5582	10.93
1995	62,389	5662	11.02
1996	64,406	5742	11.22
1997	64,803	5821	11.13
1998	64,892	5900	11.00
1999	65,118	5979	10.89
2000	66,671	6057	11.01

Source: Previous tables.

Appendix Table 10. Per-Capita Energy Consumption (P-C EC): developed countries.

Year	Energy Consumption (million BOE/yr)	Population (millions)	P-C EC (BOE/yr)
1950	11,420	834	13.69
1951	12,271	844	14.54
1952	12,452	854	14.58
1953	12,860	865	14.87
1954	13,129	876	14.99
1955	14,331	888	16.14
1956	15,067	899	16.76
1957	15,495	911	17.01
1958	15,660	922	16.98
1959	16,209	934	17.35
1960	17,135	946	18.11
1961	17,663	958	18.44
1962	18,660	970	19.24
1963	19,832	984	20.15
1964	20,881	995	20.99
1965	21,872	1006	21.74
1966	23,105	1016	22.74
1967	23,920	1026	23.31
1968	25,343	1036	24.46
1969	27,013	1045	25.85
1970	28,508	1053	27.07
1971	29,459	1062	27.74
1972	30,590	1073	28.51
1973	31,904	1082	29.49
1974	32,382	1091	29.68
1975	32,243	1100	29.31
1976	34,313	1108	30.97
1977	34,940	1116	31.31
1978	35,829	1124	31.88
1979	36,815	1131	32.55
1980	36,409	1139	31.96
1981	35,333	1146	30.83
1982	35,054	1154	30.38
1983	35,248	1161	30.36
1984	36,620	1168	31.35
1985	36,954	1176	31.42
1986	37,362	1184	31.56
1987	38,437	1191	32.27
1988	41,378	1198	34.54
1989	41,697	1206	34.57
1990	41,181	1214	33.92
1991	41,043	1221	33.61
1992	40,180	1227	32.75
1993	40,140	1233	32.55
1994	40,007	1238	32.32
1995	40,303	1244	32.40
1996	41,060	1248	32.90
1997	40,740	1253	32.51
1998	40,950	1257	32.58
1999	41,519	1260	32.95
2000	42,213	1264	33.40

Source: Previous tables.

Appendix Table 11. Per-Capita Energy Consumption (P-C EC): developing countries.

Year	Energy Consumption (million BOE/yr)	Population (millions)	P-C EC (BOE/yr)
1950	1143	1685	0.68
1951	1432	1722	0.83
1952	1581	1758	0.90
1953	1624	1794	0.90
1954	1748	1830	0.95
1955	2014	1867	1.08
1956	2227	1905	1.17
1957	2416	1945	1.24
1958	3212	1986	1.62
1959	3721	2029	1.83
1960	4249	2074	2.05
1961	3526	2121	1.66
1962	3665	2170	1.69
1963	3904	2219	1.76
1964	4173	2272	1.84
1965	4398	2328	1.89
1966	5267	2396	2.20
1967	5378	2447	2.20
1968	6060	2508	2.42
1969	6432	2572	2.50
1970	7152	2638	2.71
1971	7741	2703	2.86
1972	8294	2768	3.00
1973	9159	2835	3.23
1974	9492	2901	3.27
1975	9752	2966	3.29
1976	10,168	3031	3.35
1977	11,087	3095	3.58
1978	12,113	3159	3.83
1979	12,580	3224	3.90
1980	12,696	3291	3.86
1981	13,298	3359	3.96
1982	12,946	3429	3.77
1983	12,215	3500	3.49
1984	13,000	3574	3.64
1985	14,210	3649	3.89
1986	14,864	3725	3.99
1987	15,850	3805	4.16
1988	14,722	3884	3.79
1989	15,593	3963	3.93
1990	16,621	4041	4.11
1991	18,676	4118	4.53
1992	19,718	4194	4.70
1993	20,128	4269	4.71
1994	20,988	4343	4.83
1995	22,086	4418	5.00
1996	23,346	4493	5.20
1997	24,063	4569	5.27
1998	23,942	4643	5.16
1999	23,599	4718	5.00
2000	24,458	4792	5.10

Source: Previous tables.

Appendix Table 12. Per-Capita Energy Consumption (P-C EC): United States.

Year	Energy Consumption (million BOE/yr)	Population (millions)	P-C EC (BOE/yr)
1950	5737	158	36.31
1951	6022	160	37.64
1952	5984	163	36.71
1953	6168	165	37.38
1954	5986	168	35.63
1955	6598	171	38.58
1956	6875	174	39.51
1957	6875	177	38.84
1958	6872	180	38.18
1959	7160	183	39.12
1960	7442	186	40.01
1961	7543	189	39.91
1962	7871	192	40.99
1963	8233	195	42.22
1964	8576	197	43.53
1965	8986	200	44.93
1966	9593	202	47.49
1967	9785	204	47.97
1968	10,391	206	50.44
1969	10,976	208	52.77
1970	11,380	210	54.19
1971	11,631	212	54.86
1972	12,186	214	56.94
1973	12,620	216	58.43
1974	12,867	218	59.02
1975	12,547	220	57.03
1976	13,217	222	59.54
1977	13,505	224	60.29
1978	13,876	226	61.40
1979	13,990	228	61.36
1980	13,458	230	58.51
1981	12,776	233	54.83
1982	12,163	235	51.76
1983	12,085	238	50.78
1984	12,733	240	53.05
1985	12,747	243	52.46
1986	12,779	245	52.16
1987	13,278	247	53.76
1988	13,942	250	55.77
1989	14,131	252	56.07
1990	14,163	255	55.54
1991	14,291	257	55.61
1992	14,488	260	55.72
1993	14,813	263	56.32
1994	15,099	266	56.76
1995	15,406	269	57.27
1996	15,926	272	58.55
1997	16,035	275	58.31
1998	16,097	278	57.90
1999	16,418	280	58.64
2000	16,767	283	59.25

Source: Previous tables.

Appendix Table 13. Per-Capita Energy Consumption (P-C EC): Canada.

Year	Energy Consumption (million BOE/yr)	Population (millions)	P-C EC (BOE/yr)
1950	351	14.00	25.07
1951	370	14.00	26.43
1952	377	14.50	26.00
1953	380	15.00	25.33
1954	395	15.30	25.82
1955	407	15.70	25.92
1956	408	16.00	25.50
1957	444	16.60	26.75
1958	441	17.00	25.94
1959	466	17.50	26.63
1960	493	18.00	27.39
1961	487	18.30	26.61
1962	526	18.60	28.28
1963	543	19.00	28.58
1964	621	19.30	32.18
1965	666	19.70	33.81
1966	701	20.00	35.05
1967	739	20.50	36.05
1968	790	21.00	37.62
1969	844	21.30	39.62
1970	919	21.70	42.35
1971	983	22.00	44.68
1972	1026	22.30	46.01
1973	1072	22.60	47.43
1974	1119	22.90	48.86
1975	1129	23.10	48.87
1976	1140	23.40	48.72
1977	1179	23.70	49.75
1978	1232	24.00	51.33
1979	1276	24.20	52.73
1980	1290	24.50	52.65
1981	1237	24.80	49.88
1982	1218	25.00	48.72
1983	1211	25.30	47.86
1984	1269	25.50	49.76
1985	1291	25.80	50.04
1986	1281	26.20	48.89
1987	1319	26.50	49.77
1988	1525	27.00	56.48
1989	1557	27.30	57.03
1990	1498	27.70	54.08
1991	1505	28.00	53.75
1992	1539	28.40	54.19
1993	1606	28.70	55.96
1994	1648	29.00	56.83
1995	1665	29.30	56.83
1996	1708	29.60	57.70
1997	1738	30.00	57.93
1998	1743	30.20	57.71
1999	1794	30.50	58.82
2000	1802	30.80	58.51

Source: Previous tables.

Appendix Table 14. Per-Capita Energy Consumption (P-C EC): developed countries minus the United States and Canada.

Year	Energy Consumption (million BOE/yr)	Population (millions)	P-C EC (BOE/yr)
1950	5332	662.00	8.05
1951	5879	670.00	8.77
1952	6091	676.50	9.00
1953	6312	685.00	9.21
1954	6748	692.70	9.74
1955	7326	701.30	10.45
1956	7784	709.00	10.98
1957	8176	717.40	11.40
1958	8347	725.00	11.51
1959	8583	733.50	11.70
1960	9200	742.00	12.40
1961	9633	750.70	12.83
1962	10,263	759.40	13.51
1963	11,056	770.00	14.36
1964	11,684	778.70	15.00
1965	12,220	786.30	15.54
1966	12,811	794.00	16.13
1967	13,396	801.50	16.71
1968	14,162	809.00	17.50
1969	15,193	815.70	18.63
1970	16,209	821.30	19.74
1971	16,845	828.00	20.34
1972	17,378	836.70	20.77
1973	18,212	843.40	21.59
1974	18,396	850.10	21.64
1975	18,567	856.90	21.67
1976	19,956	862.60	23.13
1977	20,256	868.30	23.33
1978	20,721	874.00	23.71
1979	21,549	878.80	24.52
1980	21,661	884.50	24.49
1981	21,320	888.20	24.00
1982	21,673	894.00	24.24
1983	21,952	897.70	24.45
1984	22,618	902.50	25.06
1985	22,916	907.20	25.26
1986	23,302	912.80	25.53
1987	23,840	917.50	25.98
1988	25,911	921.00	28.13
1989	26,009	926.70	28.07
1990	25,518	931.30	27.40
1991	25,247	936.00	26.97
1992	24,153	938.60	25.73
1993	23,721	941.30	25.20
1994	23,260	943.00	24.67
1995	23,232	945.70	24.57
1996	23,426	946.40	24.75
1997	22,967	948.00	24.23
1998	23,110	948.80	24.36
1999	23,307	949.50	24.55
2000	23,644	950.20	24.88

Source: Previous tables.

Appendix Table 15. Per-Capita Energy Consumption (P-C EC): Australia.

Year	Energy Consumption (million BOE/yr)	Population (millions)	P-C EC (BOE/yr)
1950	122	8.20	14.88
1951	131	8.40	15.59
1952	134	8.60	15.58
1953	137	8.80	15.57
1954	149	9.00	16.55
1955	160	9.20	17.39
1956	165	9.40	17.55
1957	164	9.60	17.08
1958	174	9.80	17.75
1959	182	10.00	18.20
1960	190	10.30	18.45
1961	197	10.50	18.76
1962	205	10.70	19.16
1963	216	11.00	19.64
1964	237	11.20	21.16
1965	254	11.40	22.28
1966	261	11.60	22.50
1967	279	11.80	23.64
1968	294	12.00	24.50
1969	311	12.30	25.28
1970	326	12.50	26.08
1971	340	12.80	26.56
1972	354	13.10	27.02
1973	381	13.40	28.43
1974	383	13.70	27.96
1975	393	13.90	28.27
1976	403	14.10	28.58
1977	419	14.20	29.51
1978	417	14.30	29.16
1979	436	14.40	30.28
1980	460	14.60	31.51
1981	468	14.70	31.84
1982	488	15.00	32.53
1983	490	15.20	32.24
1984	508	15.40	32.99
1985	555	15.60	35.58
1986	542	15.90	34.09
1987	569	16.10	35.34
1988	582	16.40	35.49
1989	616	16.60	37.11
1990	652	16.90	38.58
1991	638	17.10	37.31
1992	655	17.40	37.64
1993	676	17.60	38.41
1994	696	17.80	39.10
1995	716	18.00	39.78
1996	751	18.30	41.04
1997	741	18.50	40.05
1998	774	18.70	41.39
1999	773	18.90	40.90
2000	792	19.10	41.47

Source: Previous tables.

Appendix Table 16. Per-Capita Energy Consumption (P-C EC): New Zealand.

Year	Energy Consumption (million BOE/yr)	Population (millions)	P-C EC (BOE/yr)
1950	19	1.90	10.00
1951	18	1.90	9.47
1952	20	2.00	10.00
1953	19	2.00	9.50
1954	20	2.10	9.52
1955	21	2.10	10.00
1956	22	2.20	10.00
1957	23	2.20	10.45
1958	23	2.30	10.00
1959	24	2.30	10.43
1960	26	2.40	10.83
1961	27	2.40	11.25
1962	27	2.50	10.80
1963	29	2.50	11.60
1964	30	2.60	11.54
1965	31	2.60	11.92
1966	34	2.70	12.59
1967	34	2.70	12.59
1968	34	2.70	12.59
1969	36	2.80	12.86
1970	39	2.80	13.93
1971	41	2.90	14.14
1972	44	2.90	15.17
1973	49	3.00	16.33
1974	51	3.00	17.00
1975	49	3.10	15.81
1976	53	3.10	17.10
1977	56	3.10	18.06
1978	54	3.10	17.42
1979	50	3.10	16.13
1980	51	3.10	16.45
1981	52	3.10	16.77
1982	58	3.20	18.12
1983	59	3.20	18.44
1984	65	3.20	20.31
1985	70	3.20	21.87
1986	73	3.30	22.12
1987	77	3.30	23.33
1988	89	3.30	26.97
1989	94	3.30	28.48
1990	94	3.40	27.65
1991	103	3.40	30.29
1992	106	3.40	31.18
1993	103	3.50	29.43
1994	105	3.60	29.17
1995	104	3.60	28.89
1996	111	3.60	30.83
1997	114	3.70	30.81
1998	113	3.70	30.54
1999	120	3.70	32.43
2000	123	3.80	32.37

Source: Previous tables.

Appendix Table 17. Per-Capita Energy Consumption (P-C EC): Japan

Year	Energy Consumption (million BOE/yr)	Population (millions)	P-C EC (BOE/yr)
1950	220	84.00	2.62
1951	260	85.00	3.06
1952	277	86.00	3.22
1953	307	88.00	3.49
1954	302	89.00	3.39
1955	314	90.00	3.49
1956	353	91.00	3.88
1957	397	92.00	4.31
1958	387	92.50	4.18
1959	429	93.00	4.61
1960	522	94.00	5.53
1961	589	95.00	6.20
1962	663	96.00	6.91
1963	702	97.00	7.24
1964	776	98.00	7.92
1965	852	99.00	8.61
1966	928	100.00	9.28
1967	1101	101.00	10.90
1968	1244	102.00	12.20
1969	1441	103.00	13.99
1970	1591	104.00	15.30
1971	1742	106.00	16.43
1972	1809	107.00	16.91
1973	2043	109.00	18.74
1974	2015	110.00	18.32
1975	1991	112.00	17.78
1976	2053	113.00	18.17
1977	2084	114.00	18.28
1978	2151	115.00	18.70
1979	2215	116.00	19.09
1980	2192	117.00	18.73
1981	2122	118.00	17.98
1982	2103	119.00	17.67
1983	2163	119.40	18.11
1984	2348	120.00	19.57
1985	2344	121.00	19.37
1986	2283	121.50	18.79
1987	2285	122.00	18.73
1988	2699	122.60	22.01
1989	2742	123.00	22.29
1990	2895	123.50	23.44
1991	3000	124.00	24.19
1992	3024	124.40	24.31
1993	3042	125.00	24.34
1994	3191	125.00	25.53
1995	3278	125.50	26.12
1996	3364	126.00	26.70
1997	3376	126.20	26.75
1998	3387	126.50	26.77
1999	3429	126.80	27.04
2000	3481	127.00	27.41

Source: Previous tables.

Appendix Table 18. Per-Capita Energy Consumption (P-C EC): Western Europe.

Year	Energy Consumption (million BOE/yr)	Population (millions)	P-C EC (BOE/yr)
1950	2711	283.00	9.58
1951	3022	285.00	10.60
1952	3034	287.00	10.57
1953	3029	289.00	10.48
1954	3188	291.00	10.95
1955	3434	293.00	11.72
1956	3639	295.00	12.33
1957	3671	297.00	12.36
1958	3609	300.00	12.03
1959	3603	302.00	11.93
1960	3894	305.00	12.77
1961	4027	308.00	13.07
1962	4271	311.00	13.73
1963	4616	316.00	14.61
1964	4760	319.00	14.92
1965	4938	322.00	15.33
1966	5081	325.00	15.63
1967	5193	328.00	15.83
1968	5525	330.00	16.74
1969	5949	331.00	17.97
1970	6462	333.00	19.40
1971	6545	335.00	19.54
1972	6854	338.00	20.28
1973	7251	340.00	21.33
1974	7122	342.00	20.82
1975	6879	343.00	20.05
1976	7374	345.00	21.37
1977	7262	346.00	20.99
1978	7381	347.00	21.27
1979	7858	348.00	22.58
1980	7615	349.00	21.82
1981	7333	350.00	20.95
1982	7386	351.00	21.04
1983	7479	352.00	21.25
1984	7571	352.00	21.51
1985	7772	353.00	22.02
1986	7724	354.00	21.82
1987	7788	355.00	21.94
1988	8715	356.00	24.48
1989	9030	358.00	25.22
1990	9124	360.00	25.34
1991	9841	378.00	26.03
1992	9820	380.00	25.84
1993	9746	381.00	25.58
1994	9678	383.00	25.27
1995	9892	384.00	25.76
1996	10,285	385.50	26.68
1997	10,158	386.00	26.32
1998	10,397	387.00	26.87
1999	10,491	388.00	27.04
2000	10,655	388.40	27.43

Source: Previous tables.

Appendix Table 19. Per-Capita Energy Consumption (P-C EC): Eastern Europe.

Year	Energy Consumption (million BOE/yr)	Population (millions)	P-C EC (BOE/yr)
1950	815	105.00	7.76
1951	884	106.00	8.34
1952	950	107.00	8.88
1953	1034	108.00	9.57
1954	1104	109.00	10.13
1955	1174	111.00	10.58
1956	1241	112.00	11.08
1957	1319	113.00	11.67
1958	1365	114.00	11.97
1959	1416	115.00	12.31
1960	1502	116.00	12.95
1961	1591	117.00	13.60
1962	1688	117.50	14.37
1963	1800	118.00	15.25
1964	1911	119.00	16.06
1965	1933	120.00	16.11
1966	1975	121.00	16.32
1967	2005	122.00	16.43
1968	2110	124.00	17.02
1969	2257	125.00	18.06
1970	2379	125.50	18.96
1971	2472	126.00	19.62
1972	2520	127.00	19.84
1973	2597	128.00	20.29
1974	2682	129.00	20.79
1975	2813	130.00	21.64
1976	2953	131.00	22.54
1977	3062	132.00	23.20
1978	3027	133.00	22.76
1979	3061	134.00	22.84
1980	3201	134.50	23.80
1981	3118	135.00	23.10
1982	3195	135.60	23.56
1983	3186	136.00	23.43
1984	3263	136.70	23.87
1985	3271	137.00	23.88
1986	3350	137.60	24.35
1987	3415	138.00	24.75
1988	3351	138.40	24.21
1989	3278	138.60	23.65
1990	2910	138.40	21.03
1991	2085	122.00	17.09
1992	1906	122.00	15.62
1993	1948	121.70	16.01
1994	1902	121.50	15.65
1995	1963	121.30	16.18
1996	2049	121.20	16.90
1997	1994	121.10	16.47
1998	1911	121.10	15.78
1999	1916	121.00	15.83
2000	1915	121.00	15.83

Source: Previous tables.

Appendix Table 20. Per-Capita Energy Consumption (P-C EC): Former Soviet Union.

Year	Energy Consumption (million BOE/yr)	Population (millions)	P-C EC (BOE/yr)
1950	1446	181.00	8.00
1951	1564	183.00	8.55
1952	1677	186.00	9.02
1953	1786	189.00	9.45
1954	1985	193.00	10.28
1955	2223	196.00	11.34
1956	2363	200.00	11.81
1957	2601	203.00	12.81
1958	2788	207.00	13.47
1959	2929	210.00	13.95
1960	3066	214.00	14.33
1961	3203	218.00	14.69
1962	3409	221.00	15.42
1963	3692	225.00	16.41
1964	3970	228.00	17.41
1965	4213	231.00	18.24
1966	4532	234.00	19.37
1967	4785	236.00	20.27
1968	4956	238.00	20.82
1969	5199	241.00	21.57
1970	5413	243.00	22.28
1971	5706	245.00	23.29
1972	5796	247.00	23.46
1973	5892	250.00	23.57
1974	6143	252.00	24.38
1975	6442	254.00	25.36
1976	7121	257.00	27.71
1977	7373	259.00	28.47
1978	7691	261.00	29.47
1979	7929	263.00	30.15
1980	8142	265.00	30.72
1981	8228	268.00	30.70
1982	8443	270.00	31.27
1983	8575	272.00	31.53
1984	8863	275.00	32.23
1985	9173	277.00	33.11
1986	9329	280.00	33.32
1987	9707	282.00	34.42
1988	10,475	285.00	36.75
1989	10,242	287.00	35.69
1990	9846	290.00	33.95
1991	9581	290.40	32.99
1992	8681	291.00	29.83
1993	8207	292.00	28.11
1994	7688	292.50	26.28
1995	7278	292.70	24.86
1996	6867	292.40	23.48
1997	6584	292.00	22.55
1998	6529	291.70	22.38
1999	6577	291.10	22.59
2000	6678	291.00	22.95

Source: Previous tables.

Appendix Table 21. Per-Capita Energy Consumption (P-C EC): China.

Year	Energy Consumption (million BOE/yr)	Population (millions)	P-C EC (BOE/yr)
1950	233	555	0.42
1951	268	569	0.47
1952	336	581	0.58
1953	352	591	0.60
1954	428	600	0.71
1955	509	609	0.84
1956	577	618	0.93
1957	682	627	1.09
1958	950	636	1.49
1959	1218	647	1.88
1960	1487	657	2.26
1961	1336	669	2.00
1962	1332	682	1.95
1963	1438	697	2.06
1964	1542	712	2.17
1965	1597	729	2.19
1966	1759	748	2.35
1967	1737	768	2.26
1968	1726	789	2.19
1969	1874	810	2.31
1970	2099	831	2.53
1971	2054	851	2.41
1972	2009	872	2.30
1973	1964	892	2.20
1974	1988	910	2.18
1975	2189	928	2.36
1976	2334	944	2.47
1977	2612	958	2.73
1978	2915	972	3.00
1979	2991	985	3.04
1980	2837	999	2.84
1981	2803	1012	2.77
1982	2991	1026	2.91
1983	3147	1040	3.03
1984	3448	1055	3.27
1985	3734	1070	3.49
1986	3934	1087	3.62
1987	4104	1104	3.72
1988	4463	1122	4.00
1989	4551	1139	4.00
1990	4584	1155	3.97
1991	4787	1170	4.09
1992	4992	1183	4.22
1993	5190	1196	4.34
1994	5606	1208	4.64
1995	6008	1219	4.93
1996	6290	1231	5.11
1997	6203	1243	4.99
1998	5845	1254	4.66
1999	5766	1265	4.56
2000	5898	1275	4.63

Source: Previous tables.

Appendix Table 22. Per-Capita Energy Consumption (P-C EC): India.

Year	Energy Consumption (million BOE/yr)	Population (millions)	P-C EC (BOE/yr)
1950	179	358	0.50
1951	188	364	0.52
1952	197	371	0.53
1953	202	379	0.53
1954	207	387	0.53
1955	222	395	0.56
1956	228	404	0.56
1957	256	413	0.62
1958	267	423	0.63
1959	284	432	0.66
1960	310	442	0.70
1961	336	452	0.74
1962	370	463	0.80
1963	397	473	0.84
1964	386	484	0.80
1965	420	495	0.85
1966	427	507	0.84
1967	434	518	0.84
1968	447	530	0.84
1969	448	542	0.83
1970	455	555	0.82
1971	462	568	0.81
1972	486	581	0.84
1973	494	594	0.83
1974	529	607	0.87
1975	559	621	0.90
1976	585	634	0.92
1977	617	647	0.95
1978	632	661	0.96
1979	674	675	1.00
1980	715	689	1.04
1981	751	703	1.07
1982	808	718	1.12
1983	875	733	1.19
1984	907	749	1.21
1985	985	764	1.29
1986	1059	780	1.36
1987	1119	796	1.40
1988	1219	812	1.50
1989	1325	829	1.60
1990	1381	845	1.63
1991	1486	861	1.73
1992	1561	878	1.78
1993	1632	894	1.82
1994	1772	911	1.94
1995	1847	927	1.99
1996	2057	944	2.18
1997	2086	960	2.17
1998	2169	976	2.22
1999	2221	993	2.24
2000	2338	1009	2.32

Source: Previous tables.

Appendix Table 23. Per-Capita Energy Consumption (P-C EC): Africa.

Year	Energy Consumption (million BOE/yr)	Population (millions)	P-C EC (BOE/yr)
1950	207	221	0.94
1951	219	226	0.97
1952	240	231	1.04
1953	243	236	1.03
1954	255	241	1.06
1955	283	246	1.15
1956	296	252	1.17
1957	312	258	1.21
1958	324	264	1.23
1959	328	270	1.21
1960	348	277	1.26
1961	356	283	1.26
1962	363	290	1.25
1963	383	298	1.28
1964	418	305	1.37
1965	445	313	1.42
1966	462	321	1.44
1967	468	330	1.42
1968	495	338	1.46
1969	515	347	1.48
1970	549	356	1.54
1971	602	366	1.64
1972	614	375	1.64
1973	655	385	1.70
1974	683	395	1.73
1975	737	406	1.81
1976	789	417	1.89
1977	826	429	1.92
1978	871	441	1.97
1979	927	454	2.04
1980	1008	467	2.16
1981	1069	480	2.23
1982	1122	494	2.27
1983	1185	509	2.33
1984	1187	524	2.26
1985	1209	539	2.24
1986	1250	555	2.25
1987	1285	571	2.25
1988	1319	587	2.25
1989	1416	603	2.35
1990	1446	617	2.34
1991	1475	636	2.32
1992	1458	653	2.23
1993	1463	669	2.19
1994	1525	686	2.22
1995	1623	703	2.31
1996	1668	721	2.31
1997	1708	739	2.31
1998	1743	757	2.30
1999	1789	775	2.31
2000	1830	794	2.30

Source: Previous tables.

Appendix Table 24. Per-Capita Energy Consumption (P-C EC): Latin America.

Year	Energy Consumption (million BOE/yr)	Population (millions)	P-C EC (BOE/yr)
1950	322	167	1.93
1951	367	171	2.15
1952	397	176	2.25
1953	396	181	2.19
1954	425	186	2.28
1955	468	191	2.45
1956	503	196	2.57
1957	552	201	2.75
1958	585	207	2.83
1959	611	212	2.88
1960	648	218	2.97
1961	682	224	3.04
1962	725	231	3.14
1963	746	237	3.15
1964	800	244	3.28
1965	842	250	3.37
1966	904	257	3.52
1967	966	264	3.66
1968	1042	271	3.84
1969	1127	278	4.05
1970	1228	285	4.31
1971	1329	292	4.55
1972	1398	299	4.67
1973	1521	307	4.95
1974	1575	314	5.02
1975	1613	322	5.01
1976	1683	330	5.10
1977	1789	337	5.31
1978	1936	345	5.61
1979	2047	353	5.80
1980	2161	361	5.99
1981	2163	369	5.86
1982	2209	377	5.86
1983	2219	385	5.76
1984	2254	393	5.73
1985	2290	401	5.71
1986	2361	409	5.77
1987	2464	417	5.91
1988	2559	425	6.02
1989	2614	432	6.05
1990	2721	440	6.18
1991	2841	448	6.34
1992	2964	456	6.50
1993	3067	464	6.61
1994	3258	472	6.90
1995	3330	480	6.94
1996	3415	488	7.00
1997	3619	495	7.31
1998	3593	503	7.14
1999	3651	511	7.14
2000	3782	519	7.29

Source: Previous tables.

Appendix Table 25. World population projections (billions).

Year	Low	Middle	High
1980	4.430	4.430	4.430
1985	4.820	4.820	4.820
1990	5.250	5.250	5.250
1995	5.660	5.660	5.660
2000	6.060	6.060	6.060
2005	6.400	6.475	6.500
2010	6.850	6.900	6.950
2015	7.130	7.250	7.400
2020	7.450	7.600	7.750
2025	7.700	7.900	8.150
2030	8.000	8.200	8.500
2035	8.250	8.500	8.850
2040	8.500	8.800	9.200
2045	8.700	9.075	9.500
2050	8.950	9.350	9.800
2055	9.150	9.600	10.075
2060	9.350	9.850	10.350
2065	9.500	10.050	10.600
2070	9.600	10.250	10.850
2075	9.675	10.400	11.100
2080	9.750	10.550	11.300
2085	9.820	10.675	11.475
2090	9.900	10.800	11.650
2095	9.950	10.900	11.725
2100	10.000	11.000	12.000

Appendix Table 26. Percent of the world population in the developed and developing countries.

Year	Developed	Developing Total	Africa	India	China	Latin America	Rest
1950	33.11	66.89	8.75	14.20	22.03	6.57	15.34
1955	32.22	67.78					
1960	31.32	68.68	9.16	14.64	22.20	6.75	15.93
1965	30.17	69.83					
1970	28.52	71.48	9.65	15.03	22.51	7.56	16.73
1975	27.05	72.95					
1980	25.71	74.29	10.54	15.55	22.55	7.86	17.79
1985	24.37	75.63					
1990	23.11	76.89	11.75	16.08	22.00	8.38	18.64
1995	21.97	78.03					
2000	20.87	79.13	13.10	16.66	21.05	8.59	19.73
2005	19.75	80.25					
2010	18.75	81.25	14.45	16.87	20.68	9.00	20.25
2015	17.75	82.25					
2020	16.75	83.25	15.95	16.99	20.26	9.25	20.80
2025	16.00	84.00					

Appendix Table 26. Percent of the world population in the developed and developing countries (cont.).

Year	Developed	Developing Total	Africa	India	China	Latin America	Rest
2030	15.25	84.75	17.75	17.05	19.20	9.50	21.25
2035	14.50	85.50					
2040	13.75	86.25	19.50	17.12	18.15	9.75	21.73
2045	13.25	86.75					
2050	12.75	87.25	21.30	17.20	17.00	10.00	21.75
2055	12.25	87.75					
2060	11.75	88.25	22.75	17.25	16.00	10.50	21.75
2065	11.50	88.50					
2070	11.25	88.75	24.25	17.50	15.00	10.25	21.75
2075	11.00	89.00					
2080	10.80	89.20	26.00	17.00	14.25	10.50	21.45
2085	10.60	89.40					
2090	10.40	89.60	27.50	16.75	13.40	10.60	21.35
2095	10.20	89.80					
2100	10.00	90.00	29.00	16.50	12.50	10.50	21.50

Source: Previous tables.

Appendix Table 27. Population projections: low case (billions).

Year	World Total	Developed	%	Developing	%
1980	4.43	1.14	25.70	3.29	74.30
1990	5.25	1.21	23.10	4.04	76.90
2000	6.06	1.26	20.90	4.79	79.10
2010	6.85	1.28	18.75	5.57	81.25
2020	7.45	1.25	16.75	6.20	83.25
2030	8.00	1.22	15.25	6.78	84.75
2040	8.50	1.17	13.75	7.33	86.25
2050	8.95	1.14	12.75	7.81	87.25
2060	9.35	1.10	11.75	8.25	88.25
2070	9.60	1.08	11.25	8.52	88.75
2080	9.75	1.05	10.80	8.70	89.20
2090	9.90	1.03	10.40	8.87	89.60
2100	10.00	1.00	10.00	9.00	90.00

Year	Africa	%	India	%	China	%	Latin America	%	Rest	%
1980	0.47	10.54	0.69	15.55	1.00	22.55	0.35	7.86	0.79	17.79
1990	0.62	11.79	0.84	16.08	1.15	22.00	0.44	8.38	0.98	18.64
2000	0.79	13.10	1.01	16.66	1.28	21.05	0.52	8.59	1.20	19.73
2025	1.31	16.75	1.33	17.00	1.54	19.70	0.74	9.50	1.64	21.00
2050	1.91	21.30	1.54	17.20	1.52	17.00	0.89	10.00	1.95	21.75
2075	2.45	25.25	1.67	17.25	1.42	14.60	1.00	10.30	2.12	21.90
2100	2.90	29.00	1.65	16.50	1.25	12.50	1.05	10.50	2.15	21.50

Source: Previous tables.

Appendix Table 28. Population projections: middle case (billions).

Year	World Total	Developed	%	Developing	%	Africa	%	India	%	China	%	Latin America	%	Rest	%
1950	2.52	0.83	33.11	1.68	66.89	0.22	8.75	0.36	14.20	0.55	22.03	0.17	6.57	0.39	15.34
1960	3.02	0.94	31.32	2.07	68.68	0.28	9.16	0.44	14.64	0.67	22.20	0.21	6.75	0.48	15.93
1970	3.69	1.05	28.52	2.64	71.48	0.36	9.65	0.55	15.03	0.83	22.51	0.28	7.56	0.62	16.73
1980	4.43	1.14	25.70	3.29	74.30	0.47	10.54	0.69	15.55	1.00	22.55	0.35	7.86	0.79	17.79
1990	5.25	1.21	23.10	4.04	76.90	0.62	11.79	0.84	16.08	1.15	22.00	0.44	8.38	0.98	18.64
2000	6.06	1.26	20.90	4.79	79.10	0.79	13.10	1.01	16.66	1.28	21.05	0.52	8.59	1.20	19.73
2010	6.90	1.29	18.75	5.61	81.25	1.00	14.45	1.16	16.87	1.43	20.68	0.62	9.00	1.40	20.25
2020	7.60	1.27	16.75	6.33	83.25	1.20	15.75	1.29	16.99	1.54	20.26	0.72	9.50	1.58	20.75
2030	8.20	1.25	15.25	6.95	84.75	1.45	17.75	1.40	17.05	1.57	19.20	0.78	9.50	1.74	21.25
2040	8.80	1.22	13.86	7.58	86.14	1.72	19.50	1.50	17.10	1.60	18.15	0.88	10.00	1.89	21.50
2050	9.35	1.19	12.75	8.16	87.25	1.99	21.30	1.61	17.20	1.59	17.00	0.93	10.00	2.03	21.75
2060	9.85	1.16	11.75	8.69	88.25	2.24	22.75	1.70	17.25	1.58	16.00	1.03	10.50	2.14	21.75
2070	10.25	1.15	11.25	9.10	88.75	2.49	24.25	1.79	17.50	1.54	15.00	1.05	10.25	2.23	21.75
2080	10.55	1.14	10.80	9.41	89.20	2.74	26.00	1.79	17.00	1.50	14.25	1.11	10.50	2.26	21.45
2090	10.80	1.12	10.40	9.68	89.60	2.97	27.50	1.81	16.75	1.45	13.40	1.14	10.60	2.30	21.35
2100	11.00	1.10	10.00	9.90	90.00	3.19	29.00	1.81	16.50	1.37	12.50	1.15	10.50	2.36	21.50

Source: Previous tables.

Appendix Table 29. Population projections: high case (billions).

Year	World Total	Developed	%	Developing	%
1980	4.43	1.14	25.70	3.29	74.30
1990	5.25	1.21	23.10	4.04	76.90
2000	6.06	1.26	20.90	4.79	79.10
2010	6.95	1.30	18.75	5.65	81.25
2020	7.75	1.30	16.75	6.45	83.25
2030	8.50	1.30	15.25	7.20	84.75
2040	9.20	1.26	13.75	7.94	86.25
2050	9.80	1.25	12.75	8.55	87.25
2060	10.35	1.22	11.75	9.13	88.25
2070	10.85	1.22	11.25	9.63	88.75
2080	11.30	1.22	10.80	10.08	89.20
2090	11.65	1.21	10.40	10.44	89.60
2100	12.00	1.20	10.00	10.80	90.00

Year	Africa	%	India	%	China	%	Latin America	%	Rest	%
1980	0.47	10.54	0.69	15.55	1.00	22.55	0.35	7.86	0.79	17.79
1990	0.62	11.79	0.84	16.08	1.15	22.00	0.44	8.38	0.98	18.64
2000	0.79	13.10	1.01	16.66	1.28	21.05	0.52	8.59	1.20	19.73
2025	1.36	16.75	1.38	17.00	1.60	19.70	0.77	9.50	1.71	21.00
2050	2.09	21.30	1.69	17.20	1.67	17.00	0.98	10.00	2.13	21.75
2075	2.80	25.25	1.91	17.25	1.62	14.60	1.14	10.30	2.43	21.90
2100	3.48	29.00	1.98	16.50	1.50	12.50	1.26	10.50	2.58	21.50

Source: Previous tables.

Appendix Table 30. United States population projections and percent of developed countries (millions).

Year	Low Case	%	Middle Case	%	High Case	%
1950	157.81	18.92	157.81	18.92	157.81	18.92
1955	171.07	19.27	171.07	19.27	171.07	19.27
1960	186.16	19.69	186.16	19.69	186.16	19.69
1965	199.80	19.86	199.80	19.86	199.80	19.86
1970	210.11	19.96	210.11	19.96	210.11	19.96
1975	220.16	20.02	220.16	20.02	220.16	20.02
1980	230.41	20.23	230.41	20.23	230.41	20.23
1985	242.53	20.63	242.53	20.63	242.53	20.63
1990	254.78	20.98	254.78	20.98	254.78	20.98
1995	268.74	21.61	268.74	21.61	268.74	21.61
2000	283.23	22.40	283.23	22.40	283.23	22.40
2010	305.00	23.83	308.56	23.92	312.00	24.00
2020	325.00	26.00	334.20	26.31	343.00	26.38
2030	345.00	28.28	358.49	28.68	372.00	28.61
2040	360.00	30.77	376.76	31.14	393.00	31.19
2050	375.00	32.89	397.06	33.37	419.00	33.52
2060	390.00	35.45	415.00	35.77	440.00	36.06
2070	403.00	37.31	435.00	37.83	467.00	38.28
2080	415.00	39.52	450.00	39.47	485.00	39.75
2090	425.00	41.26	465.00	41.52	505.00	41.73
2100	435.00	43.50	480.00	43.64	525.00	43.75

Source: Previous tables.

Appendix Table 31. Energy consumption projections: developed countries.

Year	Population (millions)	Case 1 P-C EC (BOE/yr)	Case 1 Consumption (million BOE/yr)	Case 2 P-C EC (BOE/yr)	Case 2 Consumption (million BOE/yr)
1990	1214	33.92	41,181	33.92	41,181
1995	1244	32.40	40,303	32.40	40,303
2000	1264	33.40	42,213	33.40	42,213
2005	1280	33.00	42,240	33.20	42,496
2010	1290	32.60	42,054	33.00	42,570
2015	1280	32.20	41,216	32.80	41,984
2020	1270	31.80	40,386	32.60	41,402
2025	1260	31.40	39,564	32.40	40,824
2030	1250	31.00	38,750	32.20	40,250
2035	1235	30.60	37,791	32.00	39,520
2040	1220	30.20	36,844	31.80	38,796
2045	1205	29.80	35,909	31.60	38,078
2050	1190	29.40	34,986	31.40	37,366
2055	1175	29.00	34,075	31.25	36,719
2060	1160	28.50	33,060	31.10	36,076
2065	1155	28.00	32,340	30.95	35,747
2070	1150	27.50	31,625	30.80	35,420
2075	1145	27.00	30,915	30.65	35,094
2080	1140	26.50	30,210	30.50	34,770
2085	1130	26.10	29,493	30.35	34,295
2090	1120	25.80	28,896	30.20	33,824
2095	1110	25.40	28,194	30.10	33,411
2100	1100	25.00	27,500	30.00	33,000

Appendix Table 32. Energy consumption projections: United States and Canada.

Year	Population (millions)	Case 1		Case 2	
		P-C EC (BOE/yr)	Consumption (million BOE/yr)	P-C EC (BOE/yr)	Consumption (million BOE/yr)
1990	282.70	55.40	15,660.66	55.40	15,660.66
1995	298.30	57.23	17,071.65	57.23	17,071.65
2000	313.80	59.17	18,568.31	59.17	18,568.31
2005	328.10	57.00	18,701.70	58.00	19,029.80
2010	341.78	56.00	19,139.68	56.90	19,447.28
2015	355.64	54.50	19,382.38	55.90	19,880.28
2020	369.80	53.00	19,599.40	54.90	19,599.40
2025	383.54	51.30	19,675.60	53.80	20,634.45
2030	396.18	49.50	19,610.91	52.60	20,839.07
2035	405.59	47.80	19,387.20	51.40	20,847.33
2040	415.95	46.10	19,175.29	50.30	20,922.28
2045	427.73	44.50	19,033.98	49.20	21,044.32
2050	437.47	43.00	18,811.21	48.20	21,086.05
2055	447.10	41.60	18,599.36	47.40	21,192.54
2060	456.60	40.40	18,446.64	46.60	21,277.56
2065	467.60	39.10	18,283.16	45.80	21,416.08
2070	477.70	37.80	18,057.06	45.00	21,496.50
2075	485.80	36.50	17,731.17	44.20	21,472.36
2080	493.80	35.20	17,381.76	43.40	21,430.92
2085	501.50	33.90	17,000.85	42.60	21,363.90
2090	510.00	32.60	16,626.00	41.80	21,318.00
2095	518.00	31.30	16,213.40	40.90	21,186.20
2100	526.00	30.00	15,780.00	40.00	21,040.00

Appendix Table 33. Energy consumption projections: developed countries minus the United States and Canada.

Year	Population (millions)	Case 1		Case 2	
		P-C EC (BOE/yr)	Consumption (million BOE/yr)	P-C EC (BOE/yr)	Consumption (million BOE/yr)
1990	931.30	27.40	25,518	27.40	25,518
1995	945.70	24.57	23,236	24.57	23,236
2000	950.34	24.88	23,645	24.88	23,645
2005	951.90	24.73	23,538	24.65	23,466
2010	948.22	24.16	22,914	24.38	23,123
2015	924.36	23.62	21,834	23.91	22,104
2020	900.20	23.09	20,787	24.22	21,803
2025	876.46	22.69	19,888	23.04	20,190
2030	853.82	22.42	19,139	22.73	19,411
2035	829.41	22.19	18,404	22.51	18,673
2040	804.05	21.97	17,669	22.23	17,874
2045	777.27	21.71	16,875	21.91	17,034
2050	752.53	21.49	16,175	21.63	16,280
2055	727.90	21.26	15,476	21.33	15,526
2060	703.40	20.77	14,613	21.04	14,798
2065	687.40	20.45	14,057	20.85	14,331
2070	672.30	20.18	13,568	20.71	13,924
2075	659.20	20.00	13,184	20.66	13,622
2080	646.20	19.85	12,828	20.64	13,339
2085	628.50	19.88	12,492	20.57	12,931
2090	610.00	20.11	12,270	20.50	12,506
2095	592.00	20.24	11,981	20.65	12,225
2100	574.00	20.40	11,720	20.83	11,960

Appendix Table 34. Energy consumption projections: developing countries.

Year	Population (millions)	Case 1 P-C EC (BOE/yr)	Case 1 Consumption (million BOE/yr)	Case 2 P-C EC (BOE/yr)	Case 2 Consumption (million BOE/yr)	Case 3 P-C EC (BOE/yr)	Case 3 Consumption (million BOE/yr)
1990	4040.6	4.11	16,621	4.11	16,621	4.11	16,621
1995	4418.1	5.00	22,090	5.00	22,090	5.00	22,090
2000	4792.4	5.00	24,458	5.10	24,458	5.10	24,458
2005	5200.0	5.00	26,000	5.25	27,300	5.50	28,600
2010	5610.0	5.00	28,050	5.50	30,855	6.00	33,660
2015	6000.0	5.00	30,000	5.75	34,500	6.50	39,000
2020	6330.0	5.00	31,650	6.00	37,980	7.00	44,310
2025	6620.0	5.00	33,100	6.25	41,375	7.50	49,650
2030	6950.0	5.00	34,750	6.50	45,175	8.00	55,600
2035	7265.0	5.00	36,325	6.75	49,039	8.50	61,752
2040	7580.0	5.00	37,900	7.00	53,060	9.00	68,220
2045	7880.0	5.00	39,400	7.25	57,130	9.50	74,860
2050	8160.0	5.00	40,800	7.50	61,200	10.00	81,600
2055	8425.0	5.00	42,125	7.75	65,294	10.50	88,462
2060	8690.0	5.00	43,450	8.00	69,520	11.00	95,590
2065	8900.0	5.00	44,500	8.25	73,425	11.50	102,350
2070	9100.0	5.00	45,500	8.50	77,350	12.00	109,200
2075	9260.0	5.00	46,300	8.75	81,025	12.50	115,750
2080	9410.0	5.00	47,050	9.00	84,690	13.00	122,330
2085	9545.0	5.00	47,725	9.25	88,291	13.50	128,857
2090	9680.0	5.00	48,400	9.50	91,960	14.00	135,520
2095	9790.0	5.00	48,950	9.75	95,452	14.50	141,955
2100	9900.0	5.00	49,500	10.00	99,000	15.00	148,500

Appendix Table 35. Energy consumption projections: world.

Year	Population (millions)	Case 1 P-C EC (BOE/yr)	Case 1 Consumption (million BOE/yr)	Case 2 P-C EC (BOE/yr)	Case 2 Consumption (million BOE/yr)
1990	5255	11.00	57,802	11.00	57,802
1995	5662	11.02	62,389	11.02	62,389
2000	6057	11.01	66,671	11.01	66,671
2005	6475	11.00	71,225	11.20	72,520
2010	6900	11.10	76,590	11.40	78,660
2015	7250	11.15	80,837	11.60	84,100
2020	7600	11.20	85,120	11.80	89,680
2025	7900	11.25	88,875	12.00	94,800
2030	8200	11.30	92,660	12.20	100,040
2035	8500	11.35	96,475	12.40	105,400
2040	8800	11.40	100,320	12.60	110,880
2045	9075	11.45	103,909	12.80	116,160
2050	9350	11.50	107,525	13.00	121,550
2055	9600	11.55	110,880	13.20	126,720
2060	9850	11.60	114,260	13.40	131,990
2065	10,050	11.65	117,082	13.60	136,680
2070	10,250	11.70	119,925	13.80	141,450
2075	10,400	11.75	122,200	14.00	145,600
2080	10,550	11.80	124,490	14.20	149,810
2085	10,675	11.85	126,499	14.40	153,720
2090	10,800	11.90	128,520	14.60	157,680
2095	10,900	11.95	130,255	14.80	161,320
2100	11,000	12.00	132,000	15.00	165,000

Appendix Table 36. Oil production and cumulative production (production in billion barrels per year).

Year	World	Cumulative Production (billion bbl)	United States	Former Soviet Union	OPEC	OPEC Production % World
1950	3.809	65.757	1.972			
1951	4.276	70.033	2.244			
1952	4.504	74.537	2.288	0.339		
1953	4.793	79.33	2.377	0.397		
1954	4.943	84.273	2.314	0.436		
1955	5.526	89.799	2.468	0.518		
1956	6.065	95.864	2.61	0.613		
1957	6.399	102.263	2.613	0.721		
1958	6.733	108.996	2.358	0.83		
1959	7.068	116.064	2.571	0.951		
1960	7.65	123.714	2.562	1.084	3.166	41.38
1961	8.149	131.863	2.621	1.219	3.409	41.83
1962	8.815	140.678	2.678	1.365	3.827	43.41
1963	9.477	150.155	2.751	1.513	4.194	44.25
1964	10.217	160.372	2.789	1.641	4.729	46.28
1965	10.946	171.318	2.83	1.783	5.222	47.71
1966	11.938	183.256	3.03	1.947	5.743	48.11
1967	12.822	196.078	3.212	2.115	6.123	47.75
1968	14.035	210.113	3.341	2.259	6.821	48.6
1969	15.062	225.175	3.35	2.407	7.593	50.41
1970	16.392	241.567	3.47	2.588	8.504	51.88
1971	17.479	259.046	3.522	2.726	9.201	52.64
1972	18.14	277.186	3.467	2.88	9.815	54.11
1973	20.325	297.511	3.367	3.086	11.18	55.01
1974	20.316	317.827	3.265	3.312	11.078	54.53
1975	19.264	337.091	3.055	3.555	9.771	50.72
1976	20.357	357.994	2.958	3.832	11.069	52.95
1977	21.75	379.744	3.008	3.997	11.259	51.76
1978	21.901	401.645	3.161	4.161	10.729	48.99
1979	22.804	424.449	3.157	4.259	11.136	48.83
1980	21.664	446.113	3.157	4.398	9.801	45.24
1981	20.359	466.472	3.135	4.409	8.207	40.31
1982	19.412	485.884	3.159	4.482	6.838	35.23
1983	19.333	505.217	3.164	4.499	6.065	31.37
1984	19.784	525.001	3.194	4.463	5.816	29.4
1985	19.703	544.704	3.255	4.475	5.448	27.65
1986	20.523	565.227	3.208	4.489	6.435	31.35
1987	20.683	585.91	3.021	4.555	6.124	29.61
1988	21.439	607.349	2.98	4.543	6.888	32.13
1989	21.85	629.199	2.802	4.431	7.465	34.16
1990	22.106	651.305	2.685	4.157	8.056	36.44
1991	21.975	673.28	2.707	3.758	8.142	37.05
1992	21.978	695.258	2.617	3.266	8.705	39.61
1993	21.986	717.244	2.499	2.867	8.844	40.22
1994	22.262	739.506	2.431	2.566	8.982	40.35
1995	22.752	762.258	2.394	2.598	8.979	39.46
1996	23.254	785.512	2.359	2.526	9.041	38.88
1997	23.977	809.489	2.355	2.594	9.283	38.72
1998	24.426	833.915	2.282	2.589	10.125	41.45
1999	24.034	857.949	2.146	2.619	9.573	39.83
2000	24.858	882.807	2.125	2.749	10.127	40.74
2001	24.803	907.61	2.117	2.961	9.808	39.54
2002	24.389	931.999	2.107	3.25	9.263	37.98

Source: *Oil & Gas Journal.*

Appendix Table 37. World natural-gas liquids (NGL) production (million barrels per year, at 22.5 bbl/mcf).

Year	Production
1950	156.55
1951	192.28
1952	208.03
1953	223.63
1954	234.36
1955	242.82
1956	261.15
1957	285.23
1958	310.09
1959	346.61
1960	380.63
1961	413.73
1962	452.07
1963	472.93
1964	515.95
1965	552.08
1966	595.03
1967	636.98
1968	706.57
1969	773.66
1970	848.81
1971	902.27
1972	943.20
1973	985.68
1974	1004.49
1975	1004.58
1976	1045.24
1977	1084.32
1978	1124.32
1979	1199.81
1980	1211.51
1981	1235.74
1982	1240.00
1983	1244.97
1984	1347.10
1985	1384.51
1986	1426.27
1987	1494.94
1988	1563.41
1989	1615.50
1990	1654.11
1991	1681.49
1992	1691.64
1993	1720.01
1994	1738.46
1995	1771.83
1996	1850.00
1997	1852.02
1998	1894.41
1999	1940.60
2000	2009.16
2001	2044.10

Appendix Table 38. Oil reserves: world and Persian Gulf region countries (billion barrels at year-end).

Year	World	Persian Gulf Region	Percent of World
1950	102.32	38.75	37.87
1951	103.63	49.82	48.07
1952	118.15	63.32	53.59
1953	135.26	76.66	56.68
1954	157.54	95.96	60.91
1955	188.82	124.71	66.05
1956	231.05	142.90	61.85
1957	264.47	168.00	63.76
1958	275.70	172.45	62.55
1959	293.04	179.94	61.40
1960	300.99	181.66	60.35
1961	309.97	186.70	60.23
1962	313.54	192.47	61.39
1963	331.04	205.87	62.19
1964	341.27	210.68	61.73
1965	353.06	213.86	60.57
1966	389.05	233.61	60.05
1967	414.34	247.21	59.66
1968	458.05	268.76	58.67
1969	539.76	315.52	58.45
1970	611.40	342.57	56.03
1971	631.86	367.39	58.14
1972	666.88	353.85	53.06
1973	627.86	348.16	55.45
1974	715.70	401.86	56.15
1975	658.69	350.00	53.13
1976	640.39	324.28	50.64
1977	645.75	364.17	56.39
1978	641.61	368.00	57.35
1979	641.92	359.95	56.07
1980	648.52	360.07	55.52
1981	670.71	360.67	53.77
1982	670.19	367.48	54.83
1983	669.30	368.24	55.02
1984	698.67	396.93	56.81
1985	700.14	395.89	56.54
1986	697.45	399.98	57.35
1987	887.35	562.37	63.38
1988	907.44	565.40	62.31
1989	1001.57	654.51	65.35
1990	999.11	656.88	65.75
1991	991.01	655.86	66.18
1992	997.04	656.08	65.80
1993	999.12	657.16	65.77
1994	999.76	653.79	65.39
1995	1007.47	653.05	64.82
1996	1018.85	669.85	65.74
1997	1019.55	670.45	65.76
1998	1034.26	667.14	64.50
1999	1016.04	669.13	65.86
2000	1031.55	677.01	65.63
2001	1031.55	679.09	65.83
2002	1038.08	679.14	65.42

Source: *Oil & Gas Journal.*

Appendix Table 39. Oil reserve increases of the Persian Gulf countries and Venezuela (billion barrels at year-end).

Year	Saudi Arabia	Iran	Iraq	Kuwait	Abu Dhabi	Venezuela
1981	164.60	57.00	29.70	64.50	30.60	20.30
1982	162.40	55.30	41.00	64.20	30.50	21.50
1983	166.00	51.00	43.00	63.90	30.40	24.85
1984	169.00	48.50	44.50	90.00	30.50	25.85
1985	168.80	47.90	44.10	89.80	31.00	25.59
1986	166.60	48.80	47.10	91.90	31.00	25.00
1987	167.00	92.80	100.00	91.90	92.20	56.30
1988	170.00	92.80	100.00	91.90	92.20	58.10
1989	255.00	92.90	100.00	94.50	92.20	58.50
1990	257.50	92.80	100.00	94.50	92.20	59.04
1991	257.80	92.90	100.00	94.00	92.20	59.10
1992	257.80	92.90	100.00	94.00	92.20	62.65
1993	258.70	92.90	100.00	94.00	92.20	63.33
1994	258.70	89.20	100.00	94.00	92.20	64.50
1995	258.70	88.20	100.00	94.00	92.20	64.50
1996	259.00	93.00	112.00	94.00	92.20	64.90
1997	259.00	93.00	112.50	94.00	92.20	71.70
1998	259.00	89.70	112.50	94.00	92.20	72.60
1999	261.00	89.70	112.50	94.00	92.20	72.60
2000	259.20	89.70	112.50	94.00	92.20	76.90
2001	259.20	89.70	112.50	94.00	92.20	77.70

Source: *Oil & Gas Journal.*

Appendix Table 40. OPEC oil reserves and percent of world reserves (billion barrels at year-end).

Year	Reserves	%
1960	218.28	72.52
1961	223.78	72.19
1962	232.07	74.02
1963	248.12	74.95
1964	254.89	74.69
1965	260.24	73.71
1966	289.87	74.51
1967	310.27	74.88
1968	332.01	72.48
1969	398.75	73.87
1970	434.33	71.04
1971	433.39	68.59
1972	465.62	69.82
1973	421.81	67.18
1974	484.97	67.76
1975	449.87	68.19
1976	440.39	68.77
1977	439.91	68.12
1978	444.94	69.35
1979	435.59	67.86
1980	434.35	66.97
1981	436.50	65.08
1982	445.16	66.42
1983	448.42	67.00
1984	476.41	68.19
1985	475.21	67.87
1986	477.52	68.47
1987	670.65	75.58
1988	676.01	74.50
1989	767.10	76.59
1990	773.82	77.45
1991	769.39	77.76
1992	770.59	77.29
1993	772.13	77.28
1994	770.25	77.04
1995	776.87	77.11
1996	788.58	77.40
1997	797.13	78.18
1998	800.48	77.40
1999	802.48	79.00
2000	818.84	79.38
2001	818.84	79.38
2002	819.01	78.90

Source: *Oil & Gas Journal.*

Appendix Table 41. World oil reserves-to-production ratios (R/PR).

Year	R/PR
1950	26.86
1951	24.23
1952	26.23
1953	28.22
1954	31.87
1955	34.17
1956	38.10
1957	41.33
1958	40.95
1959	41.46
1960	39.34
1961	38.04
1962	35.57
1963	34.93
1964	33.40
1965	32.59
1967	32.31
1968	32.64
1969	35.84
1970	37.30
1971	36.15
1972	36.76
1973	30.89
1974	35.23
1975	34.19
1976	30.64
1977	29.69
1978	29.30
1979	28.15
1980	29.93
1981	32.94
1982	34.52
1983	34.62
1984	35.31
1985	35.53
1986	33.98
1987	42.90
1988	42.33
1989	45.84
1990	45.20
1991	45.10
1992	45.36
1993	45.44
1994	44.91
1995	44.28
1997	42.52
1998	42.34
1999	42.27
2000	41.50
2001	41.59
2002	42.56

Source: Previous tables.

Appendix Table 42. Estimated oil production during the 21st century (billion barrels per year). Colors correspond to Figure 25.

	Gray		Purple		Green		Blue		Red		Brown	
Year	Production	Cumulative Production	Production	Cumulative Production	Production	Cumulative Production	Production	Cumulative Production	Production	Cumulative Production	Production	Cumulative Production
1995	22.40	760.10	22.40	760.10	24.40	760.10	22.40	760.10	22.40	760.10	22.40	760.10
2000	24.30	880.00	24.30	880.00	24.30	880.00	24.30	880.00	24.30	880.00	24.30	880.00
2005	24.60	1002.20	25.40	1004.20	25.20	1003.70	25.40	1004.20	25.40	1004.20	25.40	1004.20
2010	25.40	1127.20	27.00	1135.20	26.40	1132.70	27.40	1136.20	27.40	1136.20	27.40	1136.20
2015	25.80	1255.20	28.40	1273.70	27.40	1267.20	28.80	1276.70	28.80	1276.70	29.40	1278.20
2020	25.90	1384.40	29.40	1418.20	28.20	1406.20	30.00	1423.70	29.40	1422.20	31.20	1429.70
2025	26.00	1514.10	30.40	1567.70	28.80	1548.70	31.40	1577.20	29.80	1570.20	32.60	1589.20
2030	25.90	1643.80	30.80	1720.70	29.10	1693.40	32.20	1736.20	29.90	1719.40	33.20	1753.70
2035	25.60	1772.50	30.40	1873.70	29.20	1839.10	32.40	1897.70	30.00	1869.10	33.60	1920.70
2040	25.00	1899.00	29.60	2023.70	29.10	1984.80	32.20	2059.20	30.00	2019.10	33.60	2088.70
2045	24.40	2025.50	28.20	2168.20	28.80	2129.50	30.60	2216.20	29.90	2168.80	33.40	2256.20
2050	23.60	2145.50	25.80	2303.20	28.20	2272.00	28.80	2364.70	29.70	2317.80	33.00	2422.20
2055	22.80	2261.50	23.40	2426.20	27.40	2411.00	26.80	2503.70	29.30	2465.30	32.00	2584.70
2060	21.80	2373.00	21.00	2537.20	26.00	2544.50	25.40	2634.20	28.80	2610.50	30.20	2740.20
2065	20.20	2478.00	18.60	2636.20	24.50	2670.70	23.80	2757.20	28.00	2752.50	28.40	2886.70
2070	18.00	2573.50	16.20	2723.20	23.00	2789.40	22.20	2872.20	27.00	2890.00	26.40	3023.70
2075	16.00	2658.50	13.80	2798.20	21.20	2900.00	20.00	2977.70	26.00	3022.50	24.60	3151.20
2080	14.00	2733.50	11.40	2861.20	19.40	3001.50	18.20	3073.20	25.00	3150.00	22.60	3269.20
2085	12.00	2798.50	9.00	2912.20	17.40	3093.50	16.20	3159.20	23.60	3271.50	20.80	3377.70
2090	10.00	2853.50	6.60	2951.20	15.40	3175.50	14.20	3235.20	21.60	3384.50	19.00	3477.20
2095	8.00	2898.50	4.20	2978.20	13.40	3247.50	12.40	3301.70	19.60	3487.50	17.20	3567.70
2100	6.00	2933.50	2.60	2995.20	11.40	3309.50	10.40	3358.70	17.60	3580.50	15.20	3648.70
2105	4.20	2959.00	1.60	3005.70	9.40	3361.50	8.40	3405.70	15.60	3663.50	13.20	3719.70
2110	3.00	2977.00	1.00		7.40	3403.50	6.40	3442.70	13.60	3736.50	11.20	3780.70
2115	2.00	2989.50	0.40		5.40	3435.50	4.40	3469.70	11.60	3799.50	9.40	3832.20
2120					4.00	3459.00	3.00	3488.20	9.60	3852.50	7.80	3875.20
2125					2.60	3475.50	2.00	3500.70	7.60	3895.50	6.00	3909.70
2130					2.00	3487.00	1.20	3508.70	5.60	3928.50	4.40	3935.70

Appendix Table 43. Canadian oil sands: synthetic crude oil (SCO) and crude bitumen production (million barrels).

Year	SCO Production	SCO Cumulative Production	Crude Bitumen Production	Crude Bitumen Cumulative Production	Total Production	Total Cumulative Production
1967	0.454					
1968	5.725	6.179				
1969	10.087	16.266				
1970	12.081	28.347				
1971	15.540	43.887				
1972	18.820	62.707				
1973	18.446	81.153				
1974	16.880	98.033				
1975	15.764	113.797				
1976	17.477	131.274	2.751		20.228	134.025
1977	16.423	147.697	3.054	5.805	19.477	153.502
1978	20.375	168.072	2.850	8.655	23.225	176.727
1979	33.517	201.589	3.653	12.308	37.170	213.897
1980	46.609	248.198	3.499	15.807	50.108	264.005
1981	40.546	288.744	4.737	20.544	45.283	309.288
1982	43.769	332.513	8.057	28.601	51.826	361.114
1983	58.230	390.743	9.152	37.753	67.382	428.496
1984	48.673	439.416	12.219	49.972	60.892	489.388
1985	61.299	500.715	19.056	69.028	80.355	569.743
1986	67.477	568.192	34.053	103.081	101.530	671.273
1987	65.866	634.058	42.459	145.540	108.325	779.598
1988	73.232	707.290	47.423	192.963	120.655	900.253
1989	74.857	782.147	47.067	240.030	121.924	1022.177
1990	76.050	858.197	49.414	289.444	125.464	1147.641
1991	82.534	940.731	44.743	334.187	127.277	1274.918
1992	86.662	1027.393	46.307	380.494	132.969	1407.887
1993	88.834	1116.227	48.339	428.833	137.173	1545.060
1994	95.548	1211.775	49.125	477.958	144.673	1689.733
1995	102.704	1314.479	54.281	532.239	156.985	1846.718
1996	102.637	1417.116	59.786	592.025	162.423	2009.141
1997	105.659	1522.775	86.839	678.864	192.498	2201.639
1998	112.406	1635.181	102.928	781.792	215.334	2416.973
1999	118.045	1753.226	89.137	870.929	207.182	2624.155
2000	117.045	1870.271	105.552	976.481	222.597	2846.752
2001	127.304	1997.575	112.930	1089.411	240.234	3086.986
2002	161.017	2158.592	110.455	1199.866	271.472	3358.458

Source: Alberta Energy and Utilities Board Annual Statistical Data.

Appendix Table 44. Natural gas production and cumulative production: world, United States, and the Former Soviet Union (FSU).

	World			Production		
Year	Production (tcf/yr)	Cumulative Production (tcf)	Year	United States	FSU	United States + FSU
1950	6.958	160.07	1950	6.262	0.203	6.465
1951	8.546	168.62	1951	7.457	0.221	7.678
1952	9.246	177.86	1952	8.013	0.225	8.238
1953	9.939	187.80	1953	8.397	0.242	8.639
1954	10.416	198.22	1954	8.742	0.265	9.007
1955	10.792	209.01	1955	9.405	0.317	9.722
1956	11.621	220.63	1956	10.082	0.426	10.508
1957	12.677	233.31	1957	10.680	0.656	11.336
1958	13.782	247.09	1958	11.030	0.992	12.022
1959	15.405	262.49	1959	12.046	1.250	13.296
1960	16.917	279.41	1960	12.771	1.600	14.371
1961	18.388	297.80	1961	13.254	2.083	15.337
1962	20.092	317.89	1962	13.877	2.596	16.473
1963	21.019	338.91	1963	14.747	3.172	17.919
1964	22.931	361.84	1964	15.546	3.833	19.379
1965	24.537	386.38	1965	16.040	4.508	20.548
1966	26.446	412.82	1966	17.207	5.048	22.255
1967	28.399	441.22	1967	18.171	5.559	23.730
1968	31.403	472.63	1968	19.322	5.971	25.293
1969	34.385	507.01	1969	20.698	6.395	27.093
1970	37.725	544.74	1970	22.371	6.808	29.179
1971	40.101	584.84	1971	23.002	7.304	30.306
1972	41.920	626.76	1972	22.928	7.615	30.543
1973	43.808	670.56	1973	22.904	8.127	31.031
1974	44.644	715.21	1974	21.896	8.963	30.859
1975	44.648	759.86	1975	20.278	9.950	30.228
1976	46.455	806.31	1976	20.110	11.040	31.150
1977	48.192	854.50	1977	20.200	11.900	32.100
1978	49.970	904.47	1978	20.118	12.802	32.920
1979	53.325	957.80	1979	20.724	13.986	34.710
1980	53.845	1011.64	1980	20.552	14.969	35.521
1981	54.922	1066.57	1981	20.335	16.001	36.336
1982	55.111	1121.68	1982	18.914	17.222	36.136
1983	55.332	1177.01	1983	17.132	18.422	35.554
1984	59.871	1236.88	1984	18.590	20.204	38.794
1985	61.534	1298.41	1985	17.529	22.109	39.638
1986	63.390	1361.80	1986	17.078	23.596	40.674
1987	66.442	1428.25	1987	17.693	25.018	42.711
1988	69.485	1497.73	1988	18.168	26.488	44.656
1989	71.800	1569.53	1989	18.426	27.381	45.807
1990	73.516	1643.05	1990	18.959	28.045	47.004
1991	74.733	1717.78	1991	18.819	27.881	46.700
1992	75.184	1792.96	1992	18.971	26.853	45.824
1993	76.445	1869.41	1993	19.184	26.185	45.369
1994	77.265	1946.67	1994	19.975	24.751	44.726
1995	78.748	2025.42	1995	19.721	24.329	44.050
1996	82.222	2107.64	1996	19.991	24.665	44.656
1997	82.312	2189.96	1997	20.024	23.129	43.153
1998	84.196	2274.15	1998	20.249	23.772	44.021
1999	86.249	2360.40	1999	19.966	24.194	44.160
2000	89.296	2449.70	2000	20.089	24.874	44.963
2001	90.849	2540.55	2001	20.479	24.972	45.451

Sources: U.S. Department of Interior, *Energy Perspectives* 2, 1950–1969; *Oil & Gas Journal*, 1970–2001.

Appendix Table 45. World natural-gas reserves (trillion cubic feet at year-end).

Year	Reserves
1961	721*
1962	
1963	
1964	868
1965	898
1966	1041
1967	1185
1968	1333
1969	1498
1970	1588
1971	1725
1972	1883
1973	2033
1974	2150
1975	2232
1976	2327
1977	2520
1978	2502
1979	2574
1980	2628
1981	2911
1982	3023
1983	3200
1984	3402
1985	3484
1986	3626
1987	3797
1988	3955
1989	3989
1990	4217
1991	4378
1992	4885
1993	5016
1994	4980
1995	4934
1996	4945
1997	5086
1998	5145
1999	5146
2000	5278
2001	5451
2002	5501

*Sources: Weeks (1962, 1963); *Oil & Gas Journal.*

Appendix Table 46. World natural-gas reserves-to-production ratios (R/PR).

Year	R/PR
1961	39.2
1962	
1963	
1964	37.8
1965	36.6
1966	39.4
1967	41.7
1968	42.4
1969	43.6
1970	42.1
1971	43.0
1972	44.9
1973	46.4
1974	48.2
1975	50.0
1976	50.1
1977	52.3
1978	50.1
1979	48.3
1980	48.8
1981	53.0
1982	54.8
1983	57.8
1984	56.8
1985	56.6
1986	57.2
1987	57.1
1988	56.9
1989	55.6
1990	57.4
1991	58.6
1992	65.0
1993	65.6
1994	64.4
1995	62.6
1996	60.1
1997	61.8
1998	61.1
1999	59.7
2000	59.1
2001	60.0

Source: Previous tables.

Appendix Table 47. United States coalbed methane production and cumulative production (billion cubic feet per year).

Year	Production	Cumulative Production	% of United States Gas Production	Black Warrior basin	San Juan basin	Powder River basin	Other basins*
1982	<1			<1			
1983	5	5		5			
1984	8	13		6	2		
1985	10	23		8	2		
1986	17	40		13	4		
1987	24	64		17	7		
1988	40	104		20	20		
1989	91	195	0.5	23	68		
1990	196	391	1.0	36	159		1
1991	348	739	1.8	68	277	0.9	2
1992	539	1278	2.8	89	440	0.8	9
1993	732	2010	3.8	103	611	0.8	17
1994	851	2861	4.3	108	709	2.0	32
1995	956	3817	4.8	109	800	5.0	42
1996	1003	4820	5.0	98	849	9.0	47
1997	1090	5910	5.4	111	909	14.0	56
1998	1194	7104	5.9	123	972	30.0	69
1999	1252	8356	6.3	108	1014	57.0	73
2000	1379	9735	6.9	109	1001	147.0	122
2001	1562	11,297	7.0	111	875	245.0	331

Sources: U.S. Department of Energy, Energy Information Administration; Cairn Point Publishing (1997); Ayers (2002).
*Production mainly from Appalachian, Piceance, Raton, and Uinta basins.

Appendix Table 48. Estimates of United States coalbed methane "proved reserves" (billion cubic feet at year-end).

Year-end	Coalbed Methane "Proved Reserves"	% of United States Natural-gas Reserves
1987	200(?)	
1988	1421	0.8
1989	3676	2.2
1990	5087	3.1
1991	8163	4.8
1992	10,034	6.0
1993	10,184	6.1
1994	9712	6.0
1995	10,499	6.4
1996	10,566	6.4
1997	11,462	6.9
1998	12,179	7.3
1999	13,229	8.1
2000	15,708	9.4
2001	17,531	9.5

Sources: U.S. Department of Energy, Energy Information Administration; Ayers (2002).

Appendix Table 49. Natural-gas production and cumulative production projections. Colors correspond to Figure 35.

	Blue		Green		Red		Brown		Purple	
Year	Production (tcf/yr)	Cumulative Production (tcf)	Production (tcf/yr)	Cumulative Production (tcf)	Production (tcf/yr)	Cumulative Production (tcf)	Production (tcf/yr)	Cumulative Production (tcf)	Production (tcf/yr)	Cumulative Production (tcf)
1995	78.5	2000	78.5	2000	78.5	2000	78.5	2000	78.5	2000
2000	88.0	2416	88.0	2416	88.0	2416	85.0	2409	85.0	2409
2005	102.0	2891	102.0	2891	102.0	2891	91.0	2849	91.0	2849
2010	112.0	3426	112.0	3426	112.0	3426	98.0	3321	98.0	3321
2015	119.0	4003	119.0	4003	119.0	4003	105.0	3829	105.0	3829
2020	128.0	4621	128.0	4621	128.0	4621	111.0	4369	111.0	4369
2025	135.0	5228	135.0	5228	135.0	5228	116.0	4936	116.0	4936
2030	143.0	5923	143.0	5923	143.0	5923	120.0	5526	120.0	5526
2035	148.0	6650	148.0	6650	149.0	6653	123.0	6134	123.0	6134
2040	152.0	7400	153.0	7402	155.0	7413	125.0	6754	125.0	6754
2045	155.0	8168	157.0	8178	160.0	8200	127.0	7384	127.0	7384
2050	158.0	8950	160.0	8970	164.0	9011	129.0	8024	128.0	8021
2055	159.0	9743	163.0	9778	167.0	9838	131.0	8674	129.0	8664
2060	160.0	10,540	165.0	10,598	170.0	10,681	132.0	9331	130.0	9311
2065	161.0	11,343	167.0	11,428	173.0	11,538	133.0	9994	130.0	9961
2070	162.0	12,150	169.0	12,268	175.0	12,409	134.0	10,661	129.0	10,609
2075	161.0	12,958	169.0	13,113	177.0	13,289	135.0	11,333	128.0	11,252
2080	159.0	13,758	168.0	13,955	178.0	14,176	136.0	12,011	126.0	11,887
2085	156.0	14,545	167.0	14,793	179.0	15,069	137.0	12,693	123.0	12,509
2090	152.0	15,315	166.0	15,625	180.0	15,966	138.0	13,381	119.0	13,114
2095	147.0	16,062	164.0	16,451	180.0	16,867	138.0	14,071	114.0	13,697
2100	141.0	16,777	160.0	17,261	179.0	17,764	138.0	14,761	108.0	14,252
2105	134.0	17,465	156.0	18,051	178.0	18,657	137.0	14,948	101.0	14,774
2110	127.0	18,117	151.0	18,818	176.0	19,542	135.0	15,628	93.0	15,259
2115	119.0	18,732	145.0	19,558	172.0	20,412	133.0	16,298	85.0	15,704
2120	111.0	19,307	139.0	20,268	168.0	21,262	130.0	16,956	77.0	16,109
2125	102.0	19,840	133.0	20,948	164.0	22,092	127.0	17,598	69.0	16,474
2130	93.0	20,327	126.0	21,596	159.0	22,899	123.0	18,223	61.0	16,799
2135	84.0	20,770	119.0	22,208	154.0	23,682	119.0	18,828	53.0	17,084
2140	75.0	21,167	111.0	22,783	147.0	24,434	115.0	19,413	45.0	17,329
2145	66.0	21,515	103.0	23,318	140.0	25,152	110.0	19,976	37.0	17,524
2150	57.0	21,822	94.0	23,811	132.0	25,832	104.0	20,511	30.0	17,692
2155	48.0	22,085	85.0	24,258	123.0	26,469	96.0	21,011	24.0	17,827
2160	39.0	22,302	76.0	24,661	114.0	27,062	87.0	21,468	19.0	17,934
2165	31.0	22,477	68.0	25,021	105.0	27,609	78.0	21,881	14.0	18,017
2170	25.0	22,620	60.0	25,341	95.0	28,110	70.0	22,251	10.0	18,077
2175	20.0	22,732	53.0	25,623	86.0	28,562	63.0	22,583	7.0	18,119
2180	16.0	22,822	46.0	25,871	77.0	28,970	56.0	22,881	4.0	18,147
2185	13.0	22,895	40.0	26,086	68.0	29,332	50.0	23,146	2.0	18,162
2190	10.0	22,952	35.0	26,273	60.0	29,653	44.0	23,381	1.0	18,169
2195	7.0	22,995	30.0	26,436	52.0	29,933	38.0	23,586	0.0	18,172
2200	5.0	23,025	25.0	26,573	45.0	30,175	33.0	23,763		
2205	4.0	23,047	21.0	26,688	38.0	30,383	28.0	23,916		
2210	3.0	23,065	17.0	27,783	32.0	30,558	23.0	24,043		
2215	2.0	24,077	13.0	27,858	26.0	30,703	19.0	24,148		
2220	1.0	24,085	10.0	27,915	21.0	30,820	15.0	24,233		
2225	0.0	24,087	7.0	27,958	16.0	30,913	11.0	24,298		
2230			4.0	27,985	12.0	30,983	8.0	24,346		
2235			2.0	28,000	8.0	31,033	5.0	24,378		
2240			1.0	28,008	5.0	31,065	2.0	24,396		
2245			0.0	28,010	3.0	31,085	0.0	24,398		
2250					1.0	31,095				
2255					0.0	31,100				

Appendix Table 50. Coal production (hard coal + lignite) and cumulative production (production in million metric tons per year).

Year	World	Cumulative Production (million t)	United States	FSU	China	Germany
1950	1848.7	88,240	508.4	297.1	42.9	343.4
1951	1966.6	90,207	522.9	321.8	53.1	375.4
1952	1971.3	92,178	460.3	338.1	66.5	387.2
1953	2001.1	94,179	442.9	359.0	69.7	403.2
1954	2016.5	96,195	381.7	392.2	83.7	419.8
1955	2186.8	98,382	445.3	442.0	98.3	444.5
1956	2300.1	100,682	480.6	473.9	110.4	458.2
1957	2386.0	103,068	470.0	518.3	130.7	464.8
1958	2472.4	105,541	391.5	535.2	270.2	463.1
1959	2555.5	108,096	392.5	547.0	347.8	455.5
1960	2634.6	110,731	394.0	543.7	420.0	469.3
1961	2503.0	113,234	381.4	535.6	250.0	482.0
1962	2556.8	115,791	398.3	528.2	250.0	494.5
1963	2682.7	118,473	432.8	563.5	270.0	508.1
1964	2778.7	121,252	457.4	583.5	290.0	514.7
1965	2803.8	124,056	478.1	590.8	299.0	492.0
1966	2855.4	126,911	496.0	616.2	327.0	476.5
1967	2740.5	129,652	512.5	615.8	227.0	453.8
1968	2810.9	132,463	505.1	601.2	300.0	463.2
1969	2872.7	135,335	517.9	608.3	325.0	475.8
1970	2989.7	138,325	555.8	634.8	354.0	480.8
1971	3006.7	141,332	508.8	645.8	390.0	479.4
1972	3038.5	144,370	546.6	664.8	400.0	462.6
1973	3095.4	147,466	543.1	673.1	430.0	463.2
1974	3130.7	150,596	553.4	670.3	450.0	470.0
1975	3265.0	153,861	593.9	698.7	482.0	463.3
1976	3390.8	157,252	621.3	722.1	480.0	478.4
1977	3516.1	160,768	633.7	740.8	527.0	468.2
1978	3527.6	164,296	598.9	701.3	593.0	467.0
1979	3709.3	168,005	708.5	695.1	610.0	480.0
1980	3770.7	171,776	752.7	690.3	595.8	482.5
1981	3799.2	175,575	747.3	675.0	598.2	492.8
1982	3934.8	179,510	760.7	682.9	641.4	499.6
1983	3961.0	183,471	709.5	675.0	687.6	492.0
1984	4152.5	187,623	807.6	669.0	789.2	507.9
1985	4363.5	191,987	801.6	680.1	872.3	521.4
1986	4468.0	196,455	805.7	702.8	894.0	512.8
1987	4661.0	201,116	833.0	790.1	928.0	500.2
1988	4777.6	205,893	862.1	800.0	979.9	498.2
1989	4865.5	210,759	889.7	768.4	1054.1	488.3
1990	4735.7	215,495	933.5	657.5	1079.9	464.1
1991	4547.9	220,043	903.6	592.0	1087.4	352.3
1992	4568.9	224,611	904.9	622.8	1116.4	313.9
1993	4461.1	229,073	857.8	551.3	1149.7	286.0
1994	4594.6	233,667	937.6	489.3	1239.9	299.4
1995	4686.2	238,353	937.1	440.7	1360.7	251.5
1996	4741.9	243,095	965.2	412.7	1397.0	240.3
1997	4752.1	247,847	989.2	398.3	1372.8	228.4
1998	4617.6	252,465	1014.3	382.1	1250.0	211.3
1999	4350.3	256,815	994.4	390.2	1045.0	205.1

Sources: United Nations Department of Economic and Social Affairs (*World Energy Supplies*, 1950–1978; *Yearbook of World Energy Statistics*, 1979–1981; *Energy Statistics Yearbook*, 1982–2001).

Appendix Table 51. Hard coal production (million metric tons).

Year	World	United States	FSU	China	India	Germany
1950	1430.1	505.3	185.2	42.9	32.8	129.0
1951	1510.8	519.9	202.5	53.1	35.0	139.3
1952	1495.8	457.6	215.0	66.5	36.9	143.6
1953	1494.3	440.3	224.3	69.7	36.6	144.6
1954	1475.1	379.1	243.7	83.7	37.5	148.4
1955	1598.2	442.4	276.6	98.3	38.8	151.8
1956	1687.6	478.0	304.0	110.4	39.9	155.3
1957	1734.9	467.6	328.5	130.7	44.2	153.5
1958	1803.5	389.3	338.9	270.2	46.1	152.8
1959	1876.8	390.1	348.9	347.8	47.8	145.4
1960	1942.5	391.5	355.9	420.0	52.6	145.9
1961	1790.1	378.7	355.8	250.0	56.1	146.3
1962	1833.6	395.5	363.4	250.0	61.4	144.5
1963	1901.7	430.4	369.3	270.0	65.9	145.4
1964	1968.6	454.7	381.3	290.0	62.4	145.0
1965	2014.6	475.3	397.6	299.0	67.2	137.6
1966	2049.3	492.5	406.6	327.0	68.0	128.3
1967	1952.9	508.4	414.1	227.0	68.2	114.1
1968	2017.2	500.7	416.2	300.0	70.8	113.8
1969	2057.7	513.4	425.8	325.0	70.7	113.1
1970	2134.6	550.4	432.7	354.0	73.7	112.4
1971	2143.2	503.0	441.4	390.0	71.8	111.8
1972	2162.7	536.6	451.1	400.0	75.6	103.5
1973	2209.0	530.2	461.2	430.0	77.9	98.3
1974	2243.3	539.4	473.4	450.0	84.1	95.8
1975	2373.0	575.9	484.7	482.0	95.9	93.2
1976	2420.8	598.2	494.4	480.0	100.9	96.7
1977	2498.3	607.2	499.8	527.0	100.3	91.3
1978	2539.8	566.6	501.5	593.0	101.5	90.1
1979	2686.3	670.5	496.5	610.0	103.4	93.3
1980	2728.5	710.4	492.9	595.8	109.1	94.5
1981	2728.3	700.8	481.3	598.2	123.1	95.5
1982	2828.3	712.8	488.0	641.4	128.5	96.3
1983	2830.7	656.6	486.8	687.6	134.8	89.6
1984	2999.7	750.3	483.3	789.2	144.9	84.9
1985	3161.8	735.9	494.4	872.3	149.7	88.5
1986	3248.9	738.9	512.9	894.0	163.4	87.1
1987	3411.3	762.3	595.0	928.0	177.0	82.4
1988	3510.4	784.9	599.5	979.9	189.0	79.3
1989	3581.5	811.3	576.8	1054.1	198.7	77.4
1990	3517.5	853.6	473.9	1079.9	201.8	76.5
1991	3466.8	825.1	414.0	1087.4	226.9	72.7
1992	3531.6	823.2	464.8	1116.4	233.9	72.1
1993	3463.1	776.4	412.0	1149.7	246.0	64.2
1994	3592.4	857.7	367.8	1239.9	254.7	57.6
1995	3744.7	858.6	340.0	1360.7	265.6	58.8
1996	3794.0	885.2	300.2	1397.0	285.5	53.1
1997	3824.4	910.8	292.4	1372.8	296.2	51.2
1998	3700.5	935.7	284.3	1250.0	297.9	45.3
1999	3469.5	916.0	290.2	1045.0	296.6	43.8

Sources: United Nations Department of Economic and Social Affairs (*World Energy Supplies*, 1950–1978; *Yearbook of World Energy Statistics*, 1979–1981; *Energy Statistics Yearbook*, 1982–2001).

Appendix Table 52. Lignite production (million metric tons).

Year	World	United States	FSU	Germany
1950	418.6	3.1	111.9	214.4
1951	455.8	3.0	119.3	236.1
1952	475.5	2.7	123.1	243.6
1953	506.8	2.6	134.7	258.9
1954	541.4	2.6	148.5	271.4
1955	588.6	2.9	165.4	292.7
1956	612.5	2.6	169.9	302.9
1957	651.1	2.4	189.8	311.3
1958	668.9	2.2	196.3	310.3
1959	678.7	2.4	198.1	310.1
1960	692.1	2.5	187.8	323.4
1961	712.9	2.7	179.8	335.7
1962	723.2	2.8	164.8	350.0
1963	781.0	2.4	194.2	362.7
1964	810.1	2.7	202.2	369.7
1965	789.2	2.8	193.2	354.4
1966	806.1	3.5	209.6	348.2
1967	787.6	4.1	201.7	339.7
1968	793.7	4.4	185.0	349.4
1969	815.0	4.5	182.5	362.7
1970	855.1	5.4	202.1	368.4
1971	836.5	5.8	204.4	367.6
1972	875.8	10.0	213.7	359.1
1973	886.4	12.9	211.9	364.9
1974	887.4	14.0	196.9	374.2
1975	892.0	18.0	214.0	370.1
1976	970.0	23.1	227.7	381.7
1977	1017.8	26.5	241.0	376.9
1978	987.8	32.3	199.8	376.9
1979	1023.0	38.0	198.6	386.7
1980	1042.2	42.3	197.4	388.0
1981	1070.9	46.5	193.7	397.3
1982	1106.5	47.9	194.9	403.3
1983	1130.3	52.9	188.2	402.4
1984	1152.8	57.3	185.7	423.0
1985	1201.7	65.7	185.7	432.9
1986	1219.1	66.8	189.9	425.7
1987	1249.7	70.7	195.1	417.8
1988	1267.2	77.2	200.5	418.9
1989	1284.0	78.4	191.6	410.9
1990	1218.2	79.9	183.6	387.6
1991	1081.1	78.5	178.0	279.6
1992	1037.3	81.7	158.0	241.8
1993	998.0	81.2	139.3	221.8
1994	1002.2	79.9	121.5	207.1
1995	941.5	78.5	100.7	192.7
1996	947.9	80.0	112.5	187.2
1997	927.7	78.4	105.9	177.2
1998	917.1	78.6	97.8	166.0
1999	880.8	78.4	100.0	161.3

Sources: United Nations Department of Economic and Social Affairs (*World Energy Supplies*, 1950–1978; *Yearbook of World Energy Statistics*, 1979–1981; *Energy Statistics Yearbook*, 1982–2001).

Appendix Table 53. Coal reserves of United States, FSU, and China (billion metric tons).

		United States	FSU	China	Total of United States, FSU, and China	Percent of World
1962, WPC-SER*	pr**	81	160			
	ip***		201			
1968, WPC-SER	pr	81.4				
	ip		249			
1974, WEC-SER[†]	pr	181.8	136.6	70–80	393.4	66.5
					241–445	
	ip	363.6	273.2	(300)	936.8	66
1976, WEC-SER	pr	198	136.6			
	ip	396.1	277			
1978, WEC-Coal Resources	pr	222	137.5	123.7	483.2	60.8
(Peters and Schilling)	ip		276			
1979, Fettweis	pr	405	280	150	835	77.6
	ip	810	560	300	1670	77.6
1980, Tatsch	pr	200.4	150.6	88.2	439.2	67.5
	ip	400.7	301.2	330.7	1032.6	65.9
1980, WEC-SER	pr	223	233	99	555	62.9
	ip		276	200		
1980, Bestougeff (after 5)	pr	222	137.4	123.6	483	60.7
	ip	410.5	341.5			
1980, World Coal Study	pr	208.7	137.5	123.7	469.9	56.7
1981, Feys	pr	181.8		101.3	556.3	
	ip	363.6	273.2	200.5		
1982, USDOE-EIA[††]	pr	221	137.5	123.7	482.2	60.6
1983, WEC-SER	pr	257.1	240.3	99	596.4	63
	ip	428.8	288	200	916.8	60.3
1984, Matveev et al.	pr[†††]	397.6	280.6	102	780.2	63
	ip[‡]	1568.6	419.2	592.5	2580.3	60
1986, WEC-SER	pr	263.8	244.7	170	678.5	81
	ip	442.9	292.8	737.1	1472.8	69.7
1988, BP-SRWE[‡‡]	pr	263.8	244.7	170	678.5	66.1
1989, WEC-SER	pr	215.2	241	730.7	1186.9	74.3
	ip	430.5	287	770	1487.5	60.8
1990, BP-SRWE	pr	261.2	239.6	167.1	667.9	61.7
1991, BP-SRWE	pr	260.3	239.2	166.1	665.6	61.9
1992, WEC-SER	pr	240.6	241	114.5	596.1	57.4
	ip	427.8	287	286.4	1001.2	51.2
1992, BP-SRWE (after 20)	pr	240.6	241	114.5	596.1	57.4
1992, USDOE-EIA	pr	243.6	243.6	169.4	656.6	62
1993, BP-SRWE (after 20)	pr	240.6	241	114.5	596.1	57.4
1994, USDOE-EIA (after 20)	pr	240.4	241	114.3	595.7	57.3
1994, BP-SRWE	pr	240.6	241	114.5	596.1	57.4
1995, WEC-SER	pr	240.6	241	114.5	596.1	57.4
	ip	431.5	287	286.4	1004.9	41.9
1995, USDOE-EIA	pr	239	239	114.3	592.3	57
1995, BP-SRWE (after 26)	pr	240.6	241	114.5	596.1	57.4
1996, USDOE-EIA	pr	239	239	114.3	592.3	57
1996, BP-SRWE	pr	240.6	241	114.5	596.1	57.8
1997, USDOE-EIA	pr	248.9	238.5	114.1	601.5	58
1997, BP-SRWE	pr	240.6	241	114.5	596.1	57.8
1998, WEC-SER	pr	246.6	230.2	114.5	591.3	60.1
1998, USDOE-EIA	pr	248.6	238.3	114	600.9	58
1998, BP-SRWE	pr	240.6	241	114.5	596.1	57.8

Appendix Table 53. Coal reserves of United States, FSU, and China (billion metric tons) (cont).

		United States	FSU	China	Total of United States, FSU, and China	Percent of World
1999, BP-SRWE	pr	246.6	230.2	114.5	591.3	60.1
2000, USDOE-EIA	pr	246.7	227	118.4	592.1	60
2000, BP-SRWE	pr	246.6	230.2	114.5	591.3	60.1
2001, WEC-SER	pr	250	230	114.5	594.5	60.4
2001, BP-SRWE	pr	246.6	230.2	114.5	591.3	60.1
2002, BP-SRWE	pr	250	230	114.5	594.5	60.4
2002, USDOE-EIA	pr	247	227.2	118.5	592.7	60
2003, USDOE-EIA	pr	245.6	226	117.9	589.5	60

*WPC-SER = World Power Conference, Survey of Energy Resources.
**pr = proved recoverable reserves.
***ip = reserves in place.
†WEC-SER = World Energy Council, Survey of Energy Resources.
††USDOE-EIA = U.S. Department of Energy, Energy Information Administration.
†††Established reserves.
‡Total reserves.
‡‡BP-SRWE = British Petroleum, Statistical Review of World Energy.

Appendix Table 54. Coal reserves-to-production ratios.

Year	Production (billion t/yr)	Proved Recoverable Reserves	R/PR	In-place Reserves	R/PR
1962	2.55	800	313		
1968	2.81	730	260		
1974	3.13	591	189	1420	453
1976	3.39	607	179	1125	
1978	3.53	795	225		
1979	3.71	1076	290	2152	580
1980	3.77	882	234	1320	350
1981	3.80	695	183	1297	341
1982	3.93	796	202		
1983	3.96	946	239	1520	384
1986	4.47	827	185	2094	468
1988	4.78	1026	215		
1989	4.86	1598	329	2446	503
1990	4.73	1083	229		
1991	4.55	1079	237		
1992	4.57	1039	227	1956	428
1993	4.46	1039	233		
1994	4.59	1039	226		
1995	4.69	1032	220	2400	512
1996	4.74	1032	218		
1997	4.75	1032	217		
1998	4.62	984	213		
1999	4.35	984	226		

Source: Previous tables.

Appendix Table 55. Coal "resources" (billion [10^9] metric tons). Numbers correspond to Figure 40.

			World	United States	FSU	China	Total	Percent of world
1	1960, WPC (Brown)		4894	1673	1213	1012	3898	79.6
2	1961, Averitt	rr	4640.5	1506	1200	1011.6	3717.6	80.1
3	1962, WPC-SER	hc	7546.2	1100	4630	1011	6741	
		bc	1964	406	1350	700	2456	
			9510.2	1506	5980	1711	9197	96.7
4	1968, WPC-SER	hc	6712	1100	4122	1011	6233	
		bc	2041	406	1406	700	2512	
			8753	1506	5528	1711	8745	99.8
5	1969, Averitt		8618 m	2912 r				
			6650 u					
			15,268					
6	1972, Mel'mikov		15,763		8670			
7	1973, Averitt	i	8618	1434				
		h	6650	1491				
			15,268	2925				
8	1974, WEC-SER	hc	7065	1244	3993			
		bc	3690	1680	1720			
			10,755	2924	5713	660–1000	9297–9637	86.4–89.6
9	1975, Averitt	i	5797	1570				
		h	9281	2029				
			15,078	3599				
10	1976, WEC-SER		11,505 hc	1286 hc	1717 hr			
				2313 bc	3994 lr			
				3599	5711			
11	1977, WEC Coal Resources (Peters and Schilling) Conservation Commission		12,657	3213	6075	1797.5	11,085.5	87.6
12	1979, Fettweis	rb	8623	2114	5164	700	7978	92.5
		pgo	10,755	2924	5714	1000	9638	89.6
13	1980, WEC SER		13,476	3387.6	5926	1465	10,778.6	80
14	1980, World Coal Study		13,437.5	3212.5	6075	1797.5	11,085	82.5
15	1980, Bestougeff (after 11, 14)		12,656	3217.5	6075	1795.5	11,088	87.6
16	1980, Tatsch		11,858	3224	6298	1102	10,624	89.6
17	1981, Feys		10,781.7	2924.5	5713.7	1011.7	9649.9	89.5
18	1981, Perry (after 11)		12,657	3213	6075	1797.5	11,085.5	87.6
19	1981, Averitt	i	4890	1570	2086			
		h	10,188	2029	6532			
			15,078	3599	8618			
20	1983, WEC-SER		9643	1570	4406	1566	7542	78.2
21	1984, Matveev et al.		14,810	3600	6806	1465.5	11,871.5	80.3
22	1986, WEC-SER		11,990	1570	5501.7	2737	9808.7	81.8
23	1989, WEC-SER		10,647	1570	5489	1094	8151	76.5
24	1992, WEC-SER		10,555	1570	5487	954.3	8011.3	75.9
25	1995, WEC-SER		11,000	1570	5487	954.3	8011.3	72.9

m = mapped; hc = hard coal; u = unmapped; bc = brown coal; r = remaining resources; rb = resource base; i = identified; pgo = possible geologic occurrence; h = hypothetical; rr = remaining reserves; hr = high-ranking solid fuels; lr = low-ranking solid fuels.

Appendix Table 56. Shale-oil production and cumulative production in Brazil.

	Year	Oil Produced (bbl/yr)	Cumulative Production Since 1981 (bbl)
Before	1982	870,000–960,000	
	1982	208,141	208,141
	1983	214,101	422,242
	1984	239,002	661,244
	1985	268,232	929,476
	1986	283,716	1,213,192
	1987	226,863	1,440,055
	1988	289,672	1,729,727
	1989	292,228	2,021,955
	1990	266,161	2,288,116
	1991	260,280	2,548,396
	1992	849,660	3,398,056
	1993	1,181,766	4,579,822
	1994	1,162,714	5,742,536
	1995	1,291,989	7,034,525
	1996	1,320,838	8,355,363
	1997	1,238,500	9,593,863
	1998	1,369,417	10,963,280
	1999	1,430,816	12,394,096
	2000	1,460,000	13,854,096
	2001	1,460,000	15,314,096

Source: Petrobras.

Appendix Table 57. Shale-oil production and cumulative production in Estonia.

Year	Production (thousand bbl/yr)	Cumulative Production (thousand bbl)
1921	0.843	
1925	19.439	40
1930	73.784	273
1935	346.511	1324
1940	1275.633	5379
1941	828.290	6208
1942	557.080	6765
1943	952.900	7717
1944	1304.740	9022
1945	0.000	9022
1946	1033.530	10,056
1948	1466.000	
1950	1641.920	13,163
1955	2484.870	23,480
1960	3122.580	37,499
1965	3884.900	55,017
1966	4016.840	
1968	4031.500	66,892
1970	3738.300	
1975	3298.500	92,547
1980	3000.170	108,294
1985	2822.050	122,849
1990	2734.090	136,739
1993	2682.780	144,865
1994	2712.000	147,577
1995	2705.000	150,282
1996	2969.000	153,251
1997	2969.000	156,220
1998	2170.000	158,390
1999	1466.000	159,876
2000	2360.300	162,216
2001	2279.600	164,496
2002	2374.900	166,870

Sources: Yefimov (1993), Yefimov et al. (1994), and V. Kattai (1999, 2000, 2003, personal communication).

Appendix Table 58. Nuclear units in operation (connected to the grid at year-end).

Year	World	United States	France	Japan	Germany	FSU*	United Kingdom
1956	1						1
1957	2						2
1958	3						3
1959	8		1				7
1960	12	2	2				8
1961	13	2	2		1		8
1962	22	4	2		1		13
1963	29	7	3	1	1		14
1964	37	8	3	1	1	2	16
1965	45	8	4	2	1	2	22
1966	53	10	5	2	3	2	24
1967	61	11	7	2	4	3	26
1968	65	10	7	2	5	3	27
1969	75	13	8	3	5	4	27
1970	81	17	8	5	5	4	27
1971	96	20	9	5	5	5	29
1972	111	27	10	6	6	6	29
1973	130	37	10	8	7	9	29
1974	152	47	10	11	7	12	29
1975	167	53	10	13	7	14	30
1976	184	60	10	13	10	16	33
1977	199	63	12	16	10	17	33
1978	218	67	14	21	11	18	33
1979	223	66	16	22	12	20	33
1980	243	68	22	23	12	23	33
1981	265	72	30	25	13	27	32
1982	283	76	32	25	13	29	32
1983	305	79	36	28	14	31	35
1984	335	85	41	31	17	34	37
1985	365	93	43	33	17	39	38
1986	390	99	49	35	19	41	38
1987	408	106	53	36	19	45	38
1988	417	108	55	38	18	45	40
1989	423	109	55	39	19	47	39
1990	419	111	56	41	19	46	37
1991	417	110	56	42	19	45	35
1992	420	108	56	44	19	45	35
1993	429	109	57	48	19	46	35
1994	432	109	56	51	19	46	34
1995	435	109	56	51	19	47	35
1996	439	109	57	54	19	46	35
1997	434	107	59	54	19	46	35
1998	434	104	58	53	20	49	35
1999	433	104	59	53	19	46	35
2000	436	104	59	53	19	45	35
2001	438	104	59	54	19	43	33
2002	441	104	59	54	19	43	31

Source: International Atomic Energy Agency.
*Russia + Ukraine.

Appendix Table 59. Nuclear plants: annual construction starts and connections to the grid.

Year	Construction Starts	Grid Connections
1955	8	0
1956	4	1
1957	13	1
1958	6	1
1959	6	5
1960	9	4
1961	7	1
1962	7	9
1963	5	7
1964	9	8
1965	9	8
1966	15	8
1967	26	10
1968	35	6
1969	17	10
1970	35	6
1971	15	16
1972	24	16
1973	24	20
1974	30	26
1975	35	15
1976	33	19
1977	17	18
1978	13	20
1979	25	8
1980	20	21
1981	15	23
1982	14	19
1983	8	23
1984	7	33
1985	15	33
1986	6	26
1987	5	22
1988	5	14
1989	6	12
1990	4	10
1991	2	4
1992	3	6
1993	4	9
1994	2	6
1995	0	4
1996	1	6
1997	5	3
1998	3	4
1999	5	4
2000	5	6
2001	1	2
2002	6	6

Source: International Atomic Energy Agency.

Appendix Table 60. Nuclear electricity generation (terrawatt hour per year).

Year	World	United States	France	Japan	Germany*	FSU
1956	0.058					
1957	0.419	0.010				
1958	0.474	0.165	0.004			
1959	1.430	0.188	0.041			
1960	2.730**	0.518	0.130			
1961	4.360	1.690	0.243		0.024	
1962	6.480	2.270	0.423		0.099	
1963	10.620	3.220	0.420	0.003	0.056	
1964	14.970	3.340	0.580	0.002	0.104	
1965	24.180	3.660	0.900	0.025	0.117	
1966	34.600	5.520	1.390	0.584	0.361	1.650
1967	41.900	7.650	2.560	0.630	1.550	1.800
1968	52.230	12.530	3.160	1.044	2.160	2.500
1969	62.170	13.930	4.460	1.082	5.410	2.900
1970	78.750	21.800	5.710	4.580	6.490	3.700
1971	108.450	37.900	9.330	8.010	6.210	6.100
1972	148.120	54.030	14.590	9.480	9.520	9.200
1973	198.250	83.380	14.750	9.710	12.100	14.000
1974	255.390	113.980	14.710	19.700	14.320	18.500
1975	351.760	172.510	18.250	25.120	24.140	20.200
1976	410.140	191.110	15.760	34.080	29.530	26.360
1977	507.740	250.900	17.940	31.660	41.260	34.830
1978	594.350	276.400	29.000	59.310	43.870	39.000
1979	621.210	255.100	37.910	70.390	52.070	48.000
1980	681.200	251.110	57.950	82.590	55.590	60.000
1981	800.570	272.670	99.630	87.820	65.530	68.000
1982	870.270	282.770	103.070	102.430	74.430	84.500
1983	1006.670	293.670	136.920	114.290	78.060	109.790
1984	1221.430	327.630	181.740	134.260	103.870	142.000
1985	1448.320	383.690	213.100	159.580	138.640	167.400
1986	1550.160	414.040	241.440	168.300	130.490	160.800
1987	1704.930	455.270	265.520	187.760	141.720	186.980
1988	1837.460	526.970	275.520	178.660	156.820	216.000
1989	1897.030	529.400	303.930	182.870	158.290	212.580
1990	1984.500	576.970	313.650	202.270	158.100	212.000
1991	2022.500	612.640	331.340	213.460	147.430	212.000
1992	2060.560	618.840	321.700	217.000	150.000	193.380***
1993	2093.450	610.360	350.200	246.300	145.000	194.430
1994	2130.140	639.360	341.800	258.300	143.000	166.670
1995	2223.550	673.400	358.200	286.900	145.700	152.820
1996	2312.050	674.780	378.200	298.200	152.800	183.200
1997	2276.310	629.420	376.000	318.100	161.400	
1998	2293.730	673.700	368.400	306.940	145.200	179.820
1999	2398.200	727.700	375.000	303.260	160.400	190.200
2000	2448.400	753.900	395.000	304.870	159.600	202.290
2001	2543.680	768.830	401.300	321.940	162.300	210.380

Sources: United Nations Department of Economic and Social Affairs (*World Energy Supplies*, 1950–1978; *Yearbook of World Energy Statistics*, 1979–1981; *Energy Statistics Yearbook*, 1982–2001); International Atomic Energy Agency.
*Predominantly West Germany (more than 90% since 1985).
**Excludes FSU 1960–1965.
***Russian Federation (60–65%) + Ukraine (35–40%) from 1992.

Appendix Table 61. Hydroelectricity generation (terrawatt hour per year).

Year	World	Percent of World Electricity	United States	Canada	Brazil	FSU	China
1950	342.8	35.9	101.0	53.0	7.5	12.7	0.8
1951	374.1	34.9	104.5	59.2	8.0	13.7	1.0
1952	399.4	34.5	109.8	61.8	9.2	14.9	1.3
1953	413.0	32.6	109.8	67.1	9.2	19.2	1.5
1954	437.5	31.9	111.8	68.6	9.9	18.6	2.2
1955	472.9	30.6	116.5	76.2	10.6	23.2	2.4
1956	515.0	30.4	125.5	81.8	12.7	29.0	3.5
1957	550.0	30.5	133.6	83.4	14.9	39.4	3.5
1958	610.1	32.0	143.9	90.5	17.5	46.5	5.0
1959	634.9	30.2	141.5	97.0	17.9	47.6	8.5
1960	689.0	30.0	149.5	105.9	18.4	50.9	11.9
1961	726.2	29.6	155.7	103.9	18.9	59.1	14.3
1962	767.3	28.8	172.2	104.0	20.7	71.9	14.2
1963	802.7	27.9	169.2	103.8	20.7	75.8	16.9
1964	838.8	26.7	180.5	113.3	22.1	77.4	22.4
1965	915.9	27.1	197.2	117.1	25.5	81.4	27.6
1966	988.8	27.1	198.1	129.8	27.9	91.8	32.7
1967	1011.4	26.2	225.3	132.7	29.2	88.6	27.6
1968	1055.1	25.1	226.3	135.0	30.5	104.0	28.9
1969	1126.7	24.6	254.1	149.2	32.7	115.2	29.1
1970	1176.6	23.7	251.2	156.7	39.9	124.4	32.8
1971	1230.2	23.3	269.6	161.0	43.3	126.1	38.1
1972	1292.2	22.7	276.1	179.9	51.4	123.0	39.4
1973	1319.0	21.5	277.3	192.8	58.3	122.3	38.0
1974	1433.1	22.7	303.9	210.9	65.5	132.0	43.0
1975	1458.5	22.3	303.1	202.4	73.8	126.0	47.6
1976	1462.2	20.9	286.9	212.8	82.4	135.7	51.0
1977	1514.1	20.7	227.5	220.3	92.9	147.0	47.6
1978	1627.9	21.2	284.0	234.0	102.7	169.7	44.6
1979	1721.2	21.5	281.4	243.2	115.1	172.0	50.1
1980	1755.3	21.3	277.7	253.1	128.9	184.0	58.2
1981	1775.8	21.2	262.3	266.0	130.8	186.7	65.5
1982	1833.3	21.6	310.8	261.0	141.2	174.7	74.4
1983	1908.6	21.6	334.1	266.0	151.5	180.4	86.4
1984	1972.5	21.1	323.5	286.6	166.6	202.8	86.8
1985	1995.6	20.5	284.8	303.7	178.4	214.5	92.4
1986	2023.1	20.1	296.0	310.7	182.6	215.7	94.5
1987	2043.8	19.3	255.5	316.3	185.6	219.8	100.0
1988	2111.1	19.0	229.0	307.5	199.1	229.1	109.1
1989	2122.5	18.5	270.9	291.4	204.7	223.4	118.4
1990	2210.5	18.7	286.1	296.9	206.7	233.0	126.7
1991	2249.7	18.7	281.7	308.5	217.8	235.0	125.1
1992	2251.0	18.5	248.9	316.5	223.3	230.6	132.5
1993	2390.4	19.3	276.5	323.7	238.4	239.5	151.8
1994	2403.7	18.9	256.8	326.4	242.7	243.7	168.1
1995	2553.3	19.1	337.8	336.0	253.9	237.8	190.6
1996	2624.3	19.0	402.5	355.9	265.8	214.9	188.0
1997	2663.4	18.9	387.3	350.9	279.0	215.4	196.0
1998	2679.6	18.5	350.9	332.5	291.4	222.7	208.0
1999	2702.3	18.2	348.6	345.9	292.9	223.8	203.8

Sources: United Nations Department of Economic and Social Affairs (*World Energy Supplies*, 1950–1978; *Yearbook of World Energy Statistics*, 1979–1981; *Energy Statistics Yearbook*, 1982–2001); International Atomic Energy Agency.

Appendix Table 62. World geothermal electricity generation (terrawatt hour per year).

Year	World	Percent of World Electricity	United States	Philippines	Mexico	Italy	Indonesia	Japan	New Zealand
1950	1.28	0.13				1.28			
1960	2.52	0.10	0.03			2.10			0.38
1961	2.88								
1962	3.21								
1963	3.60								
1964	3.92								
1965	4.02								
1966	4.09								
1967	4.12								
1968	4.53								
1969	4.85								
1970	4.69	0.09	0.52		0.001	2.72			1.18
1971	4.63	0.09	0.55		0.001	2.66			1.17
1972	5.48	0.10	1.45		0.001	2.58			1.17
1973	6.05	0.10	1.97		0.150	2.48		0.27	1.16
1974	7.00	0.11	2.45		0.470	2.50		0.31	1.25
1975	7.96	0.12	3.25		0.490	2.48		0.38	1.27
1976	8.59	0.12	3.62		0.550	2.52		0.37	1.23
1977	8.93	0.12	3.58	0.001	0.570	2.50		0.58	1.28
1978	8.44	0.11	2.98	0.003	0.600	2.49		0.77	1.18
1979	10.99	0.14	4.39	0.640	1.020	2.54		0.78	1.18
1980	13.45	0.16	5.07	2.080	0.910	2.67		1.09	1.21
1981	16.03	0.19	5.69	2.770	0.960	2.66		1.15	1.09
1982	17.21	0.20	4.84	3.540	1.280	2.74	0.03	1.27	1.10
1983	19.99	0.23	6.07	4.090	1.350	2.71	0.21	1.44	1.12
1984	24.27	0.26	8.64	4.540	1.420	2.84	0.22	1.33	1.24
1985	28.44	0.29	11.45	4.940	1.640	2.68	0.21	1.46	1.14
1986	32.21	0.32	13.80	4.590	3.400	2.76	0.21	1.37	1.11
1987	35.57	0.33	15.70	4.530	4.330	2.99	0.21	1.37	1.17
1988	32.89	0.30	15.54	4.840	4.660	3.08	0.21	1.36	1.18
1989	34.45	0.30	16.40	5.320	4.670	3.15	0.21	1.38	1.24
1990	38.82	0.33	17.95	5.470	5.120	3.22	1.12	1.74	1.81
1991	41.45	0.34	18.66	5.760	5.430	3.18	1.10	1.78	2.21
1992	43.56	0.36	19.68	5.700	5.800	3.46	1.02	1.81	2.27
1993	46.13	0.37	20.68	5.760	5.880	3.67	1.09	1.80	2.16
1994	51.82	0.41	20.75	5.850	5.950	3.42	1.60	2.08	2.21
1995	51.31	0.38	18.96	5.950	7.410	3.45	2.21	3.20	2.05
1996	48.01	0.35	16.36	6.400	5.730	3.80	2.21	3.67	2.15
1997	49.51	0.35	15.81	6.750	5.470	4.03	2.44	3.76	2.27
1998	53.19	0.37	15.93	6.900	5.670	4.45	2.59	3.53	2.49
1999	59.35	0.40	19.02	7.100	5.630	4.81	2.73	3.45	2.56

Sources: United Nations Department of Economic and Social Affairs (*World Energy Supplies,* 1950–1978; *Yearbook of World Energy Statistics,* 1979–1981; *Energy Statistics Yearbook,* 1982–2001).

Appendix Table 63. Primary energy sources used in electricity generation (percentage).

Year	Coal	Oil	Gas	Hydroelectric	Nuclear	Other
1960	48.63	7.11	14.35	29.79	0.12	
1961	48.44	7.75	14.30	29.33	0.18	
1962	49.18	8.22	13.45	28.91	0.24	
1963	49.33	8.66	13.64	28.00	0.37	
1964	48.84	10.50	13.59	26.59	0.48	
1965	48.06	10.97	13.34	26.92	0.71	
1966	46.95	11.74	13.54	26.82	0.95	
1967	46.67	12.84	13.23	26.18	1.08	
1968	46.13	13.92	13.58	25.13	1.24	
1969	43.77	16.82	13.44	24.61	1.36	
1970	42.83	19.81	11.79	23.98	1.59	
1971	41.14	20.75	12.51	23.54	2.06	
1972	40.17	22.31	11.95	22.98	2.59	
1973	39.41	24.27	11.20	21.88	3.24	
1974	37.74	24.07	11.19	22.96	4.04	
1975	38.86	22.71	10.47	22.57	5.39	
1976	38.64	23.48	10.96	21.05	5.87	
1977	38.41	23.43	10.20	20.99	6.97	
1978	37.20	23.54	10.07	21.46	7.73	
1979	37.80	22.20	10.62	21.61	7.77	
1980	37.40	20.46	12.64	21.24	8.26	
1981	37.61	19.40	12.10	21.34	9.55	
1982	38.16	17.65	12.19	21.73	10.27	
1983	38.48	16.35	11.95	21.82	11.40	
1984	38.33	15.03	12.47	21.07	13.10	
1985	38.88	13.16	11.97	20.62	14.87	0.5
1986	38.87	12.85	12.01	20.35	15.42	0.5
1987	39.39	12.07	12.47	19.47	16.10	0.5
1988	39.08	11.58	13.37	18.94	16.53	0.5
1989	38.57	11.41	14.18	18.32	16.52	1.0
1990	38.18	11.08	14.49	18.42	16.83	1.0
1991	38.12	10.64	14.45	18.58	17.21	1.0
1992	37.58	10.76	14.41	18.32	17.43	1.5
1993	37.65	10.46	14.14	18.86	17.39	1.5
1994	37.89	10.35	14.44	18.49	17.33	1.5
1995	37.89	9.59	15.30	18.78	16.94	1.5
1996	38.48	9.60	15.41	18.48	16.53	1.5
1997	38.58	9.62	15.62	18.27	16.41	1.5
1998	38.44	9.98	15.61	17.91	16.56	1.5
1999	38.11	9.83	16.02	17.50	17.04	1.5

Sources: International Energy Agency (IEA-OECD) Energy balances of non-OECD countries (1999, 2000, 2001); United Nations Department of Economic and Social Affairs (*Energy Statistics Yearbook*, 1982–2001); U.S. Department of Energy, Energy Information Administration, World Energy Project System.

Appendix Table 64. Electricity generation (terrawatt hour per year).

Year	World	Developed	% of World	United States	% of World	Developing	% of World
1950	955	896	93.8	390	40.8	59	6.2
1951	1073	999	93.1	434	40.4	74	6.9
1952	1157	1079	93.2	464	40.1	78	6.8
1953	1266	1178	93.0	515	40.7	88	7.0
1954	1370	1271	92.8	546	39.8	99	7.2
1955	1544	1433	92.8	631	40.9	111	7.2
1956	1691	1561	92.3	687	40.6	130	7.7
1957	1803	1657	91.9	718	39.8	146	8.1
1958	1909	1743	91.3	727	38.1	166	8.7
1959	2100	1907	90.8	798	38.0	193	9.2
1960	2300	2077	90.3	844	36.7	223	9.7
1961	2455	2219	90.4	881	35.9	236	9.6
1962	2663	2403	90.2	946	35.5	260	9.8
1963	2879	2596	90.2	1011	35.1	283	9.8
1964	3136	2829	90.2	1084	34.6	307	9.8
1965	3380	3044	90.0	1158	34.3	336	10.0
1966	3642	3274	89.9	1249	34.3	368	10.1
1967	3863	3488	90.3	1317	34.1	375	9.7
1968	4207	3790	90.1	1436	34.1	417	9.9
1969	4571	4113	90.0	1553	34.0	458	10.0
1970	4962	4406	88.8	1640	33.0	556	11.2
1971	5273	4682	88.8	1717	32.6	591	11.2
1972	5698	5049	88.6	1853	32.5	649	11.4
1973	6126	5413	88.4	1965	32.1	713	11.6
1974	6313	5537	87.7	1967	31.1	776	12.3
1975	6527	5670	86.9	2003	30.7	857	13.1
1976	6983	6050	86.6	2123	30.4	933	13.4
1977	7299	6287	86.1	2211	30.3	1012	13.9
1978	7688	6566	85.4	2286	29.7	1122	14.6
1979	7999	6775	84.7	2319	29.0	1224	15.3
1980	8247	6907	83.5	2354	28.5	1340	16.5
1981	8384	6978	83.2	2359	28.1	1406	16.8
1982	8476	6971	82.2	2302	27.1	1505	17.8
1983	8826	7221	81.8	2368	26.8	1605	18.2
1984	9326	7581	81.3	2479	26.6	1745	18.7
1985	9739	7860	80.7	2568	26.4	1879	19.3
1986	10,055	8017	79.7	2599	25.8	2038	20.3
1987	10,587	8403	79.4	2719	25.7	2184	20.6
1988	11,117	8745	78.7	2878	25.9	2372	21.3
1989	11,483	8960	78.0	2960	25.8	2523	22.0
1990	11,788	9106	77.2	3012	25.5	2682	22.8
1991	12,017	9174	76.3	3058	25.4	2843	23.7
1992	12,142	9086	74.8	3074	25.3	3056	25.2
1993	12,394	9103	73.4	3187	25.7	3291	26.6
1994	12,697	9147	72.0	3243	25.5	3550	28.0
1995	13,385	9590	71.6	3582	26.8	3795	28.4
1996	13,846	9804	70.8	3702	26.7	4042	29.2
1997	14,163	9890	69.8	3730	26.3	4273	30.2
1998	14,515	10,072	69.4	3859	26.6	4443	30.6
1999	15,031	10,317	68.6	3976	26.4	4714	31.4

Sources: United Nations Department of Economic and Social Affairs (*World Energy Supplies*, 1950–1978; *Yearbook of World Energy Statistics*, 1979–1981; *Energy Statistics Yearbook*, 1982–2001).

Appendix Table 65. Per-capita electricity consumption.

	World			Developed Countries		
Year	Electricity Generation (TW h/yr)	Population (millions)	Per-capita Consumption (kW h/yr)	Electricity Generation (TW h/yr)	Population (millions)	Per-capita Consumption (kW h/yr)
1950	955	2519	379	896	834	1074
1951	1073	2566	418	999	844	1184
1952	1157	2612	443	1079	854	1263
1953	1266	2658	476	1178	865	1362
1954	1370	2706	506	1271	876	1451
1955	1544	2755	560	1433	888	1614
1956	1691	2804	603	1561	899	1736
1957	1803	2856	631	1657	911	1819
1958	1909	2909	656	1743	922	1890
1959	2100	2963	709	1907	934	2042
1960	2300	3020	762	2077	946	2196
1961	2455	3079	797	2219	958	2316
1962	2663	3140	848	2403	970	2477
1963	2879	3202	899	2596	984	2638
1964	3136	3267	960	2829	995	2843
1965	3380	3334	1014	3044	1006	3026
1966	3642	3402	1070	3274	1016	3222
1967	3863	3473	1112	3488	1026	3400
1968	4207	3544	1187	3790	1036	3658
1969	4571	3617	1264	4113	1045	3936
1970	4962	3691	1344	4406	1053	4184
1971	5273	3766	1400	4682	1062	4409
1972	5698	3841	1483	5049	1073	4705
1973	6126	3916	1564	5413	1082	5003
1974	6313	3991	1582	5537	1091	5075
1975	6527	4066	1605	5670	1100	5154
1976	6983	4139	1687	6050	1108	5460
1977	7299	4211	1733	6287	1116	5633
1978	7688	4283	1795	6566	1124	5842
1979	7999	4356	1836	6775	1131	5990
1980	8247	4430	1862	6907	1139	6064
1981	8384	4505	1861	6978	1146	6089
1982	8476	4583	1849	6971	1154	6041
1983	8826	4661	1893	7221	1161	6220
1984	9326	4742	1967	7581	1168	6491
1985	9739	4825	2018	7860	1176	6684
1986	10,055	4909	2048	8017	1184	6771
1987	10,587	4995	2119	8403	1191	7055
1988	11,117	5083	2187	8745	1198	7300
1989	11,483	5169	2221	8960	1206	7429
1990	11,788	5255	2243	9106	1214	7501
1991	12,017	5339	2251	9174	1221	7513
1992	12,142	5421	2240	9086	1227	7405
1993	12,394	5502	2253	9103	1233	7383
1994	12,697	5582	2275	9147	1238	7388
1995	13,385	5662	2364	9590	1244	7709
1996	13,846	5742	2411	9804	1248	7856
1997	14,163	5821	2433	9890	1253	7893
1998	14,515	5900	2460	10,072	1257	8013
1999	15,031	5979	2514	10,317	1260	8188

Source: Previous tables.

Appendix Table 65. Per-Capita electricity consumption (cont.).

	United States			Developed Countries Minus the United States			Developing Countries		
Year	Electricity Generation (TW h/yr)	Population (millions)	Per-capita Consumption (kW h/yr)	Electricity Generation (TW h/yr)	Population (millions)	Per-capita Consumption (kW h/yr)	Electricity Generation (TW h/yr)	Population (millions)	Per-capita Consumption (kW h/yr)
1950	390	158	2468	506	676	748	59	1685	35
1951	434	160	2712	565	684	826	74	1722	43
1952	464	163	2847	615	691	890	78	1758	44
1953	515	165	3121	663	700	947	88	1794	49
1954	546	168	3250	725	708	1024	99	1830	54
1955	631	171	3690	802	717	1118	111	1867	59
1956	687	174	3948	874	725	1205	130	1905	68
1957	718	177	4056	939	734	1279	146	1945	75
1958	727	180	4039	1016	742	1369	166	1986	84
1959	798	183	4361	1109	751	1477	193	2029	95
1960	844	186	4538	1233	760	1622	223	2074	107
1961	881	189	4661	1338	769	1740	236	2121	111
1962	946	192	4927	1457	778	1873	260	2170	120
1963	1011	195	5185	1585	789	2009	283	2219	127
1964	1084	197	5502	1745	798	2187	307	2272	135
1965	1158	200	5790	1886	806	2340	336	2328	144
1966	1249	202	6183	2025	814	2488	368	2386	154
1967	1317	204	6456	2171	822	2641	375	2447	153
1968	1436	206	6971	2354	830	2836	417	2508	166
1969	1553	208	7466	2560	837	3058	458	2572	178
1970	1640	210	7809	2766	843	3281	556	2638	211
1971	1717	212	8099	2965	850	3488	591	2703	219
1972	1853	214	8659	3196	859	3721	649	2768	234
1973	1965	216	9097	3448	866	3981	713	2835	251
1974	1967	218	9023	3570	873	4089	776	2901	267
1975	2003	220	9104	3667	880	4167	857	2966	289
1976	2123	222	9563	3927	886	4432	933	3031	308
1977	2211	224	9870	4076	892	4569	1012	3095	327
1978	2286	226	10,115	4280	898	4766	1122	3159	355
1979	2319	228	10,171	4456	903	4935	1224	3224	380
1980	2354	230	10,235	4553	909	5009	1340	3291	407
1981	2359	233	10,124	4619	913	5059	1406	3359	419
1982	2302	235	9796	4669	919	5080	1505	3429	439
1983	2368	238	9950	4853	923	5258	1605	3500	459
1984	2479	240	10,329	5102	928	5498	1745	3574	488
1985	2568	243	10,568	5292	933	5672	1879	3649	515
1986	2599	245	10,608	5418	939	5770	2038	3725	547
1987	2719	247	11,008	5684	944	6021	2184	3805	574
1988	2878	250	11,512	5867	948	6189	2372	3884	611
1989	2960	252	11,746	6000	954	6289	2523	3963	637
1990	3012	255	11,812	6094	959	6354	2682	4041	664
1991	3058	257	11,899	6116	964	6344	2843	4118	690
1992	3074	260	11,823	6012	967	6217	3056	4194	729
1993	3187	263	12,118	5916	970	6099	3291	4269	771
1994	3,243	266	12,192	5904	972	6074	3550	4343	817
1995	3582	269	13,316	6008	975	6162	3795	4418	859
1996	3702	272	13,610	6102	976	6252	4042	4493	900
1997	3730	275	13,564	6160	978	6298	4273	4569	935
1998	3859	278	13,881	6213	979	6346	4443	4643	957
1999	3976	280	14,200	6331	980	6460	4714	4718	999

Source: Previous tables.

References Cited

Adams, T. D., and M. A. Kirkby, 1976, Estimate of world gas reserves: Proceedings of the 9th World Petroleum Congress (Tokyo, May 1975), v. 3 (Exploration and Transportation), p. 3–9.

Adelman, M. A., and M. C. Lynch, 1997, Fixed view of resource limits creates undue pessimism: Oil & Gas Journal, v. 95, no. 14 (April 7, 1997), p. 56–60.

Alberta Department of Energy and Natural Resources, 1979, Alberta oil sands facts and figures.

Alberta Energy Resources Conservation Board, 1972, Reserves of crude oil, gas, natural gas liquids, and sulphur, province of Alberta.

Alberta Energy Resources Conservation Board, 1979–1985, Alberta's reserves of crude oil, gas, natural gas liquids, and sulphur.

Alberta Energy Resources Conservation Board, 1986–1995, Alberta's reserves of crude oil, oil sands, gas, natural gas liquids, and sulphur.

Alberta Oil and Gas Conservation Board, 1963, A description and reserves estimate of the oil sands of Alberta. The oil sands of Alberta, OGCB Report 1963, 60 p.

Alberta Oil Sands Technology and Research Authority, 1989, 1990 AOSTRA — A 15-year portfolio of achievement, 174 p.

Alcántara, J., and O. Castillo, 1982, Project focuses on Venezuelan heavy oil: Oil & Gas Journal, v. 80, no. 22 (May 31, 1982), p. 117–125.

Al-Jarri, A. S., and R. A. Startzman, 1997, Analysis of world crude oil production trends: SPE 37962 Paper presented at the 1997 SPE Hydrocarbon Economics and Evaluation Symposium (Dallas, March 1997).

Appert, O., 1998, Gas market: Forecast and challenge: Proceedings of the 15th World Petroleum Congress (Beijing, October 1997), v. 3 (Natural Gas, Reserves, etc.), p. 91–98.

Arrington, J. R., 1960, Predicting the size of crude reserves is key to evaluating exploration programs: Oil & Gas Journal, v. 58, no. 9 (February 29, 1960), p. 130–134.

Attanasi, E. D., and D. H. Root, 1993, Statistics of petroleum exploration in the Caribbean, Latin America, western Europe, the Middle East, Africa, non-communist Asia, and the southwestern Pacific: U.S. Geological Survey Circular 1096, 129 p.

Attanasi, E. D., and D. H. Root, 1994, The enigma of oil and gas field growth: AAPG Bulletin, v. 78, no. 3, p. 321–332.

Attanasi, E. D., R. F. Mast, and D. H. Root, 1999, Oil, gas field growth projections: Wishful thinking or reality?: Oil & Gas Journal, v. 97, no. 14 (April 5, 1999), p. 79–81.

Australian Gas Association, 1996, Coal seam methane in Australia: An overview: AGA Research Paper No. 2, 63 p.

Averitt, P., 1961, Coal reserves of the United States— A progress report, January 1, 1960: U.S. Geological Survey Bulletin 1136, 116 p.

Averitt, P., 1969, Coal resources of the United States, January 1, 1967: U.S. Geological Survey Bulletin 1275, 116 p.

Averitt, P., 1973, Coal, in D. A. Brobst and W. P. Pratt, eds., United States mineral resources: U.S. Geological Survey Professional Paper 820, p. 133–142.

Averitt, P., 1975, Coal resources of the United States, January 1, 1974: U.S. Geological Survey Bulletin 1412, 131 p.

Averitt, P., 1981, Coal resources, in M. A. Elliot, ed., Chemistry of coal utilization— Second supplementary volume: Hoboken, New Jersey: John Wiley & Sons, p. 55–89.

Ayers, W. B. Jr., 2002, Coalbed gas systems, resources, and production and a review of contrasting cases from the San Juan and Powder River basins: AAPG Bulletin, v. 86, no. 11, p. 1853–1890.

Ayers, W. B. Jr., and B. S. Kelso, 1989, Knowledge of methane potential for coalbed resources grows, but needs more study: Oil & Gas Journal, v. 87, no. 43 (October 23, 1989), p. 64–67.

Barthel, F., P. Kehrer, J. Koch, F. K. Miscus, and D. Weigel, 1976, Die Kunftige Entwicklung der Energienachfrage und deren beckung, in Perspektiven bis zum Jahr 2000, section 3 (Das Angebot von Energie-Rohstoffen): Hannover, Bundesanstalt für Geowissenschaften und Rohstoffe, p. 87–189.

Beck, P. W., 1994, Prospects and strategies for nuclear power: Review of the Institute for International Affairs: London, England, Earthscan, 118 p.

Beck, P. W., 1999, Nuclear energy in the twenty-first century: Examination of a contentious subject: Annual Review of Energy and the Environment, v. 24, p. 113–137.

Bender, F., 1982, Sufficient energy raw materials for everyone?, in F. C. Whitmore and M. E. Williams, eds., Resources for the twenty-first century: U.S. Geological Survey Professional Paper 1193, p. 104–114.

Bestougeff, M. A., 1980, Introduction— Summary of mondial coal resources and reserves. Geographic and geologic repartition, in P. F. Burollet and V. Ziegler, eds., Colloque C2— Energy resources: 26th International Geological Congress (Paris, 1980): Paris, France, Editions Technip, p. 353–366.

Bibler, C. J., J. S. Marshall, and R. C. Pilcher, 1998, Status of worldwide coal mine methane emissions and use: International Journal of Coal Geology, v. 35, nos. 1–4, p. 283–310.

Birdsall, N., A. C. Kelley, and S. W. Sinding, eds., 2001, Population matters— Demographic change, economic growth, and poverty in the developing world: Oxford, England, Oxford University Press, 440 p.

Bois, C., 1982, Réserves, ressources et dispossibilités mondiales en hydrocarbures: Revue de l'Institut Francais du Petrole, v. 37, no. 2, p. 135–148.

Bookout, J. F., 1989a, Two centuries of fossil fuel energy: Episodes, v. 12, no. 4, p. 257–262.

Bookout, J. F., 1989b, World faces great challenge: Geotimes, v. 34, no. 11, p. 12–13.

Borregales, C. J., 1979, Evaluation and development of the Orinoco oil belt: United Nations Institute for Training and Research Conference on Long-term Energy Resources (Montreal, November–December 1979), 8 p.

Bowman, C. W., and M. A. Carrigy, 1977, World-wide oil sand reserves: Part A of development status of Alberta oil sands— 1976, by the Alberta Oil Sands Technology and Research Authority et al., *in* R. F. Meyer, ed., Future supply of nature-made petroleum and gas— Technical reports: First United Nations Institute for Training and Research Conference on Energy and the Future and Second International Institute for Applied Systems Analysis Conference on Energy Resources (Laxenburg, Austria, July 1976): New York, New York, Pergamon Press, p. 732–738.

Boyer, C. M. II, 1992, Coalbed methane: A new international gas play?: Presentation at the Eastern Coalbed Forum, 1992, University of Alabama, Tuscalloosa.

Boyer, C. M. II, J. R. Kelafant, and D. Kruger, 1992, Coalbed gas— Conclusion: Diverse projects worldwide include mined, unmined coals: Oil & Gas Journal, v. 90, no. 50 (December 14, 1992), p. 36–41.

British Petroleum (BP), 1961, 1970, 1980, Statistical review of the world oil industry.

British Petroleum (BP), 1984–1995, Review of world gas.

British Petroleum (BP), 1988–2003, Statistical review of world energy.

Burollet, P. F., 1984, World resources of oil, *in* Colloquium 2 (Energy resources of the world): 27th International Geological Congress (Moscow, 1984) Reports, v. 2, p. 3–10.

Cairn Point Publishing, 1997, The international coal seam gas report, 218 p.

Campbell, C. J., 1991, The golden century of oil— 1950–2050— The depletion of a resource: Dordrecht, The Netherlands, Kluwer Academic Press, 345 p.

Campbell, C. J., 1995, Taking stock: Sun World, v. 19, no. 1, p. 16–19.

Campbell, C. J., 1996, Oil shock: Energy World, no. 240, p. 7–12.

Campbell, C. J., 1997, Depletion patterns show change due for production of conventional oil: Oil & Gas Journal, v. 95, no. 52 (December 29, 1997), p. 33–37.

Campbell, C. J., and J. H. Laherrère, 1998, The end of cheap oil: Scientific American, v. 278, no. 3, p. 78–83.

Carrigy, M. A., 1980, The role of bituminous sands in extending the petroleum era beyond 2000 A.D., *in* P. F. Burollet and V. Ziegler, eds., Colloque C2— Energy resources: 26th International Geological Congress (Paris, 1980): Paris, France, Editions Technip, p. 299–300 (and separate reprint).

Carrigy, M. A., 1983, Thermal recovery from tar sands: Journal of Petroleum Technology, v. 35, no. 13, p. 2149–2157.

Carrigy, M. A., 1986, New production techniques for Alberta oil sands: Science, v. 234, p. 1515–1518.

Charpentier, R. R., et al., 1993, Estimates of unconventional natural gas resources of the Devonian shales of the Appalachian basin, *in* J. B. Roen and R. C. Kepferle, eds., Petroleum geology of the Devonian and Mississippian black shale of eastern North America: U.S. Geological Survey Bulletin 1909, p. N1–N14.

Chilingarian, G. V., and T. F. Yen, eds., 1978, Bitumens, asphalts and tar sands: Developments in Petroleum Science, v. 7, 331 p.

Collett, T. S., 1993, Natural gas production from arctic gas hydrates, *in* D. G. Howell, ed., The future of energy resources: U.S. Geological Survey Professional Paper 1570, p 299–311.

Collett, T. S., 2001, Natural-gas hydrates: Resource of the twenty-first century?, *in* M. W. Downey et al., eds., Petroleum provinces of the twenty-first century: AAPG Memoir 74, p. 85–108.

Collett, T. S., 2002, Energy resource potential of natural gas hydrates: AAPG Bulletin, v. 86, no. 11, p. 1971–1992.

Congress of the United States, Office of Technology Assessment, 1980, An assessment of oil shale technologies, 517 p.

Croft, G., and K. Stauffer, 1996, Venezuelan projects advance to develop world's largest heavy oil reserves: Oil & Gas Journal, v. 94, no. 28 (July 8, 1996), p. 62–63.

Culbertson, W. C., and J. K. Pitman, 1973, Oil shale, *in* D. A. Brobst, and W. P. Pratt, eds., United States mineral resources: U.S. Geological Survey Professional Paper 820, p. 497–503.

Curtis, J. B., 1996, How well do we know the size of the U.S. natural gas resource base?: Journal of Petroleum Technology, v. 48, no. 1, p. 75–81.

Curtis, J. B., 2002, Fractured shale-gas systems: AAPG Bulletin, v. 86, no. 11, p. 1921–1938.

Danyluk, M., B. Galbraith, and R. Omaña, 1984, Toward definitions for heavy crude oil and tar sands, *in* R. F. Meyer, J. C. Wynn, and J. C. Olson, eds., The future of heavy crude and tar sands: Second International United Nations Institute for Training and Research International Conference (Caracas, Venezuela, February 1982): New York, McGraw Hill, p. 3–6.

Davidson, R. M., L. L. Sloss, and L. B. Clarke, 1995, Coalbed methane extraction: International Energy Agency Coal Research, 67 p.

Deffeyes, K. S., 2001, Hubbert's Peak— The impending world oil shortage: Princeton, New Jersey, Princeton University Press, 208 p.

DeHaan, H. J., 1995, Introduction, *in* H. J. de Haan, ed., New developments in improved oil recovery: Geological Society (London) Special Publication 84, p. 1–4.

Demaison, G. J., 1977, Tar sands and supergiant oil fields: AAPG Bulletin, v. 61, no. 11, p. 1950–1961.

Desprairies, P., 1978, Worldwide petroleum supply limits, *in* Oil and gas resources. The full reports to the Conservation Commission of the World Energy Conference: Chicago, Illinois, Independent Publishers Group (IPG) Science and Technology Press, p. 1–180.

Desprairies, P., and B. Tissot, 1980, Les limites de l'approvisionnement petrolier mondial, *in* Proceedings of the 11th World Energy Conference (Munich, 1980), Round Table A, p. 549–573.

DeWitt, W. Jr., 1986, Devonian gas-bearing shales in the Appalachian basin, *in* C. W. Spencer and R. F. Mast, eds., Geology of tight gas reservoirs: AAPG Studies in Geology 24, p. 1–8.

Donnell, J. R., 1977, Global oil-shale resources and costs, *in* United Nations Institute for Training and Research Conference, p. 843–856.

Donnell, J. R., 1980, Potential contribution of oil shale to U.S., world energy needs: Oil & Gas Journal, v. 71, no. 41 (October 13, 1980), p. 218–224.

Dowling, D. B., 1913, Introduction, *in* W. McInnes, D. B. Dowling, and W. W. Leach, eds., The coal resources of the world: 12th International Geological Congress (Toronto, Canada, 1913): Toronto, Canada, Morang & Co. Ltd., v. I, p. xvii–xxxix.

Duce, J. T., 1946, Post-war oil supply areas: The Petroleum Times (London), v. 50, no. 1271, p. 382–389.

Duncan, D. C., and V. E. Swanson, 1965, Organic rich shale of the United States and world land areas: U.S. Geological Survey Circular 523, 30 p.

Dunn, S., 2001, Hydrogen futures: Toward a sustainable energy system: Worldwatch Paper 157, 90 p.

Dutton, S. P., et al., 1993, Major low-permeability-sandstone gas reservoirs in the continental United States: The University of Texas at Austin, Bureau of Economic Geology and Gas Research Institute Report of Investigations 211, 221 p.

Eaverson, H. N., 1939, Coal through the years: New York, American Institute of Mining and Metallurgical Engineering, 129 p.

Edwards, J. D., 1997, Crude oil and alternate energy production forecasts for the twenty-first century: The end of the hydrocarbon era: AAPG Bulletin, v. 81, no. 8, p. 1292–1305.

Edwards, J. D., 2001, Twenty-first century energy: Decline of fossil fuel, increase of renewable nonpolluting energy sources, *in* M. W. Downey et al., eds., Petroleum provinces of the twenty-first century: AAPG Memoir 74, p. 21–34.

Edwards, J. D., 2002, Twenty-first century energy: Transition from fossil fuels to renewable, nonpolluting energy sources, *in* L. C. Gerhard et al., eds., Sustainability of energy and water through the 21st century: Proceedings of the Arbor Day Farm Conference (October 8–11, 2000), Kansas Geological Survey, p. 37–48.

Ehrlich, P. R., 1968, The population bomb: New York, Ballantine Books, 223 p.

Ehrlich, P. R., and A. H. Ehrlich, 1990, The population explosion: New York, Simon and Schuster, 320 p.

Elliott, M. A., and G. R. Yohe, 1981, The coal industry and coal research and development in perspective, *in* M. A. Elliott, ed., Chemistry of coal utilization—Second supplementary volume: Hoboken, New Jersey, John Wiley & Sons, p. 1–54.

Ellis, P., 1994, The significance of gas in the former Soviet Union, *in* C. L. Ruthven, ed., Global Gas Resources Workshop Proceedings, p. 37–43.

Energy Information Administration, U.S. Department of Energy, 1994–2003, International energy outlook.

Enron Corporation, 1995, The 1995 Enron Outlook, 27 p.

Enron Corporation, 1997, 1997 Enron Energy Outlook, 26 p.

Espinosa, C. E., P. Pereira, R. Guimerans, and M. E. Gurfinkel, 2001, Production and refining of heavy crude: A business opportunity in Venezuela: 18th World Energy Council Congress (Buenos Aires, October 2001).

Fearon, J. G., and Wolman, M. G., 1982, Shale oil: Always a bridesmaid? Forty years of estimating costs and prospects: Energy Systems and Policy, v. 6, no. 1, p. 63–96.

Fettweis, G. B., 1979, World coal resources: Methods of assessment and results: Developments in Economic Geology 10, 415 p.

Feys, R., 1981, Repartition des combustibles solides dans le monde: Bulletin Centre de Recherche Exploration-Production Elf-Aquitaine, v. 5, no. 2, p. 419–438.

Fiorillo, G., 1984, Exploration of the Orinoco oil belt: Review and general strategy, *in* R. F. Meyer, J. C. Wynn, and J. C. Olson, eds., The future of heavy crude and tar sands: Second International United Nations Institute for Training and Research Conference (Caracas, Venezuela, February 1982): New York, McGraw Hill, p. 304–312.

Fiorillo, G., 1987, Exploration and evaluation of the Orinoco oil belt, *in* R. F. Meyer, ed., Exploration for heavy crude oil and natural bitumen: AAPG Studies in Geology 25, p. 103–114.

Fiorillo, G. J., 1988, Exploration and evaluation of the Orinoco oil belt, final results, *in* R. F. Meyer, ed., Third United Nations Institute for Training and Research/United Nations Development Program International Conference on Heavy Crude and Tar Sands (Long Beach, California, July 1985), p. 221–237.

Fisher, W. L., 1994a, How technology has confounded U.S. gas resources estimators: Oil & Gas Journal, v. 92, no. 43 (October 24, 1994), p. 100–107.

Fisher, W. L., 1994b, The U.S. experience in natural gas: revitalization of a resource base thought exhausted, *in* C. L. Ruthven, ed., Global Gas Resources Workshop Proceedings, p. 49–63.

Galavís S., J. A., and H. Velarde Ch., 1967, Geological study and preliminary evaluation of potential reserves of heavy oil of the Orinoco Tar Belt Eastern Venezuela Basin: 7th World Petroleum Congress (Mexico, 1967), Proceedings, v. 3 (drilling and production), p. 229–234.

Galavis, J. A., and H. Velarde Ch., 1972, Estudio geológico y de evaluación preliminar de reservas potenciales de petroleo pesado en la Faja Bituminosa del Orinoco, Cuenca Oriental de Venezuela: IV Congreso Geológico

Venezolano, Memoria, tomo IV, p. 2527–2537 (Boletín de Geología, Publicación Especial No. 5).

Gautier, D. L., and R. L. Brown, 1993, National Petroleum Council Source and Supply Study—The potential for natural gas in the United States, in D. G. Howell, ed., The future of energy gases: U.S. Geological Survey Professional Paper 1570, p. 507–526.

Gayer, R., and I. Harris, 1996, Coalbed methane and coal geology: Geological Society (London) Special Publication 109, 344 p. (preface by Gayer and Harris, p. vii–viii).

Govier, G. W., 1974, Alberta's oil sands in the energy supply picture, in L. V. Hills, ed., Oil sands, fuel of the future: Canadian Society of Petroleum Geologists Memoir 3, p. 35–49.

Grace, J. D., 1994, The economics of gas resources of the former Soviet Union, in C. L. Ruthven, ed., Global Gas Resources Workshop Proceedings, p. 247–254.

Grace, J. D., 1995, FSU gas resources: Oil & Gas Journal, v. 93, no. 6 (February 6, 1995), p. 71–74; v. 93, no. 7 (February 13, 1995), p. 79–81.

Grathwohl, M., 1982, World energy supply—Resources, technologies, perspectives: Berlin, Walter de Greiyter, 450 p.

Grenon, M., 1982a, World oil resources: Their assessment and potential for the 21st century, in F. C. Witmore, Jr. and M. E. Williams, eds., Resources for the 21st century—Proceedings of the International Centennial Symposium of the U.S. Geological Survey (Reston, Virginia, 1979): U.S. Geological Survey Professional Paper 1193, p. 145–156.

Grenon, M., 1982b, A review of world hydrocarbon resource assessments: International Institute for Applied Systems Analysis, Laxenburg, Austria (Report EA-2658, Contract TPS 80-763 for Electric Power Research Institute, Palo Alto, California).

Grossling, B. F., 1976, Window on oil: A survey of world petroleum sources: The Financial Times (London), 140 p.

Gutierrez, F. J., 1981, Occurrence of heavy crudes and tar sands in Latin America, in R. F. Meyer, C. T. Steele, and J. C. Olson, eds., The future of heavy crude and tar sands: First United Nations Institute for Training and Research International Conference (Edmonton, Alberta, Canada, June 1979), p. 107–117.

Halbouty, M. T., and J. D. Moody, 1980, World ultimate reserves of crude oil: 10th World Petroleum Congress (Bucharest, 1979), Proceedings, v. 2 (exploration, supply and demand), p. 291–301.

Hendricks, T. A., 1965, Resources of oil, gas, and natural-gas liquids in the United States and the world: U.S. Geological Survey Circular 522, 20 p.

Hiller, K., 1998, Depletion mid-point and the consequences for oil supplies; 15th World Petroleum Congress (Beijing, October 1997) Proceedings, v. 3 (Natural gas, reserves, etc.), p. 184–189.

Hoffmann, P., 2001, Tomorrow's energy—Hydrogen, fuel cells, and the prospects for a clean planet: Cambridge, Massachusetts, The MIT Press, 289 p.

Holmgren, D. A., J. D. Moody, and H. H. Emmerich, 1975, The structural settings for giant oil and gas fields: 9th World Petroleum Congress (Tokyo) Proceedings, v. 2 (Geology), p. 45–54.

Houlihan, R. N., and R. G. Evans, 1989, Development of Alberta's oil sands, in R. F. Meyer and E. J. Wiggins, eds., 4th United Nations Institute for Training and Research/United Nations Development Program International Conference on Heavy Crude and Tar Sands (Edmonton, Alberta, Canada, August 1988), v. 1 (Govern., Environment), p. 95–110.

Howell, D. G., ed., 1993, The future of energy gases: U.S. Geological Survey Professional Paper 1570, 890 p.

Hubbert, M. K., 1956, Nuclear energy and the fossil fuels, in Drilling and production practices: American Petroleum Institute, p. 7–25.

Hubbert, M. K., 1962, Energy resources—A report to the Committee on Natural Resources of the National Academy of Sciences: National Research Council Publication 1000-D, 141 p.

Hubbert, M. K., 1967, Degree of advancement of petroleum exploration in United States: AAPG Bulletin, v. 51, no. 11, p. 2207–2227.

Hubbert, M. K., 1969, Energy resources, in Resources and man, a study and recommendations by the Committee on Resources and Man of the Division of Earth Sciences, National Academy of Science, National Research Council: San Francisco, W. H. Freeman and Co., p. 157–242.

Hubbert, M. K., 1973, Survey of world energy resources: The Canadian Mining and Metallurgical Bulletin, v. 66, no. 735, p. 37–53.

Hubbert, M. K., and D. H. Root, 1981, The world evolving energy system: American Journal of Physics, v. 49, p. 1007–1029.

International Atomic Energy Agency, 2001, 2002, 2003, Nuclear power reactors in the World, Reference Data Series No. 2.

International Institute for Applied Systems Analysis, World Energy Council, 1995, Global energy perspectives to 2050 and beyond, 19 p.

Ivanhoe, L. F., 1986, Oil discovery index rates and projected discoveries of the free world, in D. D. Rice, ed., Oil and gas assessment—Methods and applications: AAPG Studies in Geology 21, p. 159–178.

Janisch, A., 1981, Oil sands and heavy oil: Can they ease the energy shortage?, in R. F. Meyer, C. T. Steele, and J. C. Olson, eds., The future of heavy crude and tar sands: First International United Nations Institute for Training and Research Conference, p. 33–41.

Jardine, D., 1974, Cretaceous oil sands of western Canada, in L. V. Hills, ed., Oil sands, fuel of the future: Canadian Society of Petroleum Geologists Memoir 3, p. 50–67.

Kalisch, R., and T. Wander, 1977, The future of world natural gas supply: American Gas Association Monthly, v. 59, no. 10 (October 1977).

Kattai, V., and U. Lokk, 1998, Historical review of the kukersite oil shale exploration in Estonia: Oil Shale, v. 15, no. 2, p. 102–110.

Kelafant, J. R., S. H. Stevens, and C. M. Boyer, II, 1992, Coalbed gas 2: Vast resource potential exists in many